Springer Proceedings in Physics

Volume 281

Indexed by Scopus

The series Springer Proceedings in Physics, founded in 1984, is devoted to timely reports of state-of-the-art developments in physics and related sciences. Typically based on material presented at conferences, workshops and similar scientific meetings, volumes published in this series will constitute a comprehensive up-to-date source of reference on a field or subfield of relevance in contemporary physics. Proposals must include the following:

- name, place and date of the scientific meeting
- a link to the committees (local organization, international advisors etc.)
- scientific description of the meeting
- list of invited/plenary speakers
- an estimate of the planned proceedings book parameters (number of pages/ articles, requested number of bulk copies, submission deadline).

Please contact:

For Americas and Europe: Dr. Zachary Evenson; zachary.evenson@springer.com
For Asia, Australia and New Zealand: Dr. Loyola DSilva; loyola.dsilva@springer.com

It should be especially noted that the symposium traditionally included presentations of both purely theoretical nature as well as experimental and technological methods for studying the optical properties of various systems. On one hand, purely theoretical reports were presented on optical processes in systems such as graphene or quantum dots of the Core/Shell/Shell type. On the other hand, experimental results were discussed on far, mid and near infrared photoconductivity phenomena in *GaAs/AlGaAs* quantum wells. Also were presented the results of a detailed study of nucleation processes, characterization and optical properties of graded gap quantum dots.

Based on the results of the Symposium, 17 representative articles have been selected for publication in the present volume of Proceedings.

The 9th International Symposium "Optics & its applications" was organized by the Russian-Armenian University, the Armenian Territorial Committee of the International Commission for Optics, the Institute for Physical Research (IPR) of the NAS of Armenia, the Faculty of Physics of Yerevan State University (YSU), the A. Alikhanyan National Laboratory, the Greek-Armenian industrial company LT-PYRKAL, and the University of Wroclaw. Co-organizing student organizations were the OPTICA YSU&NAS Armenia Student Chapter, the EPS Yerevan Young Minds, the IPR Armenia OPTICA Student Chapter, the YSU SPIE Student Chapter, and the EPS Artsakh Young Minds.

The symposium was supported by the International Commission for Optics (ICO), the Russian-Armenian University, the University of Wroclaw, the A. Alikhanyan National Laboratory, the Faculty of Physics of Yerevan State University.

The official symposium website is: http://optics2022.uni.wroc.pl/.

The 1st international Symposium "Optics & its applications" was held in 2011 in Yerevan, Armenia, and by now it took place in Armenia, Germany, Italy and Poland. The upcoming 10th Symposium of this series will be held in the cities of Armenia and Cali, Colombia from 5 to 9 December 2022 as an ICTP event (https://indico.ictp.it/event/9846/).

We are grateful to Narine Gevorgyan and Hayk Ghaltaghchyan for their help in preparing this book.

Wroclaw, Poland	David Blaschke
St. Petersburg, Russia	Dmitry Firsov
Ashtarak, Armenia	Aram Papoyan
Yerevan, Armenia	Hayk A. Sarkisyan

Eduard Kazaryan—Biography

Professor Eduard Kazaryan

January 16, 2022 marked the 80th anniversary of the birth of the famous Armenian scientist, specialist in the field of physics of semiconductors and semiconductor nanostructures, Academician of the National Academy of Sciences of Armenia, Prof. Eduard Mushegh Kazaryan. (Figs. 1–3)

E. M. Kazaryan was born on January 16, 1942, in the family of the famous Armenian journalist Mushegh Kazaryan. During his school years, the future physics professor faced the problem of choosing between a sports career as a football player (in 1958 he was included in the USSR national football team for school return

Fig. 1 Professor Eduard Kazaryan

INDIRECT TRANSITIONS DUE TO THE COULOMB INTERACTIONS

E. M. KAZARIAN
Moscow State University, Moscow, U.S.S.R.

Received 30 October 1965

In this paper a new mechanism of light absorption is proposed *. It is assumed that the electron system plays the role of the third body in the indirect transitions in semiconductors. Completely degenerate semiconductors ($T = 0$) are considered, containing comparatively large number of charge carriers in one of the bands. This degeneracy may be attained either by heavy doping or just by optical or electrical injection **. To be specific we shall consider the interband transitions in the n-type sample.

We shall start from the results of ref. 1 (see also [2]), where the following formula for the absorption coefficient was obtained:

Fig. 2 First paper of Eduard Kazaryan

players) and an academic career as a physicist. Initially, the choice was in favor of football, but Eduard's father insisted on his son's scientific career, and Eduard Kazaryan was enrolled in the Physics Department of Yerevan State University in 1959. In his third year of study, Eduard Kazaryan became one of the best students and was sent to continue his studies at the leading University of the USSR at the Moscow State University, the Department of Semiconductor Physics at the Faculty of Physics. Here, at the Moscow State University, the young Eduard takes his first steps in science under the supervision of V. L. Bonch-Bruevich, one of the famous specialists of the Soviet Union in the physics of semiconductors.

Professor Bonch-Bruevich was a representative of the world-famous scientific school of Nikolai Bogolyubov, which was one of the most powerful centers of theoretical physics in the USSR. During these years, intensive research was carried out on the physical properties of semiconductors based on the methods of quantum field theory. Therefore, it was not accidental that the student Kazaryan received the task to study the plasmon absorption mechanism by a semiconductor with indirect band structure, based on the methods of Green's functions. E. Kazaryan coped with the task brilliantly. At the same time, the result turned out to be so interesting, that Prof. Dirk ter Haar, the editor of the international journal Physics Letters, invited to urgently publish it in this authoritative edition. So at 23, Eduard Kazaryan became the author (at the same time, the only author) of a scientific publication on "Indirect transitions due to the Coulomb interactions" in one of the most famous international journals.

In 1966, after graduating with honors from the Faculty of Physics of the Moscow State University, Kazaryan started the postgraduate degree under Prof. Bonch-Bruevich's supervision. After completing his postgraduate studies in 1969, he defended his Ph.D. thesis on semiconductor physics.

In 1970 began Kazaryan's active scientific and didactic career in Armenia. During these years, the physics of low-dimensional semiconductors began to develop and Eduard Kazaryan immediately understood and appreciated the prospects and importance of these studies, becoming the founder of theoretical nanophysics in Armenia. Already in 1971, the journal Soviet Physics—Semiconductors (Fizika i Tekhnika Poluprovodnikov) published an article by E. M. Kazaryan and R. Enfiajyan "To the theory of light absorption in thin semiconductor films in

Fig. 3 Eduard Kazaryan in 1970, Leningrad (now Saint-Petersburg)

the presence of size-quantized effect" dedicated to exciton absorption in quantum films—two-dimensional nanostructures, which are currently known as quantum wells.

In parallel with his active scientific activity, Kazaryan delivers lectures for students of the Yerevan University and starts his own scientific school on the physics of semiconductor nanostructures. With his young students, Prof. Kazaryan carries out active research on the optical, Coulomb, few particle properties of quantum wells, wires and of quantum dots since the mid-80s. During the period from the early 70s to the 90s, the Kazaryan School obtained a number of important results that are recognized by the scientific community. In particular:

- Comprehensive study of the interaction of semiconductors with intense laser radiation is carried out, taking into account the complex band structure of semiconductors,
- Theoretical investigation of the possibility of excitonic Bose condensation in quantum wells and quantum dots,
- Discussion of the phonon mechanism of light absorption in semiconductor quantum dots, etc.

In the early 90s, Prof. Kazaryan's attention was focused on the study of quantum dots with non-trivial geometry. Together with his scientific teams, Kazaryan conducts an extensive cycle of studies of exciton and impurity states, as well as the optical properties of quantum dots with the geometry of a strongly oblate or prolate ellipse. In this case, the calculations are carried out based on the adiabatic approximation, which allows obtaining a number of important analytical results. In particular, interband transitions in these structures were studied, including the consideration of exciton effects.

The use of the adiabatic approximation turned out to be a rather effective tool for describing not only strongly prolate and oblate ellipsoidal quantum dots but also lens-shaped and conical ones. It should be emphasized that a series of studies devoted to long-wave absorption in electron and hole gases, localized in strongly oblate lens-shaped quantum dots have been carried out. Kazaryan's team presented the possibility of realizing a generalization of Kohn's theorem in such structures. In 2016, this assumption was experimentally confirmed by a group of Prof. Dmitry Firsov from St. Petersburg Polytechnic University, for the case of a heavy hole gas, localized in strongly oblate Ge/Si lens-shaped quantum dots.

Currently, the scientific school of Prof. E. M. Kazaryan within the framework of the Russian-Armenian University is conducting a series of studies, devoted to linear and nonlinear optical, magnetic and thermodynamic properties, as well as Coulomb and spin characteristics of quantum dots of various geometric shapes and sizes.

E. M. Kazaryan is the founder of the Institute of Physics and Engineering at the Russian-Armenian University. Today, the representatives of four generations of Kazaryan's scientific school are cooperating within this University. It is very important that young researchers and students are actively involved in the study of the physical processes of quantum nanostructures, ensuring the continuity of generations of Kazaryan's scientific school.

Contents

Contributors

Aharonyan K. H. National Polytecnic University of Armenia, Yerevan, Armenia; Russian-Armenian University, Yerevan, Armenia;
Armenian State Pedagogical University, Yerevan, Armenia

Arakelyan Sh. Department of Physics, Sogang University, Seoul, Korea

Babajanyan A. Department of Radiophysics, Yerevan State University, Yerevan, Armenia

Barseghyan Manuk Department of Solid State Physics, Yerevan State University, Yerevan, Armenia;
National University of Architecture and Construction of Armenia, Yerevan, Armenia

Berthiau G. IREENA, Boulevard de L'iniversite 37, BP-406, 44602, Saint-Nazaire Cedex, France

Blaschke D. B. Institute for Theoretical Physics, University of Wroclaw, Wroclaw, Poland;
Bogoliubov Laboratory for Theoretical Physics, Joint Institute for Nuclear Research, Dubna, RU, Russia;
National Research Nuclear University (MEPhI), Moscow, RU, Russia

Chekubasheva V. Kharkiv National University of Radio Electronics, Kharkov, Ukraine

Chuluunbaatar O. Joint Institute for Nuclear Research, Dubna, MR, Russia;
Institute of Mathematics and Digital Technology, Mongolian Academy of Sciences, Ulaanbaatar, Mongolia

Derbov V. L. N.G. Chernyshevsky Saratov National Research State University, Saratov, Russia

Dmitriev V. V. Saratov State University, Saratov, RU, Russia

Firsov Dmitry Peter the Great St. Petersburg Polytechnic University, St. Petersburg, Russia

Friedman B. Department of Physics, Sam Houston State University, Huntsville, TX, USA

Gevorgyan Lilit Department of Radiophysics, Yerevan State University, Yerevan, Armenia

Gevorgyan N. T. A.I. Alikhanyan National Science Laboratory (YerPhI), Yerevan, Armenia;
Byurakan Astronomical Observatory, National Academy of Sciences, Byurakan, Armenia

Glukhov O. Kharkiv National University of Radio Electronics, Kharkov, Ukraine

Grigoryan L. Sh. Institute of Applied Problems of Physics NAS, Yerevan, RA, Armenia

Gusev A. A. Joint Institute for Nuclear Research, Dubna, MR, Russia

Hai L. L. Ho Chi Minh city University of Education, Ho Chi Minh city, Vietnam

Hakhumyan G. Institute for Physical Research, Ashtarak, Armenia

Hambaryan D. S. Department of Radiophysics, Yerevan State University, Yerevan, Armenia

Haroyan H. S. Department of Radiophysics, Yerevan State University, Yerevan, Armenia

Haroyan Hovhannes Department of Radiophysics, Yerevan State University, Yerevan, Armenia

Harutyunyan V. A. Russian-Armenian University, Yerevan, Armenia

Hayrapetyan David Russian-Armenian University, Yerevan, Armenia

Kazaryan E. M. Russian-Armenian (Slavonic) University, Yerevan, Armenia

Khachatryan H. F. Institute of Applied Problems of Physics NAS, Yerevan, RA, Armenia

Khachatryan K. S. Russian-Armenian University, Yerevan, Armenia

Kirakosyan Albert Department of Solid State Physics, Yerevan State University, Yerevan, Armenia

Kokanyan E. P. Armenian State Pedagogical University, Yerevan, Armenia

Kotanjyan A. S. Department of Physics, Yerevan State University, Yerevan, Armenia;
Institute of Applied Problems of Physics NAS, Yerevan, RA, Armenia

Kravchuk O. Kharkiv National University of Radio Electronics, Kharkov, Ukraine

Laroze David Instituto de Alta Investigación, CEDENNA, Universidad de Tarapaca, Arica, Chile

Lee K. Department of Physics, Sogang University, Seoul, Korea

Levchenko Y. Kharkiv National University of Radio Electronics, Kharkov, Ukraine

Liebing Simon JINR Dubna, Bogoliubov Laboratory of Theoretical Physics, Dubna, Russia;
Institute of Theoretical Physics, TU Bergakademie Freiberg, Freiberg, Germany

Linnyk E. Kharkiv National University of Radio Electronics, Kharkov, Ukraine

Mahato B. Institute for Theoretical Physics, University of Wroclaw, Wroclaw, Poland

Makhov Ivan National Research University Higher School of Economics, St. Petersburg, Russia

Manaselyan Aram Department of Solid State Physics, Yerevan State University, Yerevan, Armenia

Melentev Grigorii Peter the Great St. Petersburg Polytechnic University, St. Petersburg, Russia

Mkrtchyan M. A. Russian-Armenian University, Yerevan, Armenia

Mughnetsyan Vram Department of Solid State Physics, Yerevan State University, Yerevan, Armenia

Panevin Vadim Peter the Great St. Petersburg Polytechnic University, St. Petersburg, Russia

Panferov A. D. Saratov State University, Saratov, RU, Russia

Parsamyan Henrik A. Department of Radiophysics, Yerevan State University, Yerevan, Armenia;
Center for Nanoscience and Technology, Institute of Chemical Physics, Yerevan, Armenia

Pérez Laura M. Departamento de Física, FACI, Universidad de Tarapacá, Arica, Chile

Rohovets V. Kharkiv National University of Radio Electronics, Kharkov, Ukraine

Saharian A. A. Department of Physics, Yerevan State University, Yerevan, Armenia;
Institute of Applied Problems of Physics NAS, Yerevan, RA, Armenia

Sargsyan A. Institute for Physical Research, Ashtarak, Armenia

Sarkisyan D. Institute for Physical Research, Ashtarak, Armenia

Sarkisyan Hayk A. Russian-Armenian (Slavonic) University, Yerevan, Armenia;
Peter the Great Saint-Petersburg Polytechnic University, St. Petersburg, Russia

Schwalbe Sebastian Institute of Theoretical Physics, TU Bergakademie Freiberg, Freiberg, Germany

Sedova Irina Ioffe Institute, St. Petersburg, Russia

Shalygin Vadim A. Peter the Great St. Petersburg Polytechnic University, St. Petersburg, Russia

Smolyansky S. A. Saratov State University, Saratov, RU, Russia;
Tomsk State University, Tomsk, RU, Russia

Sorokin Sergey Ioffe Institute, St. Petersburg, Russia

Tonoyan A. Institute for Physical Research, Ashtarak, Armenia

Trepte Kai SUNCAT Center for Interface Science and Catalysis, Stanford University, Menlo Park, CA, USA

Tseryupa V. A. Saratov State University, Saratov, RU, Russia

Ustimenko Ratmir Peter the Great St. Petersburg Polytechnic University, St. Petersburg, Russia

Vinitsky S. I. Joint Institute for Nuclear Research, Dubna, MR, Russia;
Peoples' Friendship University of Russia (RUDN University), Moscow, Russia

Vinnichenko Maxim Peter the Great St. Petersburg Polytechnic University, St. Petersburg, Russia

Vorobjev Leonid Peter the Great St. Petersburg Polytechnic University, St. Petersburg, Russia

Chapter 1
Current-Induced Optical Activity: First Observation and Comprehensive Study

Vadim A. Shalygin

Abstract The effect of current-induced optical activity represents a circular birefringence linear in electric current which results in the rotation of the light polarization plane. It may be considered as an electro-optic analog of the magneto-optic Faraday effect. The current-induced optical activity was experimentally observed for the first time in 1978 in a tellurium single crystal. The microscopic origin of the effect in tellurium is directly associated with the current-induced spin polarization of holes. This review article includes discussion of the results of long-term studies of the current-induced optical activity; it presents details of the first experiments and some results of further research which have not been published earlier.

1.1 Introduction

The term "current-induced optical activity" was introduced in a 1979 paper devoted to a new physical effect that was experimentally found in a tellurium single crystal [1]. The effect consists of a light polarization plane rotation caused by an electric current flowing in some medium along the light propagation direction (see Fig. 1.1a). The rotation angle $\varphi(\boldsymbol{j})$ is proportional to a current density j, and a change in the polarity of the electric current leads to the inversion of the rotation direction of the polarization plane. Baranova et al. were the first to theoretically predict the possibility of the light polarization plane rotation as a result of current flow, which they called as the electrical analog of the magneto-optic Faraday effect (experimental configuration for observing the Faraday effect is shown in Fig. 1.1b) [2]. They indicated that the electrical analog of the Faraday effect may appear only in the right-left-nonsymmetrical conducting medium and estimated the magnitude of the effect for the gyrotropic liquid. Ivchenko and Pikus theoretically predicted in 1978 the same effect in tellurium, which is a gyrotropic crystal [3].

A gyrotropic medium may be defined as a medium that allows for natural optical activity, i.e., for a light polarization plane rotation that occurs in the absence of an

V. A. Shalygin (✉)
Peter the Great St. Petersburg Polytechnic University, St. Petersburg 195251, Russia
e-mail: shalygin@rphf.spbstu.ru

D. Blaschke et al. (eds.), *Optics and Its Applications*, Springer Proceedings in Physics 281, https://doi.org/10.1007/978-3-031-11287-4_1

applied magnetic or electric field. In the framework of phenomenological theory, the natural optical activity is reduced to the spatial dispersion of the dielectric permittivity and can be described in terms of linear in the light wavevector contributions to dielectric tensor components [4],

$$\varepsilon_{ik}(\omega, \boldsymbol{q}) = \varepsilon_{ik}^{0}(\omega) + i\xi_{ikl}(\omega)\, q_l, \tag{1.1}$$

where ω and $\boldsymbol{q}$ are the frequency and the wavevector of the light wave, respectively; subscripts i, k, l run through x, y, z; the diagonal tensor $\varepsilon_{ik}^{0}(\omega)$ describes the dielectric permittivity of the medium without regard for the spatial dispersion; and the gyrotropy is characterized by the third-rank tensor ξ. The tensor ξ is asymmetric with respect to the transposition of the first two subscripts: $\xi_{ikl} = -\xi_{kil}$ [5].

For the light propagating along an optical axis (we denote it as the z axis) of the optically transparent uniaxial crystal, $q \equiv |\boldsymbol{q}| = q_z$, and the Fresnel equation is reduced to the form

$$n_{\mp}^{2} = \varepsilon_{\perp}^{0} \pm \mathrm{Im}\varepsilon_{xy}(\omega, \boldsymbol{q}) = \varepsilon_{\perp}^{0} \pm \xi_{xyz}(\omega)\, q_z, \tag{1.2}$$

where n_- and n_+ are the refractive indices for the normal waves with the left-hand (σ_-) and right-hand (σ_+) circular polarization, respectively [6]. Due to the circular birefringence ($n_- \neq n_+$), the linearly polarized light undergoes the rotation of the polarization plane. Indeed, the linearly polarized wave can be considered as a superposition of two circularly polarized waves, and the polarization plane rotation angle is determined by the phase shift that occurs between the circularly polarized waves when they are transmitted through the crystal. The magnitude of the rotation angle θ is equal to half the phase shift and in the approximation of relatively weak effects of the spatial dispersion can be obtained with the following expression [6]:

$$\theta = \frac{\omega L}{2c}(n_- - n_+) = \frac{\omega L}{2c}\frac{\xi_{xyz}(\omega)}{\sqrt{\varepsilon_{\perp}^{0}}} q_z, \tag{1.3}$$

where L is the crystal length. One can conclude that the linear in the light wavevector contributions to the dielectric tensor due to spatial dispersion cause the circular birefringence which results in the natural optical activity.

The current-induced optical activity can be described phenomenologically by linear terms in the expansion of ε_{ik} with respect to the electric current $\boldsymbol{j}$, that is,

$$\varepsilon_{ik}(\boldsymbol{j}) = \varepsilon_{ik}^{0}(\omega) + i\xi_{ikl}(\omega)\, q_l + i\Theta_{ikl}(\omega)\, j_l. \tag{1.4}$$

Since the electric current density $\boldsymbol{j}$, as well as the light wave vector $\boldsymbol{q}$, is a polar vector that changes its sign due to the time reversal, one can conclude that the tensor $\Theta_{ikl}(\omega)$ is transformed in the same manner as the tensor $\xi_{ikl}(\omega)$ and differs from zero only in gyrotropic crystals. In particular, the tensor $\Theta_{ikl}(\omega)$ is asymmetric with respect to the transposition of the first two subscripts: $\Theta_{ikl} = -\Theta_{kil}$.

Let us now consider expansion (1.4) as applied to tellurium crystals. It is assumed that the electric current flows along the optical axis of the tellurium crystal (it coincides in direction with the threefold screw axis C_3, we denote this direction as z axis) and that the light propagates in the same direction. In this case the Fresnel equation can be rewritten in a following way:

$$n_{\mp}^2(\boldsymbol{j}) = \varepsilon_{\perp}^0 \pm \mathrm{Im}\varepsilon_{xy}(\boldsymbol{j}) = \varepsilon_{\perp}^0 \pm \xi_{xyz}(\omega)\, q_z \pm \Theta_{xyz}(\omega)\, j_z. \tag{1.5}$$

Thus, there is a change in the circular birefringence magnitude determined by the last term in the Eq. 1.5. Accordingly, the light polarization plane undergoes an additional rotation that is proportional to the current density

$$\varphi(\boldsymbol{j}) = \frac{\omega L}{2c}\left[n_-(\boldsymbol{j}) - n_+(\boldsymbol{j})\right] = \frac{\omega L}{2c}\frac{\Theta_{xyz}(\omega)}{\sqrt{\varepsilon_{\perp}^0}}\, j_z. \tag{1.6}$$

Therein lies the phenomenon of current-induced optical activity. When the electric current reverses its direction, the polarization plane also rotates in the opposite direction. In contrast to the Faraday effect, which is observed even in isotropic media, the current-induced optical activity occurs only in gyrotropic media and, generally speaking, is observed against the background of the polarization plane rotation θ due to the natural optical activity (see Fig. 1.1a). Among the exceptions, there are crystals with a wurtzite lattice, in which no natural optical activity occurs ($\theta = 0$) and the current-induced optical activity can be observed in the pure form.

It should be noted that when the light beam passes twice through the crystal (forward and backward), the resulting polarization plane rotation due to the current-induced optical activity doubles (as in the Faraday effect). In the case of natural optical activity, on the other hand, the resulting polarization plane rotation in the same experimental configuration turns out to be equal to zero. Thus, the effect of current-induced optical activity is not reduced to a current-induced change in the magnitude of the effect of natural optical activity. This is a new physical effect not related to spatial dispersion.

The microscopic origin of the effect of current-induced optical activity is associated with the current-induced spin polarization (CISP) of charge carriers when there arises an average (homogeneous over the sample) charge carrier spin in the region with a constant current density. The microscopic theory of the effect in tellurium was created by Ivchenko and Pikus [1, 3, 7]. Results of the experimental studies of the current-induced optical activity and CISP in tellurium were discussed in [1, 6, 8]. Later the CISP in tellurium was also experimentally investigated using nuclear magnetic resonance (NMR) technique [9, 10]. Recently ab initio calculations of the current-induced optical activity and CISP in tellurium were performed by Tsirkin et al. [11], who suggested to designate these effects as "kinetic Faraday effect" and "kinetic magnetoelectric effect", respectively.

The CISP in deformed III-V crystals was theoretically examined by Aronov and Landa-Geller [12]. Spin polarization of conduction electrons due to an electric cur-

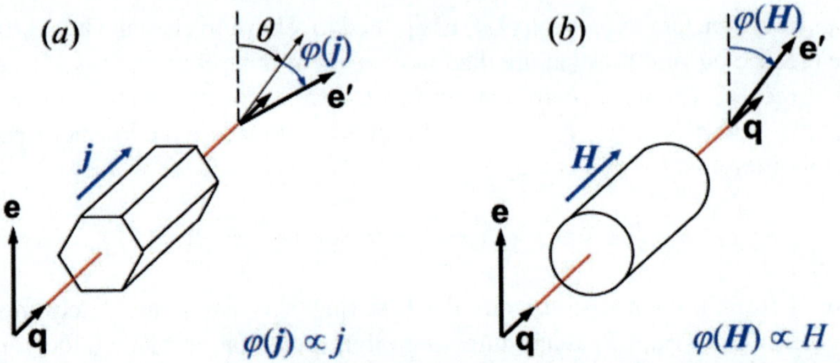

Fig. 1.1 Experimental configuration for observing the current-induced optical activity (**a**) and Faraday effect (**b**). Vectors of the current density, magnetic field strength and light wavevector are denoted as $\boldsymbol{j}$, $\boldsymbol{H}$, and $\boldsymbol{q}$, respectively; $\boldsymbol{e}(\boldsymbol{e}')$ indicates the light polarization vector at the sample input (output). The plane polarization rotation angles θ and $\varphi\,(\boldsymbol{j})$ are determined by Eqs. (1.3) and (1.6), respectively. The Faraday rotation angle is denoted as $\varphi\,(\boldsymbol{H})$

rent is also possible in semiconductor quantum well structures (see Refs. [13, 14] and references therein) and in disordered two-dimensional systems [15]. Besides bulk tellurium, to date electrically polarized spins have been experimentally observed in $GaAs$, $InGaAs$, $ZnSe$, and GaN epitaxial layers [16–18]; and single heterojunctions and quantum wells based on III-V semiconductors [19–22].

Section 1.2 of the present paper contains a story on the history of the first experiments to observe the current-induced optical activity. Microscopic theory of the effect is considered in Sect. 1.3. Then, in Sect. 1.4, comprehensive studies of this effect in tellurium are described, and the results obtained are discussed. The final Sect. 1.5 of the article summarizes the results of long-term studies of the current-induced optical activity.

1.2 First Experimental Observation of the Current-Induced Optical Activity

Electro-optical experiments on tellurium were started in Leningrad Polytechnic Institute (its current name is Peter the Great St. Petersburg Polytechnic University) in the beginning of 1978. The experiments were carried out by Shturbin, Shalygin and Vorob'ev on tellurium single crystals grown by Farbshtein and Galetskaya at Ioffe Institute.

A sketch of the original experimental setup is shown in Fig. 1.2. A self-made CO_2 laser operating in CW regime was used as a source of linearly polarized IR radiation at a wavelength of $\lambda = 10.6\ \mu$m. The laser beam was focused on the sample mounted in a vacuum-free liquid nitrogen cryostat of original design. The sample geometry used in the first experiments is illustrated in Fig. 1.4a. The samples were cut along

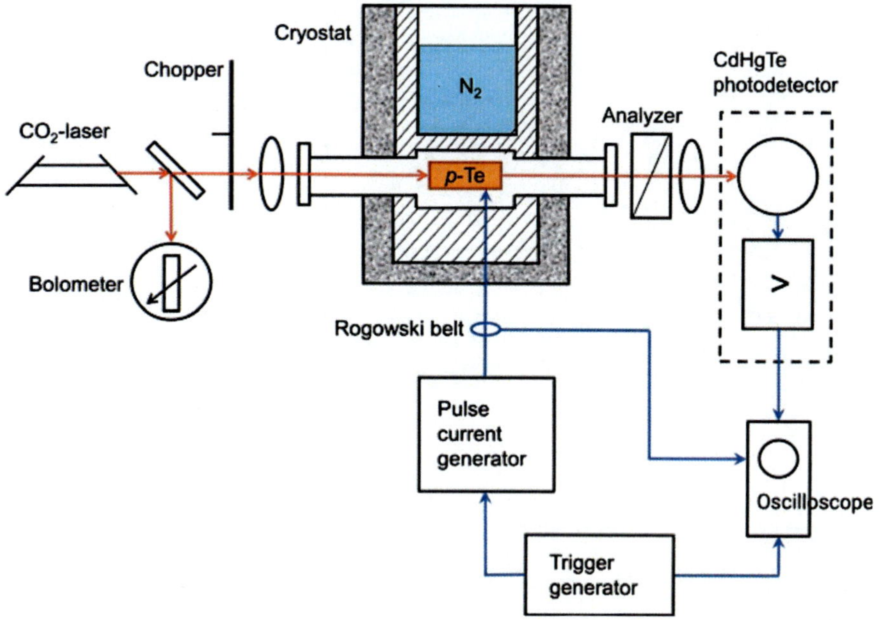

Fig. 1.2 Sketch of the original experimental setup

the C_3 axis from a tellurium single crystal using an electric spark machine and their facets were mechanically polished. Two contacts were made from a $SnBiSb$ alloy which provided an electric current flowing along the C_3 axis. The pulsed electric voltage with a pulse duration of a few μs and a repetition rate of less than 1 Hz was applied to the contacts. This regime made it possible to avoid Joule heating of the samples. To prevent the formation of acoustoelectric domains, which cause strong modulation of optical absorption, a relatively weak electric field was applied to the samples (the electric field strength in the samples did not exceed 100–200 V/cm). The Rogowski belt was used to measure the amplitude of the current. The polarization measurements were carried out using a linear polarizer mounted in front of a $CdHgTe$ photodetector. The radiation intensity modulation was recorded by an oscilloscope at various azimuth angles of the linear polarizer. Analysis of the experimental data led to the conclusion that a new physical phenomenon has been observed for the first time, namely, a radiation polarization plane rotation caused by an electric current flowing in a gyrotropic crystal along the light propagation direction. The priority has been protected by an invention application from 11.12.1978 [23].

Several months earlier, this phenomenon had been independently predicted by theoreticians Ivchenko and Pikus from Ioffe Institute [3]. In their article devoted to the circular photogalvanic effect (CPGE) in gyrotropic crystals, they also discussed a possibility to observe a reciprocal effect. In particular, they considered CPGE due to interband carrier transitions in tellurium and associated its microscopic origin with the optical spin polarization of electrons and holes that results in appearance

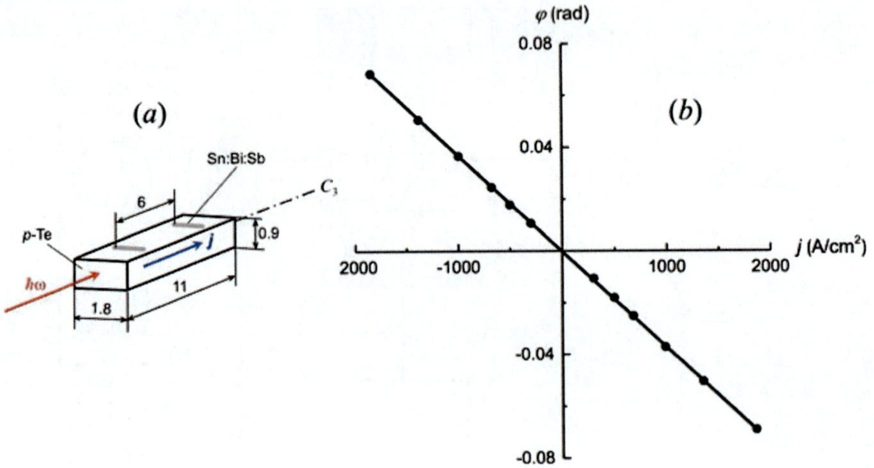

Fig. 1.3 Sample geometry (**a**) and dependence of the polarization plane rotation angle on the current density for a dextrorotatory tellurium crystal with $p = 1.5 \cdot 10^{17}$ cm^{-3} at $T = 77$ K (**b**). The electric current co-directional with the laser beam is considered as positive. Sample dimensions are indicated in millimeters. Measurements were carried out at $\lambda = 10.6$ μm

of a photocurrent owing to the spin-orbit interaction. In the reciprocal effect, an electric current leads to a partial spin polarization of free carriers, which should be accompanied by a light polarization plane rotation proportional to the current density [3].

In the beginning of 1979, the experimenters and theoreticians published a joint paper devoted to the current-induced optical activity [1]. Experimental study of the effect was carried out on a tellurium crystal of 11 mm length with a free hole concentration of $p = 1.5 \cdot 10^{17}$ cm^{-3} at a temperature of $T = 77$ K. The examined crystal was dextrorotatory. In the absence of electric current, it demonstrated a positive polarization plane rotation angle of $\theta = 102° = 1.78$ rad. At pulsed electric current co-directional with the laser beam, there was observed a negative change in the polarization plane rotation angle $\Delta\theta \equiv \varphi\,(j_z) < 0$.

In other words, in the considered case of dextrorotatory tellurium, the electric current co-directional with the radiation wavevector induced the left-handed rotation of the polarization plane. In particular, at a current density of $j_z = +1860$ A/cm^2 the current-induced rotation angle of $\varphi\,(j_z) = -4° = -0.069$ rad was determined from the experimental data on radiation intensity modulation measured at various azimuth angles of the linear polarizer. A change in the polarity of the electric current leaded to the inversion of the rotation direction of the polarization plane. The experimental dependence of the rotation angle on the electric current density was well approximated by a linear dependence over the entire range of electric current densities from -1860 to 1860 A/cm^2 (see Fig. 1.3b). Microscopic theoretical model of the current-induced optical activity in tellurium was developed by Ivchenko and Pikus and was outlined also in Ref. [1]. The model will be described below in Sect. 1.3.

1.3 Microscopic Mechanism of the Effect in Tellurium

Tellurium has a specific band structure due to strong spin-orbit interaction. It has been established that the two upper branches of the valence band (we denote them as V^+ and V^-) are anisotropic and obey the substantially non-parabolic dispersion law (see [24, 25] and references therein):

$$E_{V^\pm} = -\Delta - Ak_z^2 - Bk_\perp^2 \pm \sqrt{\Delta^2 + \beta^2 k_z^2}, \tag{1.7}$$

where k_z is the component of the electron wavevector along the C_3 axis and parameters Δ, A, B, and β are determined experimentally. Corresponding electron wave functions are superpositions of the states with the spins (i.e. angular momentum projections) $M_z = +3/2$ and $M_z = -3/2$:

$$\Psi_{V^\pm} = C^\pm_{3/2} \left|+3/2\right\rangle + C^\pm_{-3/2} \left|-3/2\right\rangle, \tag{1.8}$$

where the weight factors $C^\pm_{3/2}$ and $C^\pm_{-3/2}$ depend monotonically on the electron wavevector. As a result, in each branch of the valence band of tellurium there is a one-to-one correspondence between the electron wavevector and spin. In particular, in the V^+-branch of the valence band, the state of the electron with the wavevector projection k_z is characterized by the average spin of

$$M_z\left(k_z\right) = \left(+\frac{3}{2}\right)\left|C^+_{3/2}\right|^2 + \left(-\frac{3}{2}\right)\left|C^+_{-3/2}\right|^2 = \frac{3}{2}\frac{\beta k_z}{\sqrt{\Delta^2 + \beta^2 k_z^2}}. \tag{1.9}$$

As it was shown in Ref. [24], the positive sign of the constant β in Eq. (1.9) corresponds to levorotatory tellurium crystals, while its negative sign is related to dextrorotatory crystals. For the latter case the dependence $M_z\left(k_z\right)$ is shown in Fig. 1.4b. The monotonic dependence of the electron spin M_z on the electron wavevector k_z (which is a direct consequence of the spin-orbit interaction in tellurium) underlies the microscopic mechanism of the current-induced optical activity.

Let us consider propagation of light in tellurium crystal along the z axis. If the photon energy $\hbar\omega$ is less than the band gap energy E_g, then the magnitude of the refractive indices for normal waves σ_- and σ_+ is determined by virtual transitions of electrons from the valence band V^+ to the conduction band C. In Fig. 1.4a, the virtual transitions excited by the σ_- circularly polarized wave are shown by blue vertical arrows, and red arrows correspond to the transitions excited by the σ_+ circularly polarized wave. We don't take into account the spatial dispersion, i.e. we put the light wavevector q equal to 0 and depict the optical transitions with vertical (rather than tilted) arrows. In the valence band V^+, the electron states with a spin of $M_z = -3/2$ dominate at $k_z > 0$ (Fig. 1.4b). Electrons from these states can be exited to the conduction band C by σ_+ photons only (the latter have an angular momentum of $+1$). As a result, electrons with spin $-1/2$ appear in metastable states in the C-band. The σ_- photons with angular momentum of -1 cannot excite electron

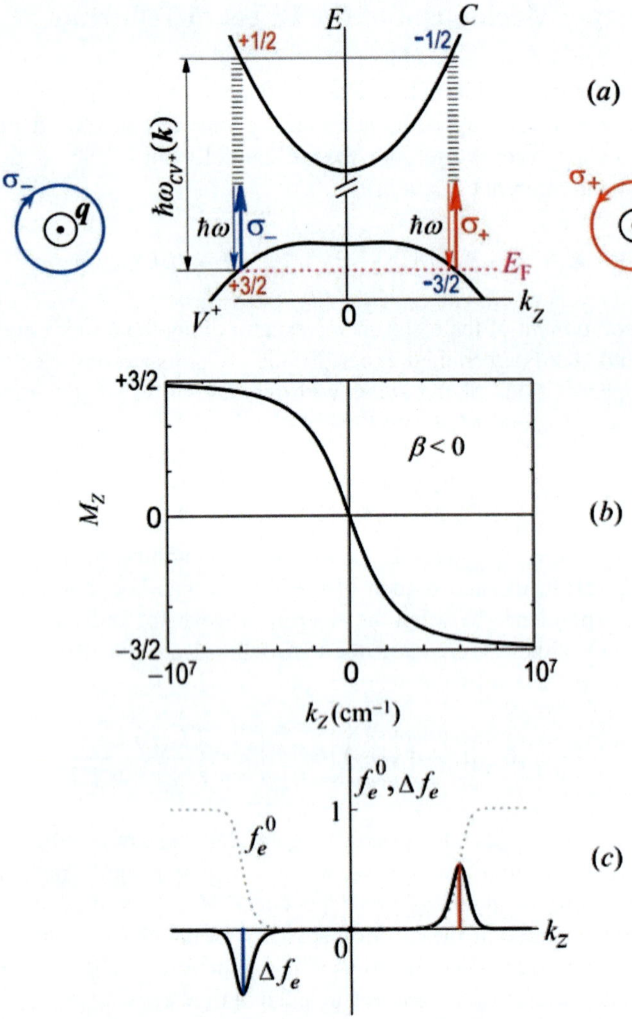

Fig. 1.4 Microscopic mechanism of the current-induced optical activity in dextrorotatory tellurium ($\beta < 0$). Illustration for a degenerated p-type crystal (E_F denotes the Fermi energy). (**a**) Diagram of virtual interband transitions of electrons excited from the states around the Fermi energy by circularly polarized photons with the energy $\hbar\omega < E_g$. (**b**) Dependence of the average spin M_z for states of the valance subband V^+ on the electron wave vector component k_z. (**c**) Equilibrium electron distribution function f_e^0 and its variation Δf_e due to an electric current flow (for the electron states in the V^+-band with $k_x = 0$ and $k_y = 0$)

transitions from these states to metastable states in the C-band because in this case the electron spins in those states should be equal to $-5/2$ due to the angular momentum conservation law but such states are absent in the conduction band. There are the states with spins of $-1/2$ and $+1/2$ only. Similarly, the electron states at $k_z < 0$ are mainly excited by σ_- circularly polarized wave. In this case electrons undergo virtual transitions from the valence band states with spin of $+3/2$ to metastable C-band states with spin of $+1/2$ (see Fig. 1.4a). The virtual transitions excited by σ_+ and σ_- photons with the same energy deficit give the same contribution to a refractive indices n_+ and n_-, respectively. In the absence of an electric current, the electron distribution function f_e^0 in the valence band V^+ is an even function of k_z (see dashed curve in Fig. 1.4c). Consequently, $n_+ = n_-$, i.e. there is no circular birefringence (if neglecting the spatial dispersion).

When an electric current $\boldsymbol{j}$ co-directional with the light beam flows through the tellurium crystal, the electron distribution function in the V^+-band shifts in $\boldsymbol{k}$-space against the current $\boldsymbol{j}$. Therefore, a non-equilibrium addition to the distribution function Δf_e is an odd function of k_z (it is plotted in Fig. 1.4c by a solid line). The positive Δf_e at positive values of k_z increases the refractive index for the σ_+-polarized wave, while the negative Δf_e at negative values of k_z decreases the refractive index for the σ_--polarized wave (in the considered case of $\beta < 0$). Both the factors lead to a current-induced circular birefringence, namely, $\left[n_-(\boldsymbol{j}) - n_+(\boldsymbol{j})\right] < 0$. In accordance with the phenomenological formula (1.6), it results in a negative rotation of the polarization plane of linearly polarized light: $\varphi(\boldsymbol{j}) < 0$. This qualitative description explains the existence and sign of the current-induced optical activity observed in the experiment (see Fig. 1.3).

Ivchenko and Pikus performed a microscopic simulation of the effect [1]. They considered probabilities of the virtual interband electron transitions excited in tellurium by σ_+- and σ_--polarized light in the conditions of electric current flow. Non-equilibrium distribution function of holes in the V^+-band was calculated using the Boltzmann kinetic equation in the approximation of relaxation time. It was found that the specific current-induced rotation angle $\gamma = \varphi(j_z)/(j_z l)$, where l is a distance between the electrical contacts, is proportional to the term

$$\frac{\beta E_g}{E_g^2 - (\hbar\omega)^2}. \tag{1.10}$$

The theoretically calculated value of $|\gamma|$ is $9 \cdot 10^{-5}$ rad · cm/A, which is rather close to the value of $\gamma = -6 \cdot 10^{-5}$ rad · cm/A found from the experimental data presented in Fig. 1.3.

1.4 Comprehensive Study of the Effect and Discussion of Its Results

For studying the microscopic mechanism of the current-induced optical activity, it was of particular interest to examine its spectral and temperature dependences. Previously, the sample preparation technique was improved.

Pieces with a length of 11−16 mm along the C_3 axis were cut from single tellurium crystals (4−8 mm diameter) using an electric spark machine or chemical cutting technique. The end faces were subjected to optical polishing. Some samples had the shape of a hexagonal prism, in which the lateral surface was a natural facet of the single crystal (Fig. 1.5b). Other samples were made by cleaving the hexagonal prism at $T = 77$ K and had the form of a triangular prism (Fig. 1.5a). To provide a more homogeneous current distribution in a sample, ring electrical $SnBiSb$ contacts were made on the lateral surface of every crystal near its ends. Moreover, to increase additionally the homogeneity of electric current in the samples with a cross section area S exceeding 6 mm^2 (see Table 1.1), gold from a solution of chloroauric acid was deposited on the sample ends. In the central parts of the sample ends, regions with a diameter of 1−3 mm were left uncovered by gold for the passage of the laser beam (see Fig. 1.5b).

A crucial issue was the optical alignment of the light beam relative to the sample. In the case when the light beam in the crystal is not strictly parallel to the C_3 axis, the strong linear birefringence of tellurium causes noticeable distortions in the natural/current-induced optical activity [26]. By moving the sample in two direc-

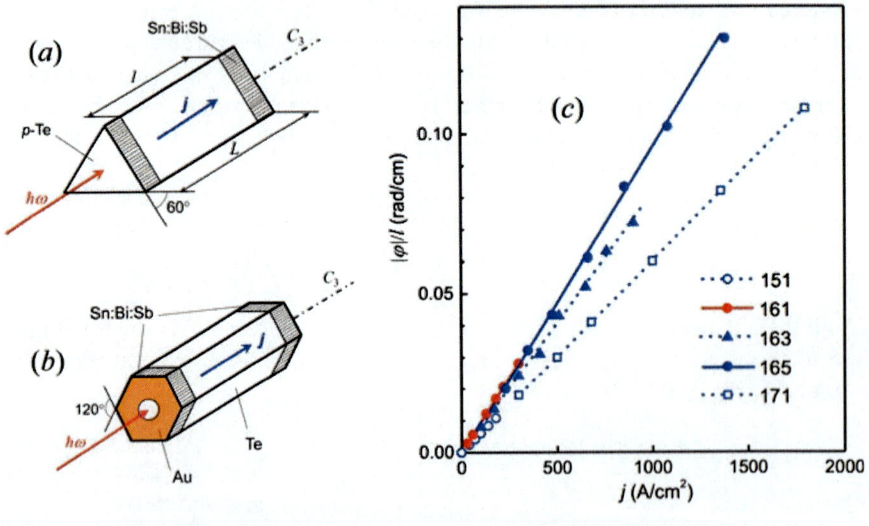

Fig. 1.5 Two variants of the sample design (**a**, **b**) and current dependence of the induced polarization plane rotation in p-tellurium samples with different hole concentrations (**c**). Measurements were carried out using CO_2 laser ($\lambda = 10.6$ μm) at $T = 77$ K

Table 1.1 Sample parameters at $T = 77K$ [26]

Sample number	151	161	163	165	166	171
signθ	+1	−1	+1	+1	+1	+1
p (cm^{-3})	$5 \cdot 10^{15}$	$1 \cdot 10^{16}$	$3.2 \cdot 10^{16}$	$4 \cdot 10^{16}$	$4.8 \cdot 10^{16}$	$1.5 \cdot 10^{17}$
Geometry	Figure 1.5b	Figure 1.5b	Figure 1.5b	Figure 1.5b	Figure 1.5a	Figure 1.3a
L (mm)	12.4	15	15	12.5	12	11
l (mm)	10.5	13	13	9	9.5	6
S (mm^2)	25	37	19	6.8	0.6	1.6
γ (rad · cm/A)	$-5.7 \cdot 10^{-5}$	$+9 \cdot 10^{-5}$	$-8.5 \cdot 10^{-5}$	$-9.5 \cdot 10^{-5}$	$-6.9 \cdot 10^{-5}$	$-6 \cdot 10^{-5}$

tions and rotating it around two axes, a linearly polarized light beam from a CO_2 laser was adjusted so that the degree of linear polarization of light at the exit from the crystal P_{lin} was maximum. The quite high values of $P_{lin} = 0.95 - 0.99$ were obtained for different samples. This ensured the propagation of the light along the optical axis of the crystal even if the normal to the input face formed a small angle with the optical axis (the deflection of the laser beam relative to the optical axis did not exceed 0.5°). Note, that in the first experiment (Fig. 1.3a) quality of alignment was lower ($P_{lin} = 0.83$) [1].

The current-induced optical activity was investigated in p-type tellurium crystals with various hole concentrations in the range of $5 \cdot 10^{15} - 1.5 \cdot 10^{17}$ cm^{-3} (at $T = 77$ K), see Table 1.1. The electron mobility at this temperature varied insignificantly from sample to sample and ranged from 1700 to 3300 cm^2/(V · s). The studies were carried out on both dextrorotatory (sign$\theta = +1$) and levorotatory (sign$\theta = -1$) tellurium crystals. Current dependence of the induced polarization plane rotation experimentally measured with a CO_2 laser at $T = 77$ K is presented in Fig. 1.5c. For all samples, this dependence turned out to be linear. The values of specific current-induced rotation $\gamma = \varphi(j)/(jl)$ are negative for the dextrorotatory crystals and positive for the levorotatory ones (see Table 1.1). These facts are consistent with the theory of the current-induced optical activity created by Ivchenko and Pikus (See Ref. [1]) and with ab-initio calculations made recently by Tsirkin et al. [11].

Studies of the effect on the crystals with $p < 5 \cdot 10^{15}$ cm^{-3} are difficult because of their high resistivity: acoustoelectric domains arise in them at relatively low current densities [27]. As the hole concentration increases, the threshold of the generation of acoustoelectric domains moves to the region of higher currents, and, accordingly, higher values of the current-induced rotation angles $\varphi(j)$ become observable (see Fig. 1.5c). On the other hand, it is practically impossible to measure current-induced polarization plane rotation on the crystals with $p > 2 \cdot 10^{17}$ cm^{-3} due to their low optical transmission (less than 1% at $\lambda = 10.6$ μm).

Note, that the specific current-induced rotation angle γ does not show a strong concentration dependence in the examined range of hole concentrations, namely, when p changed by one and a half orders of magnitude, the value of γ changed by no more than 1.5 times (see Table 1.1). Such a small difference in the magnitude of the effect seems to be associated not so much with the different concentration of holes, but with the varying degrees of perfection of the crystal structure and of the electric field homogeneity in different samples. Indeed, the maximum values of $|\gamma| = (8-9) \cdot 10^{-5}$ rad · cm/A were observed in the samples with a large cross section ($S = 7-37$ mm^2), the lateral surface of which was a natural facet of the single tellurium crystal and was not subjected to any mechanical processing (see Table 1.1, samples 161, 163, 165). It is obvious that electric field homogeneity in the sample 165 (Fig. 1.5b) is higher than in the sample 171 (Fig. 1.3a) due to special design of the electrical contacts. With this we associate a lower value of $|\gamma|$ obtained in the first experiments (Fig. 1.3).

In order to obtain data on the spectral dependence of the current-induced optical activity in tellurium, additional experiments were carried out using a carbon monoxide (CO) laser, which generates radiation in a shorter wavelength range than a CO_2 laser. The optical scheme of the experimental setup was modified as follows. Since the CO laser simultaneously generates a whole series of lines in the $5.2-6.2$ μm wavelength range, it was used in tandem with a monochromator. A Ge:Au resistor was utilized as a photodetector.

The experimental results on the specific value of the current-induced optical activity in $p-Te$ (sample 165) for several wavelengths at a fixed temperature are presented in Fig. 1.6. For comparison, the experimental spectral dependence of the specific value of the natural optical activity (θ/L) is shown in the same figure [6, 28]. As the threshold of interband transitions ($\lambda_g = 2\pi c\hbar/E_g = 3.7$ μm) is approached with decreasing wavelength, the magnitudes of both the effects increase monotonically, but their rate of rise differs significantly. Spectral dependence of the current-induced optical activity is well described by the theoretical expression (1.10), while the natural optical activity has much stronger spectral dependence in accordance with a theoretical model developed by Natory [29]:

$$\frac{\theta}{L} \propto \frac{1}{\sqrt{E_g - \hbar\omega}} + \frac{1}{\sqrt{E_g + \hbar\omega}} - \frac{2}{\sqrt{E_g}}, \qquad (1.11)$$

This circumstance confirms the idea that the current-induced optical activity is not a simple consequence of the modulation of the natural optical activity by the electric field. The latter effect, which is referred to as the electro-optical activity or the electrogyration effect, is associated with spatial dispersion and is not possible in tellurium in the considered geometry of the experiment [1]. As it was already mentioned in Sect. 1.1, the phenomenon of current-induced optical activity is qualitatively different from the effects associated with the spatial dispersion.

Experimental study of the temperature dependence of the effect in p-tellurium (sample 163) revealed a 6-fold decrease of its magnitude upon increasing the temperature from 77 to 350 K (see Fig. 1.7). Theoretical simulation of the $\gamma(T)$ dependence in the framework of the model from Ref. [1] was performed in Ref. [8]. The equi-

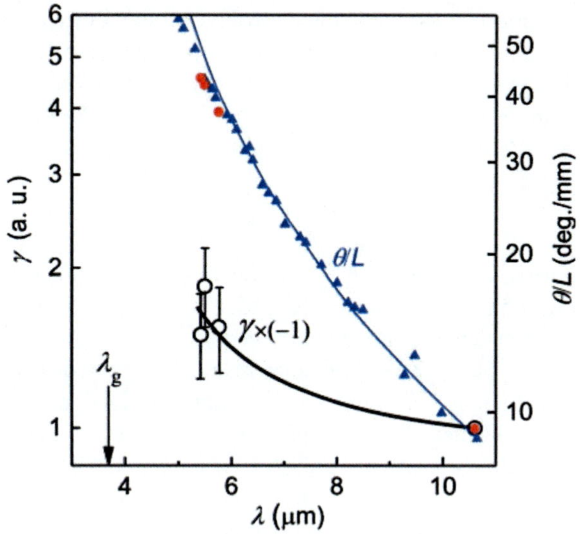

Fig. 1.6 Spectral dependences of the current-induced optical activity (γ) and natural optical activity (θ/L) in the dextrorotatory tellurium. Circles demonstrate results of our experiments at $T = 77$ K [6], and triangles represent experimental data from Ref. [28]. Results of the theoretical simulation of the $\gamma(\lambda)$ dependence according to formula (1.10) are shown by thick black line. The theoretical spectral dependence of θ/L was obtained using the model from Ref. [29] and is shown by thin blue curve. The threshold wavelength for the interband electron transitions is indicated by an arrow

Fig. 1.7 Temperature dependence of the current-induced optical activity in p-tellurium at $\lambda = 10.6$ μm. Circles demonstrate experimental results. Four curves represent results of simulation according to the theoretical model form Ref. [1]. Numbers near the curves indicate different values of the scattering parameter n (see Eq. (1.12))

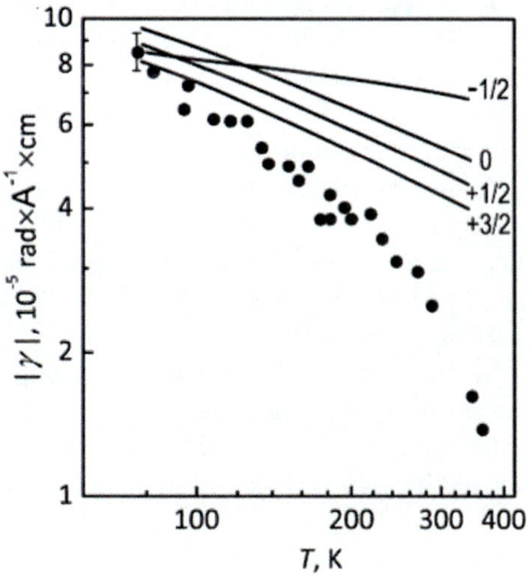

librium hole distribution in $\boldsymbol{k}$-space was described by the Boltzmann distribution function and a non-equilibrium hole distribution function was calculated using the Boltzmann kinetic equation in the approximation of relaxation time. It was assumed that the momentum relaxation time τ has a power-law dependence on the hole energy E_h:

$$\tau(E_h) \propto E_h^n, \tag{1.12}$$

where n is a parameter that depends on the scattering mechanism of holes. The calculation was carried out for various values of n in the range from $-1/2$ to $+3/2$, which obviously overlaps the limits of variation of this parameter in tellurium in the examined temperature range. The simulation results are plotted in Fig. 1.7 as four lines. The figure shows that the specific current-induced polarization plane rotation angle at $T = 77$ K weakly depends on the parameter n, and its calculated values are in good agreement with the experimental value of $|\gamma| = (8.5 \pm 0.7) \cdot 10^{-5}$ rad · cm/A. However, the experimentally observed dependence of γ on temperature is much stronger than that which is calculated for any possible value of n (see Fig. 1.7).

The sharp decrease in the absolute value of the current-induced optical activity with increasing temperature was explained in the theoretical work of Averkiev [30]. The contributions to the effect from the virtual optical transitions between three bands (V^+, V^-, and C) were considered. A change in the occupations of these bands with an increase in the temperature was taken into account. It was shown, that at higher temperatures, the electron contribution to the total electric current begins to dominate over the hole contribution. On the other hand, it was found that asymmetry in the distribution of electrons in the C band (caused by the electric current flow) makes a much smaller contribution to the current-induced optical activity compared to that of holes in the V^+ band. This is precisely the additional reason for the decrease in γ upon increasing temperature.

As discussed above, the microscopic origin of the effect of current-induced optical activity is associated with the CISP of charge carriers when there arises an average charge carrier spin in the region with a constant current density. The dependence of the degree of spin polarization of holes in tellurium on the electric current density was considered in Ref. [6].

In the absence of an electric current, the distribution of holes in $\boldsymbol{k}$-space is described by the Fermi-Dirac function with the Fermi level E_F that can be determined by the normalization to hole concentration p. This equilibrium distribution function, denoted below as $f_h^0(\boldsymbol{k})$, is an even function of k_z, consequently, the average spin of the holes $\langle M_z \rangle$ is equal to zero, because $M_z(k_z)$ is an odd function. It should be noted that the dependences $M_z(k_z)$ in the electron band $E_{V^+}(\boldsymbol{k})$ and in the hole band $E_h(\boldsymbol{k})$ are described by the same relationship, Eq. (1.9) [6]. When the electric current is flowing through the sample, the addition to the distribution function of the holes, $\Delta f_h(\boldsymbol{k})$, appears, which is an odd function of k_z, and the average spin the holes $\langle M_z \rangle$ ceases to be equal to zero:

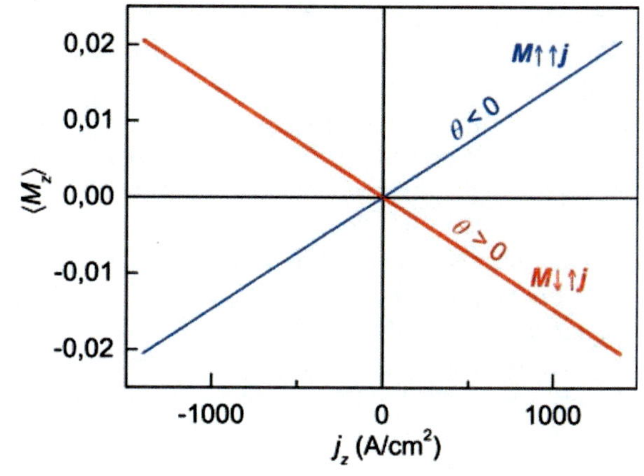

Fig. 1.8 Dependences of the average hole spin on the current density for the dextrorotatory (red line) and levorotatory (blue line) tellurium crystals. Simulations were made for tellurium crystals with a hole concentration of $p = 4 \cdot 10^{16}\ \mathrm{cm}^{-3}$ at $T = 77$ K under condition of $\boldsymbol{j} \parallel z \parallel C_3$

$$\langle M_z \rangle = \frac{1}{p} \int M_z(k_z)\, \Delta f_h\left(k_x, k_y, k_z\right) \frac{d^3k}{(2\pi)^3}. \tag{1.13}$$

The non-equilibrium addition to the distribution function of the holes can be found using the Boltzmann kinetic equation in the approximation of relaxation time τ [6]:

$$\Delta f_h(\boldsymbol{k}) = f_h^0(\boldsymbol{k})\left[1 - f_h^0(\boldsymbol{k})\right] \frac{eE_z}{k_B T} \frac{1}{\hbar} \frac{\partial E_h}{\partial k_z} \tau(E_h), \tag{1.14}$$

where $\boldsymbol{j} \parallel \boldsymbol{E} \parallel z \parallel C_3$ is assumed, and k_B is the Boltzmann constant.

Figure 1.8 presents the calculated dependence of the average hole spin on the electric current in tellurium for the range of electric current densities corresponding to experiments on the current-induced optical activity (see Fig. 1.8). The calculations were performed for the dextrorotatory ($\mathrm{sign}\theta = -\mathrm{sign}\beta = +1$) and levorotatory ($\mathrm{sign}\theta = -\mathrm{sign}\beta = -1$) tellurium crystals using Eqs. (1.9), (1.13), and (1.14). Results of the simulations for a p-tellurium crystal ($p = 4 \cdot 10^{16}\ \mathrm{cm}^{-3}$) at $T = 77$ K are presented in the figure. In the framework of the considered model, the average spin of the holes and the current density are related by a linear expression. It can be seen from Fig. 1.8 that, in the case of the dextrorotatory tellurium crystal, the direction of the average hole spin is opposite to the direction of the electric current. And vice versa, in the levorotatory crystal, the electric current and the average spin of the holes have the same direction.

Since the holes in tellurium are in a superposition of states with the z-projections of the angular momentum $\pm 3/2$, the degree of spin polarization of the holes is conveniently determined using the expression

$$\rho_h = \frac{\langle M_z \rangle}{3/2}. \tag{1.15}$$

In accordance with our simulations, the maximum absolute value of the spin polarization of the holes in considered experiments on the current-induced optical activity in p-tellurium reaches $1.4 \cdot 10^{-2}$ (at $p = 4 \cdot 10^{16}$ cm^{-3}, $T = 77$ K, and $j_z = 1400$ A/cm^2). The spin polarization of the holes is spatially uniform in the region of current flow and is characterized by a very high spin density of 560 spins/μm^3 [6]. This agrees surprisingly well with the value of 561 spins/μm^3 obtained by Tsyrkin et al. from ab initio calculations at the same conditions [11].

A significantly weaker CISP was observed in other bulk materials. The value of the volume-uniform spin density under lateral electric current flow in the 1.5 μm-thick epitaxial layer of the n-$ZnSe$ gyrotropic semiconductor at $T = 20$ K did not exceed 12 spins/μm^3 [16]; in this case the degree of electron spin polarization ρ_e was only $1.3 \cdot 10^{-6}$. The CISP, uniform over the volume of semiconductor, was also observed in strained n-$InGaAs$ epitaxial layers grown on a $GaAs$ substrate [17]. The maximum value of the electron spin density detected at $T = 5$ K was 8.1 spins/μm^3 ($\rho_e = 2.7 \cdot 10^{-4}$). In contrast to tellurium, where spins are oriented along the direction of the electric current, in the n-$ZnSe$ and n-$InGaAs$ semiconductors, when the electric current flows along the epitaxial layer, the spin polarization of electrons has been observed in the layer plane in the direction perpendicular to the current [16, 17]. As the authors noted, the microscopic origin of the observed effects is not quite clear. The measurements of CISP in Refs. [16, 17] were carried out using magnetooptical methods. The Faraday and Kerr rotation angles detected in these studies did not exceed $4 \cdot 10^{-5}$ rad, which is four orders of magnitude less than the maximum current-induced rotation angle of the polarization plane observed in tellurium [6].

A few years ago there was published a conference abstract titled as "Current induced optical activity in topological insulator Bi_2Te_2Se" [31]. Unfortunately, this publication does not contain any experimental data on the magnitude of the observed effect. It seems that in terms of the magnitude of the current-induced polarization plane rotation, tellurium is a unique semiconductor. Moreover, tellurium is the only material for which the CISP has been also detected using NMR technique [9, 10]. Namely, a current-induced shift of ^{125}Te NMR spectrum was measured which was attributed to the appearance of a magnetic field that is parallel to the applied current. It was shown that the current-induced magnetization has opposite polarities in dextra- and levorotatory tellurium crystals. For a crystal of given chirality, the magnetization is linear in current, and a reversal of the current leads to a reversal of the magnetization. It is important to mention, that comparison of the current-induced NMR shift [10] with the experimental data on CISP from current-induced optical activity measurements [6] enables one to determine a constant of the hyperfine coupling of ^{125}Te in elemental trigonal tellurium.

Current-induced optical activity can play a significant role in the optical noise spectroscopy. As it was shown in Refs. [32, 33], in addition to the spin noise, in gyrotropic systems, the spectrum of the Faraday and Kerr rotation fluctuations contains information about the electric current fluctuations. In tellurium, for instance, stochastic fluctuations of electric current give rise to a stochastic rotation of the light polarization plane as it passes through a gyrotropic system due to the effect of current-induced optical activity. The microscopic theory of the noise of the Faraday or Kerr rotation effects has been recently developed for bulk tellurium, ensembles of carbon nanotubes, and two-dimensional systems, such as $GaAs$ quantum wells grown along different crystallographic directions [32]. The estimates showed that among the studied systems the effect is the strongest in quantum well structures. It was found for zinc-blende lattice quantum wells grown along the [110] axis, that contribution from current fluctuations dominates near the optical resonance, while contribution from spin fluctuations dominates far away from it [33].

1.5 Conclusion

This review paper summarizes results of the experimental and theoretical investigations of the current-induced optical activity by a number of researchers in 1978–2021. This effect is also designated as the electrical analog of the Faraday effect, or kinetic Faraday effect.

Phenomenologically, the effect is associated with linear terms in the expansion of dielectric tensor components with respect to the electric current. In a gyrotropic uniaxial medium, when an electric current is flowing along the optical axis, it may result in a circular birefringence for the light propagating co-directionally with the electric current. In the case of linearly polarized light, the circular birefringence leads to rotation of the light polarization plane. This is precisely the effect of current-induced optical activity.

So far, p-tellurium is the only material for which comprehensive studies of the current-induced optical activity have been performed. The spectral, temperature and concentration dependences of the effect have been experimentally examined. The microscopic theory of the effect has been developed and ab initio calculations have been carried out. They are in good agreement with the experimental results.

The microscopic origin of the effect in tellurium is associated with the spin-orbit interaction that leads to a very particular kind of the electron wave functions in the topmost valence subband. Namely, the valence band electron spin orientation is locked with its wavevector, hence, the electric current magnitude in the valence band is directly related to the average spin of the holes (and therefore, to the degree of hole spin polarization).

One of the possible applications of the effect is an infrared modulator. Its main advantages are high speed and linear dependence of the modulation magnitude on current [23, 26]. Unlike hot electron infrared modulators, the modulator based on the effect of current-induced optical activity does not distort the signal waveform.

Acknowledgements The author would like to thank H.A. Sarkisyan for useful discussions. The work was supported in part by the Ministry of Science and Higher Education of the Russian Federation (state assignment No. 075-03-2021-050/5).

References

1. Vorob'ev, L.E., Ivchenko, E.L., Pikus, G.E., Farbshtein, I.I., Shalygin, V.A., Shturbin, A.V.: Optical activity in tellurium induced by a current. JETP Lett. **29**(8), 441–445 (1979)
2. Baranova, N.B., Bogdanov, Yu.V., Zel'dovich, B.Ya.: Electrical analog of the Faraday effect and other new optical effects in liquids. Opt. Commun. **22**(2), 243–247 (1977)
3. Ivchenko, E.L., Pikus, G.E.: New photogalvanic effect in gyrotropic crystals. JETP Lett. **27**(11), 604 (1978)
4. Agranovich, V.M., Ginzburg, V.L.: Spatial Dispersion in Crystal Optics and the Theory of Excitons. Wiley, New York (1966)
5. Sturman, B.I., Fridkin, V.M.: The Photovoltaic and Photorefractive Effects in Non-centrosymmetric Materials. Gordon and Breach, New York (1992)
6. Shalygin, V.A., Sofronov, A.N., Vorob'ev, L.E., Farbshtein, I.I.: Current-induced spin polarization of holes in tellurium. Phys. Solid State **54**(12), 2362–2373 (2012)
7. Ivchenko, E.L., Pikus, G.E.: Photogalvanic effects in optically active crystals. Ferro-Electr **43**, 131–136 (1982)
8. Shalygin, V.A., Shturbin, A.V.: Temperature dependence of current-induced optical activity of tellurium. Sov. Phys. Semicond. **15**(11), 1309–1310 (1981)
9. Furukawa, T., Shimokawa, Y., Kobayashi, K., Itou, T.: Observation of current-induced bulk magnetization in elemental tellurium. Nat. Commun. **8**, 954 (2017)
10. Furukawa, T., Watanabe, Y., Ogasawara, N., Kobayashi, K., Itou, T.: Current-induced magnetization caused by crystal chirality in nonmagnetic elemental tellurium. Phys. Rev. Res. **3**(2), 023111 (2021)
11. Tsirkin, S.S., Puente, P.A., Souza, I.: Gyrotropic effects in trigonal tellurium studied from first principles. Phys. Rev. B **97**(3), 035158 (2018)
12. Aronov, A.G., Lyanda-Geller, Y.B.: Nuclear electric resonance and orientation of carrier spins by an electric field. JETP Lett. **50**, 431–434 (1989)
13. Edelstein, V.M.: Spin polarization of conduction electrons induced by electric current in two-dimensional asymmetric electron systems. Solid State Commun. **73**, 233–235 (1990)
14. Golub, L.E., Ivchenko, E.L.: Spin orientation by electric current in (110) quantum wells. Phys. Rev. B **84**, 115303 (2011)
15. Smirnov, D.S., Golub, L.E.: Electrical spin orientation, spin-galvanic, and spin-Hall effects in disordered two-dimensional systems. Phys. Rev. Lett. **118**, 116801 (2017)
16. Kato, Y.K., Myers, R.C., Gossard, A.C., Awschalom, D.D.: Current-induced spin polarization in strained semiconductors. Phys. Rev. Lett. **93**(17), 176601 (2004)
17. Stern, N.P., Ghosh, S., Xiang, G., Zhu, M., Samarth, N., Awschalom, D.D.: Current-induced polarization and the spin Hall Effect at room temperature. Phys. Rev. Lett. **97**(12), 126603 (2006)
18. Koehl, W.F., Wong, M.H., Poblenz, C., Swenson, B., Mishra, U.K., Speck, J.S., Awschalom, D.D.: Current-induced spin polarization in gallium nitride. Appl. Phys. Lett. **95**(7), 072110 (2009)
19. Silov, A.Y., Blajnov, P.A., Wolter, J.H., Hey, R., Ploog, K.H., Averkiev, N.S.: Current-induced spin polarization at a single heterojunction. Appl. Phys. Lett. **85**(24), 5929 (2004)
20. Sih, V., Myers, R.C., Kato, Y.K., Lau, W.H., Gossard, A.C., Awschalom, D.D.: Spatial imaging of the spin Hall effect and current-induced polarization in two-dimensional electron gases. Nat. Phys. **1**, 31–35 (2005)
21. Yang, C.L., He, H.T., Ding, L., Cui, L.J., Zeng, Y.P., Wang, J.N., Ge, W.K.: Spectral dependence of spin photocurrent and current-Induced spin polarization in an In-GaAs/InAlAs two-dimensional electron gas. Phys. Rev. Lett. **96**(18), 186605 (2006)
22. Ivchenko, E.L., Ganichev, S.D.: Spin photogalvanics. In: Dyakonov, M.I. (ed.) Spin Physics in Semiconductors, 2nd edn., pp. 281–328. Springer, Berlin (2018)
23. Shturbin, A.V., Shalygin, V.A., Vorobjev, L.E., Farbstein, I.I., Galetskaya, A.D., Stafeev V.I.: Light modulator. Inventor's certificate No. 776180 (USSR), invention application from 11.12.1978 [in Russian]

24. Ivchenko, E.L., Pikus, G.E.: Natural optical activity of semiconductors (tellurium). Sov. Phys. Solid State **16**(7), 1261 (1975)
25. Averkiev, N.S., Asnin, V.M., Bakun, A.A., Danishevskii, A.M., Ivchenko, E.L., Pikus, G.E., Rogachev, A.A.: Circular photogalvanic effect in tellurium. Sov. Phys. Semicond. **18**(4), 402 (1984)
26. Shalygin, V.A.: Optical and photogalvanic effects in bulk semiconductors and two-dimensional structures, St. Petersburg Polytechnic University (2013) [dissertation in Russian]. https://elib.spbstu.ru/dl/2/4754.pdf/download/4754.pdf. Accessed 30 Nov 2021
27. Müller, K.H., Nimtz, G.: Acoustoelectric domain induced transparency in tellurium at 11 μm. Appl. Opt. **16**(11), 2961–2967 (1977)
28. Fukuda, S., Shiosaki, T., Kawabata, A.: Infrared optical activity in tellurium. Phys. Stat. Sol. (b) **68**(2), K107–K110 (1975)
29. Natory, K.: Band theory of the optical activity of crystals. J. Phys. Soc. Jpn. **39**(4), 1013–1021 (1975)
30. Averkiev, N.S.: On the problem of the temperature dependence of the current-induced optical activity in tellurium. Sov. Phys. Semicond. **18**(4), 450 (1984)
31. Mandal, N., Mitkowski, I., Glazov, M., Chen, Y.: Current induced optical activity in topological insulator Bi_2Te_2Se. In: APS March Meeting 2016 Abstracts, Bulletin of the American Physical Society, vol. 61(2), P29. 011 (2016)
32. Smirnov, D.S., Glazov, M.M.: Stochastic Faraday rotation induced by the electric current fluctuations in nanosystems. Phys. Rev. B **95**(4), 045406 (2017)
33. Smirnov, D.S., Mantsevich, V.N., Glazov, M.M.: Theory of optically detected spin noise in nanosystems. Phys. Usp. **64**(9), 923–946 (2021)

Chapter 2
Optically Pumped Terahertz Radiation Sources Based on Impurity Carrier Transitions in Quantum Wells

Dmitry Firsov, Ivan Makhov, Vadim Panevin, Hayk A. Sarkisyan, and Leonid Vorobjev

Abstract A review of the results of studies of terahertz radiation associated with impurity electron transitions in n-doped $GaAs/AlGaAs$ quantum wells under conditions of interband optical excitation of nonequilibrium charge carriers is presented. The principles of radiation generation and methods of controlling its intensity are described: a decrease in the lifetime of electrons at impurity levels due to stimulated interband radiation and the introduction of a compensating acceptor impurity.

2.1 Introduction

Terahertz (THz) radiation (wavelength 30–300 μm) has a wide range of applications. High penetrating power for dry non-metallic objects allows its use in the field of non-destructive testing and in security systems. In the terahertz frequency range, many organic molecules have characteristic absorption bands, which makes it possible to create systems for the diagnosis of materials. Terahertz radiation is non-ionizing, unlike X-rays, which are used in medical diagnostics. It makes terahertz radiation safer for living organisms. Despite the vast areas of possible application of terahertz radiation, its use is significantly limited due to the difficulties with the creation of affordable compact and efficient sources of terahertz radiation.

To date, the most efficient and compact source of terahertz radiation is a quantum cascade laser, the operation principle of which is based on intersubband electron tran-

D. Firsov (✉) · V. Panevin · H. A. Sarkisyan · L. Vorobjev
Peter the Great St. Petersburg Polytechnic University, St. Petersburg 195251, Russia
e-mail: dmfir@rphf.spbstu.ru

I. Makhov
National Research University Higher School of Economics, St. Petersburg 194100, Russia

H. A. Sarkisyan
Russian-Armenian University, Yerevan 0051, Armenia

D. Blaschke et al. (eds.), *Optics and Its Applications*, Springer Proceedings in Physics 281, https://doi.org/10.1007/978-3-031-11287-4_2

sitions during their vertical transport in nanostructures with tunnel-coupled quantum wells [1]. The most important factor limiting the widespread use of terahertz quantum cascade lasers is the extreme technological complexity of these devices. Therefore, the problem of finding and developing alternative mechanisms and schemes for generating terahertz radiation is urgent.

In n-type semiconductors, under conditions of optical interband excitation of nonequilibrium charge carriers, terahertz radiation can arise when nonequilibrium electrons are captured from the conduction band to donor impurity states, the depopulation of which occurs due to spontaneous electron-hole recombination during electron transitions from the ground donor state to the valence band. Terahertz radiation of this type was previously observed in bulk semiconductors [2–6]. Placing an impurity center in a quantum well opens up the possibility of controlling its energy spectrum due to the size quantization effect by changing the parameters of doped quantum wells and the possibility of controlling the frequency of impurity terahertz luminescence.

In this paper, we review the results of studying the terahertz impurity luminescence under the conditions of interband photoexcitation of nonequilibrium charge carriers in nanostructures with doped $GaAs/AlGaAs$ quantum wells. An increase in the intensity of the observed radiation can be obtained due to the effective depopulation of the ground donor state. To this end, two approaches were used: the fast depopulation of the ground impurity states by stimulated near-infrared (NIR) emission and a substantial compensation of impurities in quantum wells. In the latter case, the compensation of impurities leads to the appearance of an additional recombination channel of the "donor-acceptor" type, and also decreases the equilibrium population of impurity states.

2.2 Terahertz Photoluminescence Under Interband Photoexcitation of Quantum Wells Doped with Shallow Donors

2.2.1 Mechanism of Terahertz Emission

Let us consider the mechanism of impurity terahertz photoluminescence in quantum wells doped with shallow donors under interband optical excitation. A diagram illustrating optical transitions of charge carriers in doped quantum wells under interband photoexcitation is shown in Fig. 2.1. At a low temperature of the crystal lattice, the donor is neutral, i.e. in equilibrium, an electron is located in the ground state of the donor $1s$. Under interband optical pumping, depending on the photon energy of the exciting radiation, the generation of electron-hole pairs will occur either directly in the quantum wells or also in the barrier layers of the structure (this optical transition is marked with a solid arrow labeled "optical pumping" in Fig. 2.1). Then, nonequilibrium photoexcited electrons and holes, being thermalized, will descend to the lower subbands of the size quantization of electrons $e1$ and heavy holes $hh1$. After this,

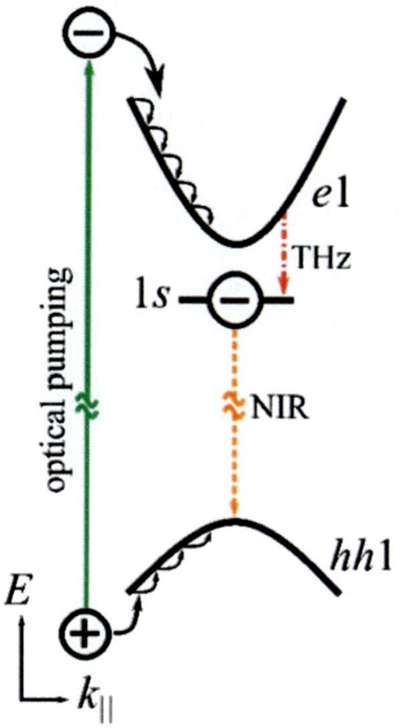

Fig. 2.1 Diagram of optical transitions of charge carriers in donor-doped quantum wells under interband optical excitation

the recombination of an electron in the ground state $1s$ of the donor impurity with a nonequilibrium hole from the $hh1$ subband is possible, accompanied usually by spontaneous emission of a photon corresponding to the near-infrared range. After the $1s$ donor ground state is depopulated, a nonequilibrium electron from the first electron subband $e1$ can be captured by an ionized donor with a spontaneous emission of a terahertz photon.

This section presents the results of the first studies of low-temperature impurity terahertz photoluminescence in nanostructures with $GaAs/AlGaAs$ quantum wells doped with shallow donors under conditions of interband optical excitation of nonequilibrium charge carriers.

2.2.2 *Experimental Samples and Techniques*

Investigations of impurity terahertz photoluminescence were performed on samples with quantum wells of various widths. Here, we present the results for a nanostructure containing 50 periods of $GaAs$ quantum wells of 30 nm, separated by tunnel-nontransparent $Al_{0.3}Ga_{0.7}As$ barriers with a thickness of 7 nm. Each quantum well of the nanostructure was doped with silicon in a narrow layer 4 nm wide, shifted from the center of the quantum well by 6 nm. The surface concentration of the dopant was

$3 \cdot 10^{10}$ cm^{-2}. The nanostructure had a 20 nm thick $GaAs$ cover layer doped with donors with a volume concentration of $5 \cdot 10^{17}$ cm^{-3}.

For measurements in the temperature range from 4 to 320 K, a Janis PTCM-4-7 optical closed-cycle cryostat was used. Interband photoexcitation of nonequilibrium charge carriers was carried out by modulated radiation of a continuous-wave solid-state Nd:YAG laser with frequency doubling in a lithium iodate crystal ($LiIO_3$) with a wavelength of 532 nm.

The terahertz photoluminescence emission of the sample was collected by an off-axis parabolic mirror of a Bruker Vertex 80v vacuum infrared Fourier spectrometer operating in the step-scan mode. In order to prevent the scattered pumping radiation from entering the measuring channel of the setup, a black polyethylene filter 100μm thick was installed in front of the entrance window of the Fourier spectrometer. An Infrared Laboratories, Inc., liquid helium-cooled silicon bolometer was used as a terahertz radiation detector. The photoresponse signal from the bolometer preamplifier was measured with a phase-sensitive detector lock-in amplifier SR-830 at the modulation frequency of the pumping radiation. The measured signal interferogram was converted into a spectrum using the OPUS software.

We also studied photoluminescence spectra in the near-infrared frequency range using an experimental setup based on a Horiba JobinYvon FHR 640 monochromator with a holographic grating consisted of 1200 groves per mm and a liquid nitrogen cooled silicon CCD matrix.

2.2.3 *Experimental Results*

When the structure is illuminated, optical generation of electron-hole pairs can occur not only in low-dimensional layers, but also in a semi-insulating GaAs substrate, which can contain residual impurities. In addition, nonequilibrium photoexcited charge carriers can reach the substrate due to diffusion from the active layers of the nanostructure, since the total thickness of the active layers of the nanostructure is comparable with the diffusion lengths of charge carriers in $GaAs$ [7]. As a result, two bands were found in the terahertz photoluminescence spectrum of the nanostructure, one of which is associated with impurities in the quantum wells. The second band, which was also observed in the sample, which is a substrate without epitaxial layers, is associated with optical transitions of charge carriers in the semi-insulating substrate of the structure. Note that the first band of terahertz photoluminescence cannot be caused by intersubband optical transitions of charge carriers in quantum wells. This is due to the fact that, in accordance with the selection rules, such optical transitions are allowed only for light polarized along the growth axis of the structure, while in the geometry of the experiment performed, radiation polarized in the plane of quantum wells was recorded.

The terahertz photoluminescence spectra of the nanostructure with quantum wells for various lattice temperatures are shown in Fig. 2.2 [11, 12].

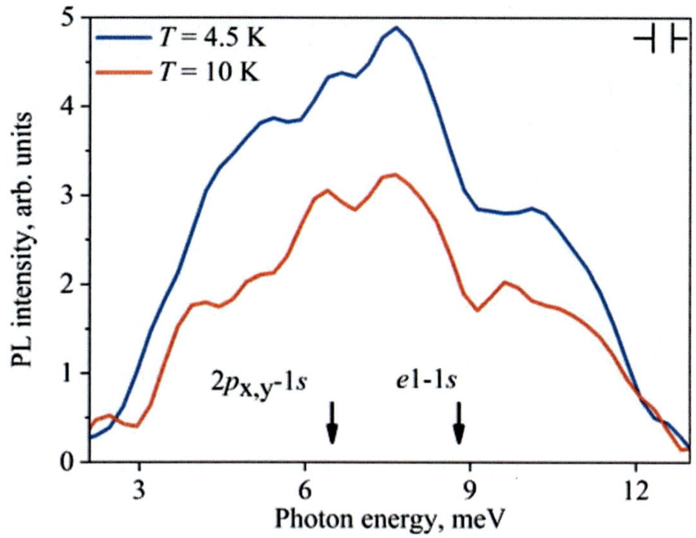

Fig. 2.2 Terahertz photoluminescence spectra of 30 nm $GaAs/AlGaAs$ quantum wells at various temperatures of the crystal lattice

In accordance with the calculation of the energy spectrum of shallow donors in $GaAs/Al_{0.3}Ga_{0.7}As$ QWs 30 nm wide, the binding energy of the donor impurity is about 9 meV [8]. Since the spectral position of the terahertz photoluminescence band of the nanostructure is close to the calculated value of the donor binding energy in quantum wells, the observed photoluminescence can be associated with optical transitions of nonequilibrium electrons with the participation of shallow donor states.

The observed radiation can be associated with several optical transitions of electrons. Firstly, this band can be caused by intracenter optical transitions of electrons between the excited $2p_{x,y}$ and ground $1s$ donor states. The calculated value of the transition energy is about 6.6 meV and is marked with an arrow $2p_{x,y}$-$1s$ in Fig. 2.2. Secondly, this terahertz photoluminescence band can be associated with optical transitions of electrons from the first electron subband $e1$ to the $1s$ ground state of the donor impurity. The calculated value of the optical transition energy $e1 - 1s$ is 8.8 meV and is indicated by the arrow $e1 - 1s$ in Fig. 2.2. The significant width of the spectrum is caused by the broadening of impurity states due to the high concentration of donor impurity centers [9], as well as by the broadening of the donor binding energy due to the finite width of the doping region of quantum wells. Narrow lines of impurity terahertz luminescence are observed in semiconductors with an impurity concentration much lower than in our case [10].

An increase in the lattice temperature from 4.5 K to 10 K leads to a 1.5-fold decrease in the integral luminescence intensity. Such quenching of impurity terahertz photoluminescence with increasing temperature has already been observed for the case of doped bulk semiconductors [3] and was attributed to a decrease in the probability of capture of nonequilibrium charge carriers by ionized donor centers with increasing lattice temperature. In addition, the decrease in the intensity of impurity terahertz photoluminescence with an increase in the lattice temperature can be caused by the expansion of the distribution function of charge carriers in the QW subband with temperature.

The photoluminescence spectra of the near-infrared range were also investigated. In accordance with the mechanism of the observed terahertz impurity luminescence, these spectra exhibit features associated with radiative electron-hole recombination between the ground donor state and the first hole subband of the quantum well.

Similar measurements were performed for a nanostructure with narrower quantum wells 16.1 nm wide [13]. The calculated value of the binding energy of the donor impurity of silicon in such quantum wells is about 10 meV, which led to a short-wavelength shift of the impurity terahertz photoluminescence band relative to the impurity luminescence band of wide quantum wells and to the corresponding modification of the photoluminescence spectra of the near-infrared range.

2.3 Influence of Stimulated Near-Infrared Radiation on Terahertz Photoluminescence

2.3.1 Introduction

According to the proposed mechanism of terahertz impurity photoluminescence, the depopulation of the donor ground state occurs due to spontaneous radiative or nonradiative recombination during transitions of charge carriers between the donor ground state and the states of the first hole subband. There are a number of approaches that make it possible to increase the efficiency of impurity terahertz photoluminescence in doped quantum wells. One of them is an increase in the rate of depletion of the final state for the impurity terahertz transition of an electron (i.e., the ground donor state), since the capture of nonequilibrium electrons to the ground impurity state, accompanied by the emission of photon in the terahertz range, in particular, is determined by the population of the ground donor state. An increase in the rate of depopulation of the ground donor state can be achieved due to the organization of stimulated emission in the near-infrared range in the same nanostructure, namely, the organization of stimulated optical transitions of charge carriers from the ground state of the donor to the hole subband.

A similar mechanism has already been demonstrated earlier in [14, 15] when observing mid-infrared radiation from laser diode structures with vertically coupled self-organized $InGaAs/AlGaAs$ quantum dots and $InGaAs/GaAs$ quantum wells. Laser generation, realized in the near-infrared range of the spectrum in structures with quantum dots during transitions of charge carriers between the lower electron and hole levels, led to an increase in the intensity of spontaneous intraband transitions of charge carriers between the levels of quantum dots, accompanied by the emission of photons in the mid-infrared range. At the same time, in the absence of lasing in the near-infrared range of the spectrum, spontaneous emission of the mid-infrared region was not observed at all in structures with quantum dots. In structures with quantum wells, spontaneous emission of mid-infrared radiation associated with intersubband transitions of nonequilibrium charge carriers in the wells was also observed, but it did not have a threshold character as in structures with quantum dots.

This section presents the results of studies of the spectra of spontaneous terahertz photoluminescence in laser nanostructures with doped quantum wells with a waveguide for near-infrared radiation. When lasing in the near-infrared range occurs in such a nanostructure, the rate of depopulation of the ground donor state increases due to stimulated transitions from the ground donor state to the first hole subband, which in turn should be reflected in an increase in the intensity of terahertz radiation when electrons are captured from the first electron subband and excited donor states to the ground donor state.

2.3.2 Experimental Samples and Techniques

We studied a nanostructure with doped quantum wells with waveguide layers, as well as a semi-insulating $GaAs$ substrate, which did not contain epitaxial layers. A sample of a nanostructure with quantum wells was grown by molecular beam epitaxy on a semi-insulating $GaAs$ substrate. The quantum wells were formed by 7.6 nm thick $GaAs$ layers separated by 5 nm thick $Al_{0.3}Ga_{0.7}As$ barriers. Doping of quantum wells with a surface concentration of $5 \cdot 10^{10}$ cm^{-2} was carried out with silicon in the central region of each quantum well with a width of 2.6 nm. In total, the nanostructure contained 10 quantum wells, which were placed in a symmetric gradient waveguide for near-infrared radiation, formed by wide-gap gradient $Al_xGa_{1-x}As$ layers (composition x varied from 0.4 to 0.6) 0.6 μm thick. The nanostructure had a 10 nm thick $GaAs$ cover layer doped with silicon with a volume concentration of $5 \cdot 10^{17}$ cm^{-3}.

To get lasing in the near-infrared range, the $GaAs$ semi-insulating substrate was grinded down to a thickness of about 100 μm, and then high-Q cavity of total internal reflection with sides of about $600 \cdot 600$ μm^2 was cleaved out of it.

During the experiment, the samples were placed on a copper holder in a Janis PTCM-4-7 closed cycle optical cryostat, which made it possible to change the sample temperature in the range of $4 - 320$ K. The photoluminescence studies of both terahertz and near-infrared ranges were carried out simultaneously. The pump radiation was directed to the sample in a cryostat using a system of mirrors and a spherical lens. In this case, the lens was installed in a defocused position to illuminate the entire surface of the laser nanostructure resonator. Terahertz photoluminescence radiation was collected by an off-axis parabolic mirror of a Fourier spectrometer from the sample surface. Spontaneous and stimulated near-infrared radiation was collected from the facet of the nanostructure cavity using a system of lenses and mirrors and directed to the entrance slit of the grating monochromator.

To achieve lasing in the near-infrared spectral range, high-power optical excitation of the samples was carried out by a pulsed solid-state Nd:YAG laser with frequency doubling in a $LiIO_3$ crystal. The pumping radiation parameters are as follows: the radiation wavelength is 532 nm, the pulse duration is 250 ns, and the repetition rate is 8 kHz. The intensity of the polarized pumping laser radiation was varied using an adjustable attenuator consisting of a combination of a half-wave plate and a Glan-Taylor prism, and was additionally modulated by a mechanical chopper at a frequency of 87 Hz. The need for additional modulation of the pulsed pump radiation at a low frequency is caused by the limited bandwidth of the bolometer, which acts as a photodetector of terahertz radiation. The photoresponse signal of bolometer was measured with a phase-sensitive lock-in amplifier SR-830 at a frequency of 87 Hz.

Except for the details described above, the technique for studying the terahertz and near-infrared photoluminescence spectra of samples does not differ from the technique described in Sect. 2.2.2.

2.3.3 Experimental Results, Their Analysis and Discussion

The spectral dependence of the terahertz photoluminescence intensity measured at a low pulsed optical excitation power of 40 W/cm^2 for a nanostructure with doped quantum wells and a waveguide for near-infrared radiation, prepared in the form of a total internal reflection cavity, is shown in Fig. 2.3 [21]. In this case, in the spectra of spontaneous near-infrared photoluminescence, an emission band is observed, associated with radiative electron-hole recombination through donor states. As can be seen, in the terahertz photoluminescence spectrum, one emission band is observed with a maximum intensity near the photon energy of 20 meV. Note, that, under the same conditions, terahertz radiation from a semi-insulating $GaAs$ substrate was not observed.

In accordance with the calculation of the energy spectrum of donor states [16], the binding energy of a donor impurity located in the center of a $GaAs/Al_{0.3}Ga_{0.7}As$ quantum well with a width of 7.6 nm is approximately 13.5 meV. Therefore, the observed band of terahertz radiation with a maximum intensity near the photon energy of 20 meV (see Fig. 2.3) is associated with optical transitions of nonequilibrium electrons from the first electron subband $e1$ to the ground state of the donor impurity $1s$.

A significant shift of the spectral position of the intensity maximum of the terahertz photoluminescence band from the binding energy of the donor impurity in the quantum well can be explained as follows. It is known that the wave functions of electrons on donor impurity states are formed mainly by the wave functions of the conduction band. Therefore, optical transitions of electrons between the $1s$ donor ground state and the first electron subband e1 are forbidden in the standard dipole approximation, since the wave functions of the initial and final states in this case have the same symmetry. However, this condition is strictly fulfilled only for the zero value of the wave vector of the electron. Theoretical calculations of the spec-

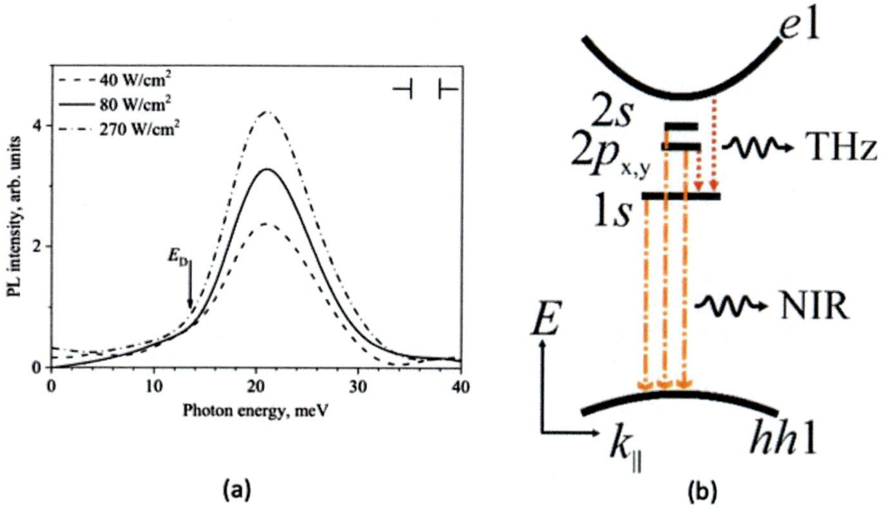

Fig. 2.3 Terahertz photoluminescence spectra of a nanostructure with doped quantum wells, measured at a temperature $T = 5$ K and various optical excitation powers (**a**); Diagram of impurity optical transitions of charge carriers in quantum wells. Dotted and dash-dotted arrows show optical transitions in the terahertz and near-infrared ranges of the spectrum, respectively (**b**)

tral dependence of the photoionization cross section (which is proportional to the photodeionization cross section [17]) of shallow donors in quantum wells performed in [18] showed that the maximum value of the donor photoionization cross section in quantum wells is located at photon energies exceeding the binding energy of the donor impurity. Thus, the maximum of the photodeionization cross section, which determines the shape of the spectrum of impurity photoluminescence associated with the $e1$-$1s$ transitions, should be shifted to the short-wavelength region of the spectrum relative to the binding energy of the donor impurity, which agrees with the observed spectrum of terahertz photoluminescence (see Fig. 2.3a).

The next step was to increase the optical excitation power of the nanostructure, which should lead to the appearance of stimulated near-infrared radiation during donor impurity transitions of charge carriers. This, in turn, should lead to a more efficient depopulation of the ground donor state, which is the final state for terahertz electron transitions. Figure 2.4 shows the dependence of the intensity of the near-infrared radiation, integrated over the entire photoluminescence spectrum, on the optical excitation power. This dependence has a threshold character, which indicates the appearance of lasing in the near-infrared range of the spectrum. The threshold value of the pumping power for the appearance of lasing is about 120 W/cm^2.

In the near-infrared photoluminescence spectra, when the pumping power reaches 120 W/cm^2, a high-intensity narrow emission line appears against the background of a broad photoluminescence spectrum. Based on the results of calculating the energy spectrum of donor states in quantum wells, the arising stimulated emission line is

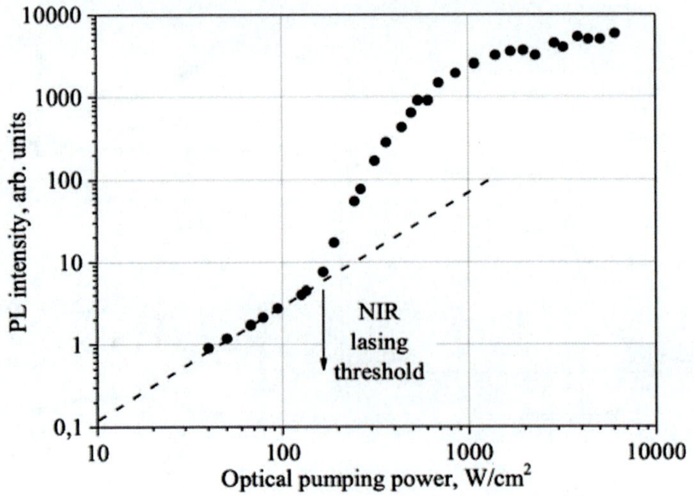

Fig. 2.4 Dependence of the integrated intensity of the near-infrared photoluminescence of a nanostructure with doped quantum wells on the optical excitation power

associated with the radiative electron-hole recombination between the excited state of the donor $2s$ and the first hole subband $hh1$ (see Fig. 2.3 b).

Lasing in the near-infrared range initially occurs during optical transitions of charge carriers with the participation of an excited rather than a ground donor state. This may be due to the different gain for optical transitions with the participation of the excited and ground donor states.

The dependence of the integrated intensity of terahertz radiation on the power of optical excitation of the nanostructure is shown in Fig. 2.5.

It can be seen from this dependence that at low excitation levels, i.e. at pumping powers not exceeding the near-infrared lasing threshold (the threshold value is indicated by the arrow $2s$-$hh1$ in Fig. 2.5), the integral intensity of the impurity terahertz photoluminescence is proportional to the square root of the optical pumping power (the square root dependence is shown by the dashed line in Fig. 2.5). As follows from the terahertz photoluminescence spectra measured at low pumping powers (see Fig. 2.3a), an increase in the terahertz photoluminescence intensity in this range of photoexcitation powers is associated with an increase in the intensity of the $e1$-$1s$ band. This dependence of the terahertz impurity photoluminescence intensity on the pump power is in a good agreement with the theoretical and experimental results for terahertz impurity radiation upon interband optical pumping of bulk semiconductors [4].

When the pumping power corresponding to the onset of near-infrared lasing at $2s - hh1$ transitions is reached, the rate of increase in the terahertz photoluminescence intensity associated with optical transitions $e1$-$1s$ begins to decrease (see Fig. 2.5). The intensity of such terahertz photoluminescence I_{THz}^{e1}, associated with

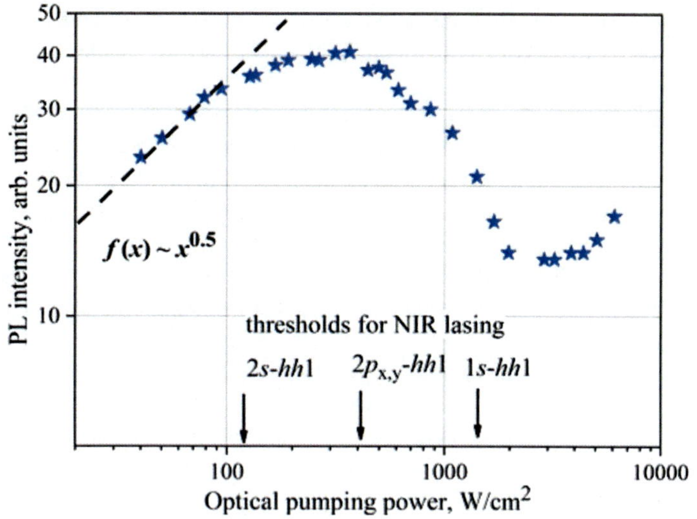

Fig. 2.5 Dependence of the integrated intensity of terahertz photoluminescence of a nanostructure with doped quantum wells on the level of interband optical excitation, measured at a temperature of $T = 5$ K

electron transitions $e1$-$1s$, in the first approximation depends on the concentration of electrons $n\,(e1)$ in the $n^+\,(1s)$ subband and the concentration of free donor states $1s$:

$$I_{THz}^{e1} \propto n\,(e1) \cdot n^+\,(1s)\,. \tag{2.1}$$

Without taking into account the compensation of impurities, the concentration $n^+(1s)$ of free donor $1s$ states can be determined as follows:

$$n^+\,(1s) = N_D - n\,(1s) - n\left(2p_{x,y}\right) - n\,(2s)\,, \tag{2.2}$$

where N_D is the total donor concentration in quantum wells, and $n\,(1s)$, $n\left(2p_{x,y}\right)$, and $n\,(2s)$ are the quasi-equilibrium electron concentrations in donor states $1s$, $2p_{x,y}$ and $2s$, respectively. The presence of the last two terms in expression (2.1) is caused by the fact that the considered donor impurity centers are singly charged, i.e. one donor impurity center cannot capture more than one electron [19]. The concentrations $n\,(e1)$ and $n^+\,(1s)$ themselves are determined by the rates of capture of electrons from the first electron subband $e1$ to excited and ground donor states and the rates of recombination of electrons from donor states with holes from the valence subbands. Probably, the arising stimulated emission line caused by stimulated impurity transitions $2s$-$hh1$ leads to a faster depopulation of the excited donor state $2s$ in the QW compared to depletion due to spontaneous transitions, which leads to the corresponding more efficient nonradiative capture of electrons from the first electron subband $e1$ into an excited donor state $2s$. It follows that, with the onset of stimulated $2s$-$hh1$ transitions, the rate of increase in the electron concentration in the subband

n ($e1$) with increasing pumping power can decrease, as well as a change in the concentration n^+ ($1s$) of free donor states $1s$ capable of capturing an electron from the $e1$ subband.

A subsequent increase in the optical excitation power to a level of 400 W/cm^2 leads to a decrease in the integrated intensity of the impurity terahertz photoluminescence with an increase in the level of optical excitation of the nanostructure (see Fig. 2.5). As follows from the corresponding terahertz photoluminescence spectra measured at photoexcitation powers exceeding 270 W/cm^2, the decrease in the integrated terahertz photoluminescence intensity is accompanied by a decrease in the intensity of the terahertz luminescence band associated with optical transitions of electrons $e1$-$1s$.

The decrease in the integrated intensity of impurity terahertz photoluminescence at pumping powers above 400 W/cm^2 is accompanied by the appearance of an additional line of impurity stimulated near-infrared emission, observed in the corresponding spectra. The appearance of the second line of impurity stimulated emission is associated with stimulated transitions of charge carriers $2p_{x,y}$-$hh1$. Such transitions probably lead to a faster nonradiative capture of nonequilibrium electrons from the first electron subband $e1$ to excited donor states $2p_{x,y}$, which in turn leads to a decrease in the intensity of impurity terahertz electron transitions $e1$-$1s$. This is due to a change in the concentration of electrons in the first electron subband and free ground donor states included in expression (2.1), by analogy with the effect of stimulated transitions $2s$-$hh1$ on the intensity of terahertz photoluminescence at transitions $e1$-$1s$.

A further increase in the optical excitation power above 1 kW/cm^2 leads to the appearance of a third line of stimulated near-infrared radiation, which is associated with radiative recombination of electrons from the ground donor state with nonequilibrium holes from the first subband of heavy holes $hh1$.

It is the onset of stimulated optical transitions $1s$-$hh1$ that should lead to the effective depopulation of the $1s$ donor ground state and, accordingly, to an increase in the intensity of impurity terahertz photoluminescence. Indeed, the dependence of the integral intensity of the impurity terahertz photoluminescence on the optical excitation power shows an increase in the terahertz photoluminescence intensity with increasing pumping at photoexcitation powers exceeding 2 kW/cm^2 (see Fig. 2.5), i.e. exceeding the threshold for lasing at impurity transitions $1s$-$hh1$.

It is noteworthy that in the terahertz photoluminescence spectra, the appearance of stimulated emission in the near-infrared range during the $1s$-$hh1$ transitions leads to the appearance of a new terahertz radiation line near the photon energy of 9 meV. In accordance with the results of calculating the energy spectrum of donor impurity states in our quantum wells [16], a new terahertz radiation line at a photon energy of 9 meV is associated with intracenter optical transitions of nonequilibrium electrons between the excited $2p_{x,y}$ and ground $1s$ donor states in quantum wells. Probably, at pump powers exceeding 2 kW/cm^2, the ground donor states $1s$ begin to depopulate by near-infrared stimulated radiation faster than excited states $2p_{x,y}$, which leads to the appearance of radiative transitions $2p_{x,y}$-$1s$, accompanied by the emission of terahertz photons.

2.4 Terahertz Photoluminescence in Compensated Quantum Wells

2.4.1 Introduction

There is another way to increase the intensity of terahertz photoluminescence associated with impurity transitions of electrons in quantum wells. The intensity of the impurity terahertz photoluminescence depends on the population of the donor ground state, which is final for terahertz electron transitions. In this section, we consider the possibility of increasing the intensity of impurity terahertz radiation due to additional doping of quantum wells with acceptors, i.e., due to compensation of donors with acceptors. Firstly, such compensation of donor impurities by acceptor impurities will significantly lower the equilibrium population of donor states. Secondly, an additional recombination channel will arise for electrons from the $1s$ ground donor state due to recombination of the donor-acceptor type. The above factors should increase the intensity of impurity terahertz photoluminescence for compensated quantum wells as compared to uncompensated ones. In compensated quantum wells, emission bands associated with donor-acceptor recombination and recombination of the "first electron subband-acceptor" type can appear in the photoluminescence spectra of the near-infrared range. In the terahertz photoluminescence spectra, an emission band associated with the capture of holes from the first hole subband into acceptor states can also be detected.

2.4.2 Experimental Samples and Techniques

On semi-insulating $GaAs$ substrates, two structures were grown, each containing 50 periods of quantum wells formed by $GaAs$ layers 7.6 nm wide and separated by 5 nm wide $Al_{0.3}Ga_{0.7}As$ barriers. Quantum wells in the first nanostructure were doped only with a donor impurity (Si) with a surface concentration of $3 \cdot 10^{10}$ cm^{-2}. The quantum wells in the second nanostructure, in addition to the same donor impurity, contained a compensating acceptor (Be) impurity with the same concentration. Doping was carried out into the central region of the quantum wells 2.6 nm wide. Both nanostructures had a 5 nm thick $GaAs$ cover layer doped with silicon with a volume concentration of $5 \cdot 10^{17}$ cm^{-3}.

The technique for studying the photoluminescence of the near-infrared and terahertz spectral ranges in the described nanostructures is completely analogous to the technique described in Sect. 2.2.2.

In order to compare the terahertz photoluminescence intensities of two nanostructures, the dependences of the integrated terahertz photoluminescence intensity on the optical excitation power were measured. For these studies, a Ge:Ga photoresistor sensitive to terahertz radiation (sensitivity range $8 - 30$ meV) was installed opposite to the surface of the sample placed on a copper holder in a closed-cycle

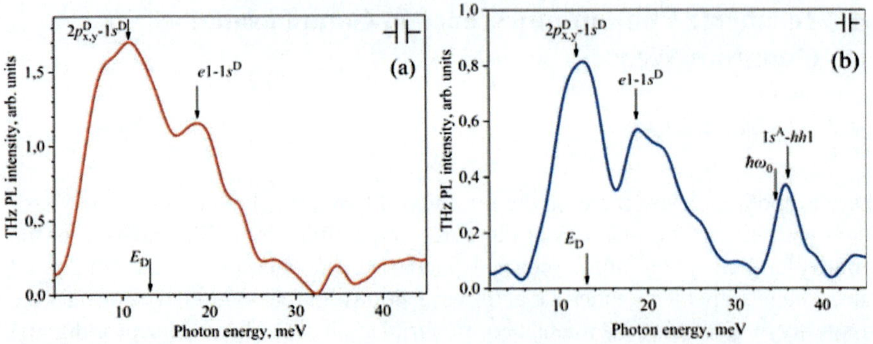

Fig. 2.6 Terahertz photoluminescence spectra of nanostructures with quantum wells doped with donors (**a**) and donors and acceptors (**b**), measured at a temperature of $T = 8$ K. The scales of the vertical axes in figures (**a**) and (**b**) differ

cryostat. The distance between the sample and the $Ge{:}Ga$ photodetector was about 12 mm. Thus, upon optical pumping of the sample, the integrated intensity of terahertz radiation from the sample surface was recorded by a $Ge{:}Ga$ photodetector. In order to prevent penetration of the reflected and scattered pumping radiation, as well as the near-infrared radiation of the photoluminescence of the sample, to the $Ge{:}Ga$ photodetector, cold filters made of black polyethylene (approximately 100 μm thick) and high-resistive germanium compensated with antimony and gold (1 mm thick) were installed at the input of the photodetector.

2.4.3 Experimental Results, Their Analysis and Discussion

In the spectra of near-infrared photoluminescence for both nanostructures with doped quantum wells, an emission band appears, associated with optical transitions of charge carriers through donor impurity states [25]. In addition, for a nanostructure with compensated quantum wells, the spectrum contains emission bands associated with impurity optical transitions of charge carriers between the first electron subband and acceptor states, as well as between the states of donors and acceptors in quantum wells.

The terahertz photoluminescence spectrum of a nanostructure with $GaAs/AlGaAs$ quantum wells doped only by donors is shown in Fig. 2.6a.

In the photoluminescence spectrum, two emission bands are observed with intensity maxima near the photon energies of 10 and 21 meV. No terahertz photoluminescence signal from the semi-insulating $GaAs$ substrate was observed under similar experimental conditions; therefore, the observed emission bands are caused by optical transitions of nonequilibrium charge carriers in low-dimensional doped layers. The detected bands of terahertz radiation cannot also be caused by intersubband transitions of charge carriers in quantum wells, since they are forbidden for radiation polarized in the plane of quantum wells (it is this polarization of radiation that was used in experiments).

In accordance with the results of calculating the energy spectrum of the states of shallow donors in quantum wells $GaAs/Al_{0.3}Ga_{0.7}As$ [16], the binding energy of the ground donor state in quantum wells 7.6 nm wide is about 13.5 meV (this value is indicated by the arrow E_D in Fig. 2.6 a). Thus, the terahertz photoluminescence band with a maximum intensity near the quantum energy of 21 meV can be associated with optical transitions of nonequilibrium electrons from the first electronic subband $e1$ to the ground donor state $1s$. The corresponding luminescence band is indicated by the arrow $e1$-$1s^D$ in Fig. 2.6a.

The second band of terahertz photoluminescence with a maximum intensity near a quantum energy of 10 meV is probably associated with intracenter optical transitions of electrons between the excited $2p_{x,y}$ and ground $1s$ states of donors in quantum wells. This luminescence band is marked with an arrow $2p^D_{x,y}$-$1s^D$ in Fig. 2.6a. The spectral position of the observed band caused by intracenter transitions of electrons is in a good agreement with the results of calculating the energy spectrum of shallow donors in quantum wells from [16], according to which the energy gap between the ground $1s$ and excited $2p_{x,y}$ donor states is about 10 meV.

In the terahertz photoluminescence spectra of a nanostructure with quantum wells doped with both donors and acceptors, emission bands associated with optical transitions of charge carriers with the participation of both donor and acceptor impurity states should appear. Such a spectrum is shown in Fig. 2.6 b on the same line for the crystal lattice temperature of the nanostructure $T = 8$ K. Since, in the $6 - 28$ meV photon energy range, the emission spectrum of this nanostructure exhibits terahertz luminescence bands similar to those found for a nanostructure with quantum wells doped only by donors (see Fig. 2.6a), then emission in this spectral range is also associated with optical transitions of nonequilibrium photoexcited electrons from the first electron subband $e1$ and excited donor states $2p_{x,y}$ to the ground donor state $1s$. These bands of impurity terahertz photoluminescence are marked with arrows $2p^D_{x,y}$-$1s^D$ and $e1$-$1s^D$ in Fig. 2.6b.

The terahertz photoluminescence band, located near the photon energy of 36 meV, is observed only for a nanostructure with quantum wells, in which donors are compensated by acceptors, which makes it possible to associate this terahertz radiation band with the presence of acceptor impurities in the quantum wells. The binding energy of a beryllium acceptor located in the center of a $GaAs/Al_{0.3}Ga_{0.7}As$ quantum well 7.6 nm wide is about 35 meV [22]. Therefore, the terahertz photoluminescence band near a photon energy of 36 meV is associated with optical transitions of nonequilibrium electrons between the ground acceptor state and the first subband of size quantization of heavy holes $hh1$. This photoluminescence band is marked with an arrow $1s^A$-$hh1$ in Fig. 2.6b on the same line. It should be noted that the true shape of the terahertz photoluminescence band associated with the presence of acceptor impurity states in quantum wells may differ from that presented. This is caused by the possible influence of lattice absorption in the $GaAs$ layers, since the energy of the transverse optical phonon in $GaAs$ is about 33.8 meV [23], which is very close to the binding energy of the beryllium impurity in quantum wells. The energy of a transverse optical phonon is marked with an arrow $\hbar\omega_0$ in Fig. 2.6b.

Next, we compared the intensities of terahertz photoluminescence associated with donor impurity transitions of nonequilibrium electrons in quantum wells of both types. Comparative studies of the integrated intensity of terahertz photoluminescence for two nanostructures were performed using $Ge{:}Ga$ photoresistor, which is sensitive to terahertz radiation in the 8–30 meV photon energy range [24] and does not detect short-wavelength radiation associated with optical transitions with participation of acceptors.

The dependences of the integrated terahertz photoluminescence intensity on the optical pumping power in both studied nanostructures, measured in a wide range of photoexcitation powers at a grating temperature $T = 8$ K look similar. However, a nanostructure in which donors in quantum wells are compensated by acceptors demonstrates an order of magnitude greater integrated intensity of terahertz photoluminescence associated with donor impurity electron transitions, compared to a nanostructure in which shallow donors in quantum wells are not compensated. This is observed in the sufficiently wide range of photoexcitation powers.

2.5 Conclusion

This paper presents the results of experimental studies of low-temperature terahertz photoluminescence associated with impurity transitions of nonequilibrium charge carriers in quantum wells under interband optical excitation.

The spectral dependences of the intensity of spontaneous terahertz photoluminescence under interband optical pumping of nanostructures with donor-doped $GaAs/AlGaAs$ quantum wells of various widths are studied for the first time. Two approaches have been implemented to increase the efficiency of terahertz impurity luminescence:

- Increase in the rate of depletion of the ground donor state due to stimulated radiation in the near infrared range, implemented in the same nanostructure.
- Compensation of donor with acceptors, which leads to a decrease in the equilibrium population of donor states and to an additional depopulation of the donor ground state due to electron transitions from the donor ground state to acceptor states.

Acknowledgements Financial support from the Ministry of Science and Higher Education of the Russian Federation (state assignment) is gratefully acknowledged. IM also acknowledges a support from the Basic Research Program of the National Research University Higher School of Economics.

References

1. Köhler, R., Tredicucci, A., Beltram, F., Beere, H.E., Linfield, E.H., Davies, A.G., Ritchie, D.A., Iotti, R.C., Rossi, F.: Terahertz semiconductor-heterostructure laser. Nature **417**, 156–159 (2002)
2. Andrianov, A.V., Zakhar'in, A.O., Ivanov, Y.L., Kipa, M.S.: Terahertz impurity luminescence under the interband photoexcitation of semiconductors. JETP Lett. **91**(2), 96–99 (2010)
3. Zakhar'in, A.O., Andrianov, A.V., Egorov, AYu., Zinov'ev, N.N.: Terahertz photoluminescence from GaAs doped with shallow donors at interband excitation. Appl. Phys. Lett. **96**(21), 211118 (2010)
4. Zakhar'in, A.O., Bobylev, A.V., Andrianov, A.V.: Terahertz emission upon the interband excitation of GaN layers. Semiconductors **46**(9), 1135–1139 (2012)
5. Andrianov, A.V., Zakhar'in, A.O., Zhukavin, R.K., Shastin, V.N., Abrosimov, N.V., Bobylev, A.V.: Terahertz intracenter photoluminescence of silicon with lithium at interband excitation. JETP Lett. **100**(12), 771–775 (2015)
6. Andrianov, A.V., Zakhar'in, A.O., Petrov, A.G.: Intraexciton and intracenter terahertz radiation from doped silicon under interband photoexcitation. JETP Lett. **107**(9), 540–543 (2018)
7. Casey, H.C., Miller, B.I., Pinkas, I.: Variation of minority-carrier diffusion length with carrier concentration in GaAs liquid-phase epitaxial layers. J. Appl. Phys. **44**(3), 1281–1287 (1973)
8. Firsov, D.A., Shalygin, V.A., Panevin, V.Yu., Melentyev, G.A., Sofronov, A.N., Vorobjev, L.E., Andrianov, A.V., Zakhar'in, A.O., Mikhrin, V.S., Vasil'ev, A.P., Zhukov, A.E., Gavrilenko, L.V., Gavrilenko, V.I., Antonov, A.V., Aleshkin, V.Ya.: Terahertz emission and photoconductivity in n-type GaAs/AlGaAs quantum wells: the role of resonant impurity states. Semiconductors **44**(11), 1394–1397 (2010)
9. Kogan, Sh.M., Lifshits, T.M.: Photoelectric spectroscopy—a new method of analysis of impurities in semiconductors. Phys. Status Solidi A **39**(1), 11–39 (1977)
10. Melngailis, I., Stillman, G.E., Dimmock, J.O., Wolfe, C.M.: Far-infrared recombination radiation from impact-ionized shallow donors in GaAs. Phys. Rev. Lett. **23**(19), 1111–1114 (1969)
11. Vorobjev, L.E., Firsov, D.A., Panevin, V.Y., Sofronov, A.N., Balagula, R.M., Makhov, I.S.: Near- and far-infrared emission from GaAs/AlGaAs quantum wells under interband optical excitation. St. Petersburg State Polytechnical Journal. Phys. Math. **4**(182), 109–114 (2013)
12. Firsov, D.A., Vorobjev, L.E., Panevin, V.Y., Sofronov, A.N., Balagula, R.M., Makhov, I.S., Kozlov, D.V., Vasil'ev, A.P.: Terahertz radiation associated with the impurity electron transition in quantum wells upon optical and electrical pumping. Semiconductors **49**(1), 28–32 (2015)
13. Vinnichenko, M.Ya., Makhov, I.S., Panevin, V.Yu., Sofronov, A.N., Firsov, D.A., Vorobjev, L.E., Sadofev, Y.G., Vasiliev, A.P.: Terahertz radiation related to the electron relaxation after interband optical pumping in doped quantum wells. In: International Conference on Infrared, Millimeter, and Terahertz Waves 2016, 7758782 (2016)
14. Vorob'ev, L.E., Firsov, D.A., Shalygin, V.A., Tulupenko, V.N., Shernyakov, Yu.M., Ledentsov, N.N., Ustinov, V.M., Alferov, Zh.I.: Spontaneous far-IR emission accompanying transitions of charge carriers between levels of quantum dots. JETP Lett. **67**(4), 275–279 (1998)
15. Vorob'ev, L.E., Firsov, D.A., Shalygin, V.A., Tulupenko, V.N., Ledentsov, N.N., Kop'ev, P.S., Ustinov, V.M., Shernyakov, Yu.M., Alferov, Zh.I.: The outlook for the development of radiation sources in the middle-IR range based on the intraband transitions between the energy levels of charge carriers in injection laser heterostructures with quantum dots and wells. Physics-Uspekhi **42**(4), 391–405 (1999)
16. Mailhiot, C., Chang, Y.-C., McGill, T.C.: Energy spectra of donors in $GaAs - Ga_{1-x}Al_xAs$ quantum well structures in the effective-mass approximation. Phys. Rev. B **26**(8), 4449–4457 (1982)
17. Ridley, B.K.: Quantum processes in semiconductors, 5th edn. Oxford University Press, New York (2013)
18. El-Said, M., Tomak, M.: Photoionization of impurities in infinite-barrier quantum wells. J. Phys. Chem. Solids **52**(4), 603–606 (1991)

19. Blakemore, J.S.: Semiconductor statistics. Pergamon Press, Oxford (1962)
20. Makhov, I.S., Panevin, V.Yu., Sofronov, A.N., Firsov, D.A., Vorobjev, L.E., Vinnichenko, M.Ya., Vasil'ev, A.P., Maleev, N.A.: The effect of stimulated interband emission on the impurity-assisted far-infrared photoluminescence in GaAs/AlGaAs quantum wells. Superlattices Microstruct. **112**, 79–85 (2017)
21. Makhov, I.S., Panevin, VYu., Firsov, D.A., Vorobjev, L.E., Vasil'ev, A.P., Maleev, N.A.: Terahertz photoluminescence of the donor doped GaAs/AlGaAs quantum wells controlled by the near-infrared stimulated emission. J. Lumin. **210**, 352–357 (2019)
22. Zheng, W.M., Halsall, M.P., Harmer, P., Harrison, P., Steer, M.J.: Acceptor binding energy in δ-doped GaAs/AlAs multiple-quantum wells. J. Appl. Phys. **92**(10), 6039–6042 (2002)
23. Blakemore, J.S.: Semiconducting and other major properties of gallium arsenide. J. Appl. Phys. **53**(10), R123–R181 (1982)
24. Haller, E.E., Hueschen, P.L., Richards, P.L.: Ge: Ga photoconductors in low infrared backgrounds. Appl. Phys. Lett. **34**(8), 495–497 (1979)
25. Makhov, I.S., Panevin, VYu., Firsov, D.A., Vorobjev, L.E., Klimko, G.V.: Impurity-assisted terahertz photoluminescence in compensated quantum wells. Appl. Phys. Lett. **126**(19), 175702 (2019)

Chapter 3
Broadband Absorption of Microwaves in Periodic Cylindrical Structures

Lilit Gevorgyan, Henrik A. Parsamyan, and Hovhannes Haroyan

Abstract The absorption efficiency of a subwavelength conductive wire can be essentially increased in the broad microwave spectrum from 4 to 12 GHz by the appropriate choice of the radius and height of a wire. Such functional dependence ensures matching between the configuration and the incident field oriented by the wire axis. Theoretical results obtained within the limits of electrostatic approximation and numerical calculations reveal that the absorption cross-section of a wire can exceed the geometrical one by about 10 times, whereas the scattering efficiency is negligibly small. Such properties allow one to achieve relatively high absorption of the incident wave by a system consisting of the wires periodically distributed on a flat surface.

3.1 Introduction

For many years, electromagnetic absorbers have been widely used in the field of electromagnetic compatibility, sensors, bolometers, solar energy harvesting, heat emitters and new passive cooling technologies [1–3]. Effective materials that shield and absorb microwave radiation, particularly in the 2–18 GHz radar frequency range, are required for different defense and aerospace applications, for example creation of "stealth" aircraft [4, 5] camouflaging military ground devices and units from radar surveillance, and the design of anechoic chambers. In this context, the research and development of radio-absorbing materials have become essential.

In order to control electromagnetic waves to eliminate interference between different devices in a wide frequency band from RF to millimeter-wave spectrum, high-efficiency electromagnetic wave absorbers are required. This has an important role in future 5G wireless networks [6].

L. Gevorgyan · H. A. Parsamyan · H. Haroyan (✉)
Department of Radiophysics, Yerevan State University, A. Manoogian 1, 0025 Yerevan, Armenia
e-mail: hharoyan@ysu.am

H. A. Parsamyan
Center for Nanoscience and Technology, Institute of Chemical Physics, NAS RA, P. Sevak 5/2, 0014 Yerevan, Armenia

D. Blaschke et al. (eds.), *Optics and Its Applications*, Springer Proceedings in Physics 281, https://doi.org/10.1007/978-3-031-11287-4_3

As wireless communication technology is advancing, electronic equipment becomes widely used in many fields, thus leading to an increase in electromagnetic pollution [7]. In this regard, microwave absorbers effectively absorb electromagnetic waves, lowering the effects of electromagnetic pollution.

Commonly, the absorbing principles in absorbent materials are mainly linked to the dielectric loss or the magnetic loss. Nevertheless, because of the impedance mismatching at broad frequency ranges, these absorbers hardly achieve ultra-wideband absorption [8, 9].

Many absorbers have been developed, however the main drawbacks include thickness and difficulty of broadband impedance matching to free space. An ideal absorber should enjoy the benefits of a lightweight, low thickness, cost-effectiveness, wide bandwidth and good processability. The operating bandwidth is one of the most challenging problems, because different applications mostly demand wide-bandwidth absorbers. Multilayer structures and lossy materials with a tapered shape are commonly used to obtain wider operating bandwidths. However, this may result in a bulk volume and high cost [4, 10].

Many attempts have focused on developing metamaterial/metasurface-based absorbers from the microwave band to optical band. Limited by the resonance property, the operating bandwidth of the perfect metamaterial absorber (PMA) is usually narrow. In microwave band PMAs [11], typically comprised of dielectric thin-films sandwiched by a metallic split-ring resonator and cutting wire, can show near 100% absorbance at the resonant frequency, but they generally exhibit a narrow absorption bandwidth.

In recent years, all-dielectric metamaterial absorbers, composed of traditional microwave absorbing materials have shown their unique potential in improving impedance matching in a wide frequency band. One can achieve impedance matching with the surrounding by varying specific structures in these absorbers. Otherwise, material properties such as plasma frequency can be modified according to the targeted operating frequency. It is challenging to improve the absorbance and the bandwidth simultaneously because of the fundamental trade-off between the operational bandwidth and the attainable absorption. Therefore, most of the broadband perfect absorption structures are based on near to 90% absorption, and by trying to obtain higher absorbance, the bandwidth will drop dramatically. Consequently, further research is needed for obtaining the broadband absorption structure with higher absorbance.

3.2 Theory

Let ε_1 be the dielectric permittivity of the medium of a prolate spheroid with semi-axes a, b and c (so that $a > b = c$) placed in the environment with a real dielectric constant ε_2. Within the limits of electrostatic approximation (the dimensions of the spheroid are much smaller than the incident wavelength in the surrounding), the total dipole moment of the prolate spheroid is determined by the relation [12]:

$$P_x = \frac{ab^2}{3} \frac{\varepsilon_1 - \varepsilon_2}{\varepsilon_2 + (\varepsilon_1 - \varepsilon_2) n^{(x)}} E_x, \tag{3.1}$$

where

$$n^{(x)} = \frac{1 - e^2}{e^3} (\text{arctanh}\, e - 1), \, e = \sqrt{1 - \frac{b^2}{a^2}}. \tag{3.2}$$

In the investigated case, when $b \ll a$, we have [13]:

$$n^{(x)} \approx \frac{b^2}{a^2} \ln \frac{2a}{2.7b}. \tag{3.3}$$

Henceforth, we will assume that the medium of the spheroid is a lossy material with the complex dielectric permittivity $\varepsilon_1 = \varepsilon_{1r} + i \cdot \varepsilon_{1i}$. Within the range of electrostatic approximation, the absorption and scattering cross-sections of the prolate spheroid are as follows:

$$\sigma_{abs} = \frac{8\pi^2}{\lambda} \frac{a^3}{3 \ln \frac{2a}{2.7b}} \frac{\varepsilon_2 \eta_i}{(\varepsilon_2 + \eta_r)^2 + (\eta_i)^2}, \tag{3.4}$$

$$\sigma_{scat} = \frac{2^7 \pi^5}{3^4} \frac{a^6}{\lambda^4} \frac{1}{\left(\ln \frac{2a}{2.7b}\right)^2} \frac{(\eta_r)^2 + (\eta_i)^2}{(\varepsilon_2 + \eta_r)^2 + (\eta_i)^2}. \tag{3.5}$$

Here

$$\eta_r = (\varepsilon_{1r} - \varepsilon_2) \frac{b^2}{a^2} \ln \frac{2a}{2.7b}, \, \eta_i = \varepsilon_{1i} \frac{b^2}{a^2} \ln \frac{2a}{2.7b}. \tag{3.6}$$

and λ is the wavelength in the surrounding medium.

In order to quantitatively describe the absorption and scattering properties of the investigated configuration, we employ the absorption and scattering efficiency factors:

$$Q_{abs} = \frac{\sigma_{abs}}{S} \text{ and } Q_{scat} = \frac{\sigma_{scat}}{S}. \tag{3.7}$$

We consider the case when $\lambda \gg a \gg b$. Therefore under specific conditions (the real and imaginary parts of dielectric constant of wires materials are the same order ~103 for the fixed geometry: $r = 20\,\mu\text{m}$, $L = 5\,\text{mm}$ within considered frequency range (4–12 GHz)), the scattering efficiency is negligible compared to the absorption.

3.3 Results and Discussion

Full-wave numerical analysis based on the finite element method (FEM) was conducted to reveal the dependence of the absorption properties of the geometrical parameters. To simplify the simulated geometry, prevent simulation model from memory overflow due to the high density of the meshes in narrow regions, the prolate spheroid was replaced by an equivalent cylindrical wire having radius r=b and height $L = 2a$, schematically illustrated in Fig. 3.1a. The wire is illuminated by ax-polarized plane wave propagating along the y-axis. Within all simulations, we will assume that the wire height $L = 5$ mm.

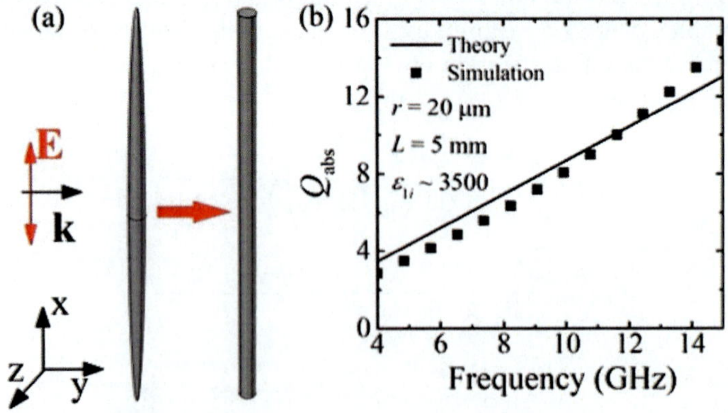

Fig. 3.1 **a** Schematic sketch of the prolate spheroid and the equivalent wire and **b** absorption efficiency spectra of the prolate spheroid obtained according to Eqs. (3.4) and (3.7) (solid line) and simulated (symbol). The sizes of the wire are for $r = 20\,\mu$m, $L = 5$ mm, wire dielectric constant $\varepsilon_1 = \varepsilon_{1r} + i \cdot \varepsilon_{1i} = 1000 + i \cdot 3500$. The surrounding medium is air with $\varepsilon_2 = 1$

Figure 3.1b shows the theoretical (lines) and simulated (symbols) efficiency factor of the absorption for $r = 20\,\mu$m, $L = 5$ mm, wire material dielectric constant is $\varepsilon_1 = \varepsilon_{1r} + i \cdot \varepsilon_{1i} = 1000 + i \cdot 3500$, which can be obtained by composite materials like graphite mixtures. The surrounding medium is air with $\varepsilon_2 = 1$. One sees that the theoretical results of the absorption efficiency for a prolate spheroid and simulations for an equivalent wire are rather in good agreement. Both numerical and theoretical analyses show that the absorption efficiency of the structures linearly depends on the frequency and increases for higher frequencies reaching to about 13 (Fig. 3.1b). However, the small mismatch of the numerical and theoretical results appearing at higher frequencies is the result of neglecting the quadrupole effects in the theoretical analysis.

To realize the absorption metasurface the periodical structure of appropriate distributed rods (wires) can be used. In this case, a trade off rod spacing should be chosen to minimize interaction, thereby reducing incident wave reflections without a large transition coefficient. In Fig. 3.2 is shown the geometry of the simulation, consisting

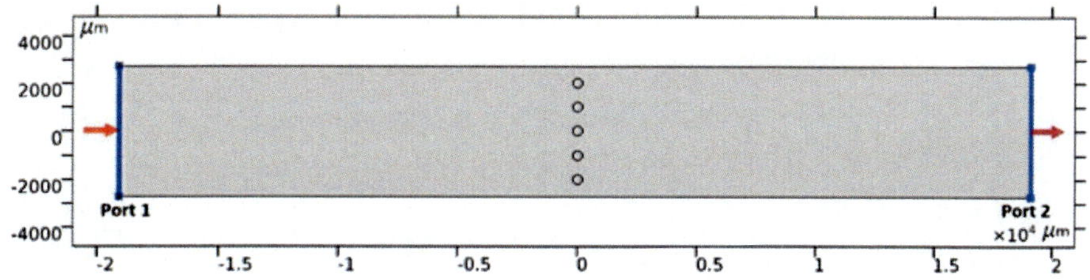

Fig. 3.2 Modeling geometry for one layer

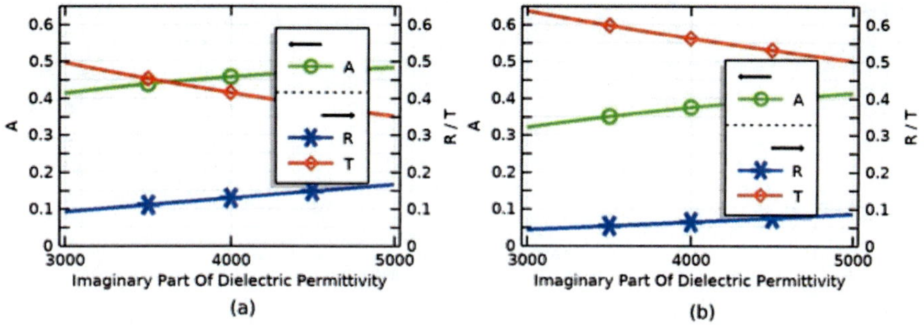

Fig. 3.3 Dependency of the reflection (asterisk), absorption (circle) and transmission (diamond) coefficients from the imaginary part of the dielectric permittivity, consisting of one layer of infinite rods, with parameters $r = 20\,\mu\text{m}$, $\varepsilon_r = 1000$, $f = 11$ GHz. The distance between the two elements were **a** 1 mm and **b** 2 mm

of one layer of periodically distributed rods at distance of 1000 μm from each other. The grey rectangle illustrates the simulation domain composed of orderly arranged cylinders placed in free space. The incident plane wave is excited and received using the port boundary condition (Port1 and Port2, respectively). Arrows show the propagation direction of the incident electromagnetic waves.

Figure 3.3 shows the change in the coefficients of reflection (asterisk), absorption (circle) and transmission (diamond) (respectively R, A and T) as a function of the imaginary part of dielectric constant of the rods with parameters $\varepsilon_{1r} = 1000$, $r = 20\,\mu\text{m}$, $L = 5$ mm, $f = 11$ GHz with a distance between the two elements of 1 mm (Fig. 3.3a), and 2 mm (Fig. 3.3b). The absorption coefficient is calculated using the following expression $A = 1 - R_2 - T_2$, where $R = |S_{11}|^2$ and $T = |S_{21}|^2$ are the reflection coefficient and the transmission coefficients, respectively. It is seen that the distance between the elements affects the reflection and absorption coefficient of the absorber: in the case of 1 mm $A \sim 0.48$, $R \sim 0.16$, $T \sim 0.36$, and in the case of 2 mm, $A \sim 0.41$, $R \sim 0.07$, $T \sim 0.52$.

It should be noted that the weak dependence of A, R, T parameters on the imaginary part of the dielectric constant of a wire enables do not to have precision quantities of the mixture components of graphite mixture which makes easier the fabrication of such structures.

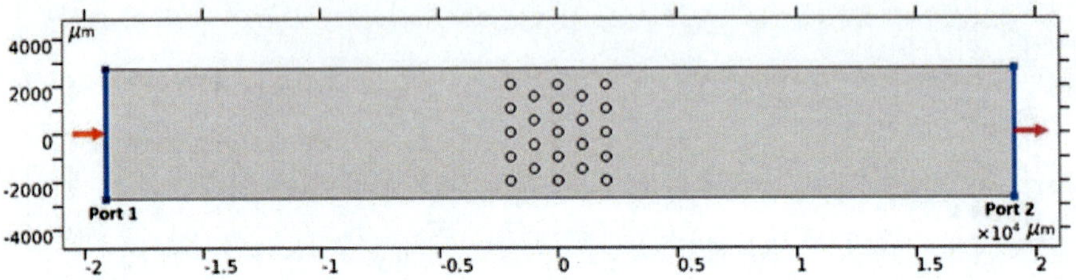

Fig. 3.4 Modeling geometry for five layers

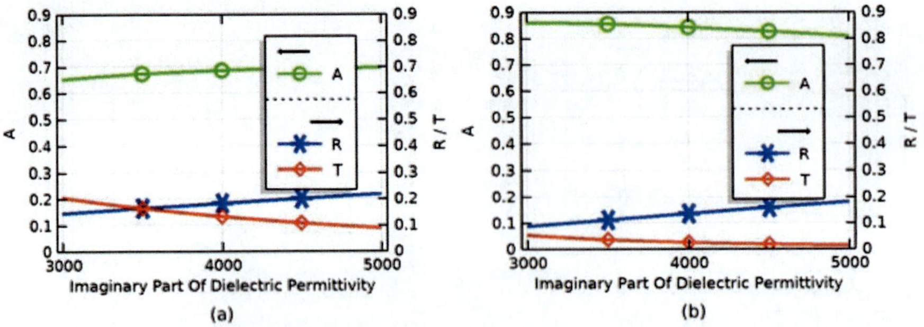

Fig. 3.5 Dependency of the reflection (asterisk), absorption (circle) and transmission (diamond) coefficients from the imaginary part of the dielectric permittivity of the absorber layers, the distance between the two elements of the layers were 1 mm. The distance between the layers were 1 mm (**a**), and 2 mm (**b**)

The efficiency of the absorption can be increased by creating an absorbing system consisting of multiple layers of such wires (Fig. 3.4).

The improvement of absorption feature of multilayer structure is clearly seen in Fig. 3.5. The dependence of the reflection (asterisk), absorption (circle) and transmission (diamond) coefficients on the imaginary part of the dielectric permittivity of the absorber wire, in the case of a five-layer structure presented in Fig. 3.5. The distance between two elements in one layer were 2 mm, $r = 20\,\mu\text{m}$, $\varepsilon_r = 1000$, $f = 11\,\text{GHz}$ **a** the distance between the layers 1 mm, **b** distance between layers 2 mm.

Further improvement of absorption can be reached by optimizing the distance between two nearby layers. To find the optimal distance between the layers, numerical analyses were carried out at different distances between the layers.

Figure 3.6 shows that increasing the distance between layers respectively increases the absorption coefficient and decreases the reflectance. However, at some point (in our case, from a distance of 10 mm), the absorption coefficient decreases.

To describe the dependency of absorption on the number of layers, the simulations were carried out by regularly adding the layers. Figure 3.7 shows that an increase in the number of layers leads to an increase in the absorption coefficient, but at some point (in our case from the 9th layer) the absorption coefficient becomes almost constant.

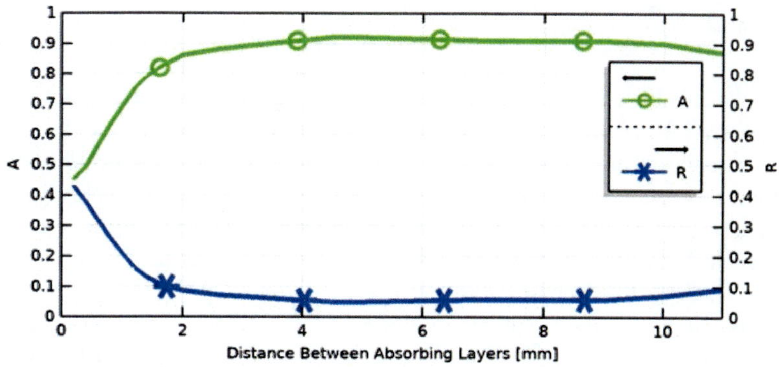

Fig. 3.6 Dependency of the reflection and absorption coefficients from the distance between the layers of a five-layer system

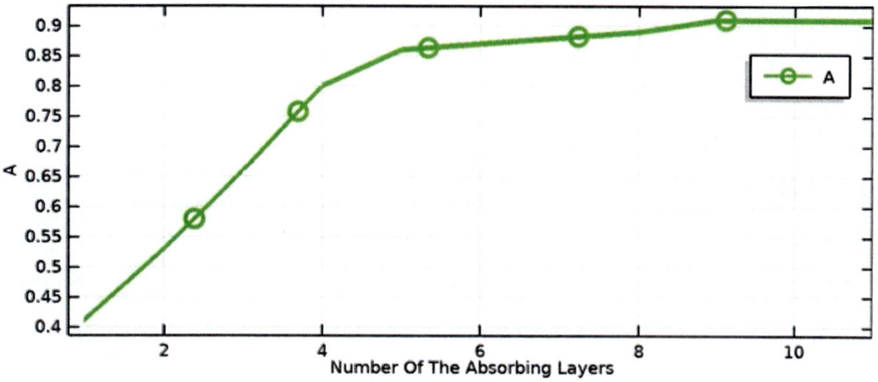

Fig. 3.7 Dependency of the absorption coefficient on the number of layers

As shown in Fig. 3.7, there is a significant increase in absorptance (0.85 and more) in the number of layers up to 5–7, after which the growth rate decreases sharply. This suggests that in practice, it is possible to not use too many layers, especially since an increase in the number of layers leads to an increase in the total thickness of the absorber. For example, in the case of having 5 layers, if the distance between the layers is 2 mm, the thickness of the system will be 1 cm, which, for example, at an operating frequency of 10 GHz is three times less than the wavelength.

The proposed structure is polarization-sensitive and it can absorb only linearly polarized incident field parallel to wires. This issue can be solved in a multilayer system by applying additional through one layers perpendicular to the initial ones.

Thus, depending on the formulation of the problem, it is possible to achieve different rates of absorption depending on the dielectric permittivity of the rods, their radius, the distance between them, and the distance between the layers.

3.4 Conclusion

Conductive subwavelength wires oriented to the polarization of the incident wave can serve as an effective basic element of broadband absorbers in the 4–12 GHz microwave range. The absorption capacity of such absorbing rods can be significantly increased due to the correct choice of the geometrical and electrodynamics parameters of the structure. To obtain absorbing metasurfaces, periodic systems of absorber rods can be used, with which it is possible to provide surfaces with an absorption coefficient near 0.9. In particular, the results of numerical calculations show that the corresponding periodic arrangement of rods with a radius of $r = 20\,\mu\text{m}$, length $L = 5\,\text{mm}$, $\varepsilon_1 = 1000 + i \cdot 3500$ can lead to broadband absorption of the order of 0.9, which is nearly uniform in a broad spectrum from 4 to 12 GHz.

Acknowledgements This work was supported by a scientific research grant through the Science Committee of MESCS of Armenia (20DP-1C05, 20APP-1C009,21AA-2B035and 21AG-1C061) and by a faculty research funding program 2021 implemented by Enterprise Incubator Foundation with the support of PMI Science.

References

1. Grant, J., Escorcia-Carranza, I., Li, C., McCrindle, I.J.H., Gough, J., Cumming, D.R.S.: A monolithic resonant terahertz sensor element comprising a metamaterial absorber and micro-bolometer. Laser Photon. Rev. **7**(6), 1043–1048 (2013)
2. Matsuno, Y., Sakurai, A.: Perfect infrared absorber and emitter based on a large-area metasurface. Opt. Mater. Express **7**(2), 618 (2017)
3. Raman, A.P., Anoma, M.A., Zhu, L., Rephaeli, E., Fan, S.: Passive radiative cooling below ambient air temperature under direct sunlight. Nature **515**(7528), 540–544 (2014)
4. Munk, B.: Frequency Selective Surfaces: Theory and Design, 1st edn. Wiley-Interscience, New York (2000)
5. Knott, E.F., Shaeffer, J.F., Tuley, M.T.: Radar Cross Section (Radar, Sonar and Navigation), 2nd edn. Scitech Publishing, New York (2004)
6. Rappaport, T.S., Xing, Y., MacCartney, G.R., Molisch, A.F., Mellios, E., Zhang, J.: Overview of millimeter wave communications for fifth-generation (5G) wireless networks - with a focus on propagation models. IEEE Trans. Antennas Propag. **65**(12), 6213–6230 (2017)
7. Liu, T., Pang, Y., Zhu, M., Kobayashi, S.: MicroporousCo@CoO nanoparticles with superior microwave absorption properties. Nanoscale **6**(4), 2447 (2014)
8. Peng, Z., Hwang, J.-Y., Andriese, M.: Absorber impedance matching in microwave heating. Appl. Phys. Express **5**(7), 077301 (2012)
9. Chen, X., Wu, Z., Zhang, Z., Heng, L., Wang, S., Zou, Y.: Impedance matching for omnidirectional and polarization insensitive broadband absorber based on carbonyl iron powders. J. Magn. Magn. Mater. **476**, 349–354 (2019)
10. Cui, Y., Fung, K.H., Xu, J., Ma, H., Jin, Y., He, S., Fang, N.X.: Nano Lett. **12**, 1443 (2012)
11. Tao, H., Landy, N.I., Bingham, C.M., Zhang, X., Averitt, R.D., Padilla, W.J.: Opt. Express **16**, 7181 (2008)
12. Landau, L.D., Pitaevskii, L.P., Lifshitz, E.M.: Electrodynamics of Continuous Media, 2nd edn. Elsevier Science & Technology, Oxford, United Kingdom (1984)
13. Solivérez, C.E. (eds.): Electrostatics and Magnetostatics of Polarized Ellipsoidal Bodies: The Depolarization Tensor Method. Free Scientific Information (2016)

Chapter 4
Dielectric Confinement Affected Exciton-Polariton Properties of the Semiconductor Nanowires

K. H. Aharonyan, E. M. Kazaryan, and E. P. Kokanyan

Abstract The nonlocal susceptibility with the cylindrical symmetry in semiconductor quantum wires embedded in the dielectric barrier environment (DQWW) is calculated. The strong dependence from the wire radius R due to the dielectric mismatch effect is established ($\sim R^{-8/3}$). It has been received that the oscillator strength of the one-dimensional (1D) excitons in DQWW increase strongly with decreasing R. The Maxwell's equations are solved in presence with the nonlocal excitonic response to the dielectric polarization for the DQWW and in a result the effective boundary conditions just considering a quantum wire presence in the structure have been established. The scattering coefficient of the light incident on the DQWW is obtained which strongly depends on the DQWW material parameters such as: (a) quantum wire radius R and dielectric constant ε_w, (b) barrier environment dielectric constant ε_b. This made it possible to obtain the dispersion spectrum and lifetime for the 1D exciton-polaritons of the DQWW near the exciton resonance with the opportunities of the valuable manipulations along with the magnitudes of the DQWW material parameters.

4.1 Introduction

The strong coupling of the low-dimensional excitons and localized photon states (exciton-polaritons) has been a subject of considerable interest for a long time in the optics of semiconductor nanostructures [1–3]. A number of unique physical phenomena were discovered (a Rabbi splitting, a large decrease of the exciton resonance lifetime [4, 5]) which can be directly aimed at promising applications in semiconductor optoelectronics. In this regard, intense theoretical efforts [6–10] have been carried

K. H. Aharonyan (✉)
National Polytecnic University of Armenia, Yerevan 0009, Armenia
e-mail: ahkamo@yahoo.com

K. H. Aharonyan · E. M. Kazaryan
Russian-Armenian University, Yerevan 0051, Armenia

K. H. Aharonyan · E. P. Kokanyan
Armenian State Pedagogical University, Yerevan 0010, Armenia

D. Blaschke et al. (eds.), *Optics and Its Applications*, Springer Proceedings in Physics 281, https://doi.org/10.1007/978-3-031-11287-4_4

out over the past time to realize the photonic structures such as planar and cylindrical multilayered microcavities, wide band-gap semiconductor photonic structures which offer unique optical properties.

To provide a substantial enhancement of light-matter interaction with the quantum-confined excitons and thus to support high polariton stability, the photonic structures should provide both the large exciton binding energy and the oscillator strength values [1, 2, 9]. Due to this a search for ways to enhance these physical values in photonic structures is a decisive factor in this area [11–15].

In the bulk semiconductors a role of Wannier-Mott excitons to optical response is insignificant compared to interband transitions since the exciton line dimensionless oscillator strength (per unit crystal cell) is the order of $(d_0/a_{0ex})^3 \approx 10^{-4} \div 10^{-5}$, where d_0 is the lattice constant and a_{0ex} is the excitons effective radii taking the values as $a_{0ex} \gg d_0$.

Such an estimate follows from the physical condition that the dimensionless oscillator strength is of the order of exciton radiative recombination probability (i.e. localization probability of electron and hole in the same crystal lattice unit cell) proportional to the$|\phi_{ex}(0)|^2$ value as compared to the intensity of the absorption spectrum of the semiconductor, where ϕ_{ex} is the exciton wave function. Here the localization probability is inversely proportional to the exciton effective volume $V_{ex} \sim a_{ex}^3$ and the absorption intensity is characterized by the localized Wannier functions linear combinations of band states just within the crystal unit cell volume V_0 and estimating as proportional to $V_0^{-1} \sim d_0^{-3}$.

In turn, the exciton effects in semiconductor low-dimensional structures-quantum wells (QW), quantum wires (QWW) are much more prominent and accessible for experimental detection than in bulk samples both in the absorption and in the emission. This, first of all, found the confident confirmation in experiments on optical absorption and photoluminescence [16, 17]. Theoretically, this is explaining by an increase in the exciton binding energy and in the oscillator strength of the corresponding exciton transition with a decrease in the spatial dimensions of the semiconductor quantum sample when takes place the compression of the exciton wave function in the spatial confinement direction (quantum confinement (QC)). In particular, in the thin QW there is a correspondingly fourfold and eightfold increase of the exciton binding energy and the exciton transition oscillator strength [18, 19].

These results refers the physical situation, when the dielectric constants mismatch between semiconductor quantum sample (semiconductor/dielectric quantum wells (DQW) and quantum wires (DQWW)) and surrounding dielectric barrier environment (ε_w, ε_b) is small or neglected. The difference between ε_w and ε_b leads to an unhomogeneous polarization of the quantum structure. In result for the case of $\varepsilon_w \gg \varepsilon_b$ and the distances between the charges as large as the QC spatial size (DQW thickness d or DQWW radius R) the charges produced field in the barrier begins to play a perceptible role and force electron and hole into the middle of the quantum sample and, so, to modify and enhance their interaction (dielectric confinement effect (DC)) [20–22]. As correspondingly are shown in Refs. [23, 24] and [22] the effective exciton volume in DQW and DQWW might radically reduced since in the direction normal to the DQW the exciton dimension is $\sim d$ and the in-plane effective radius

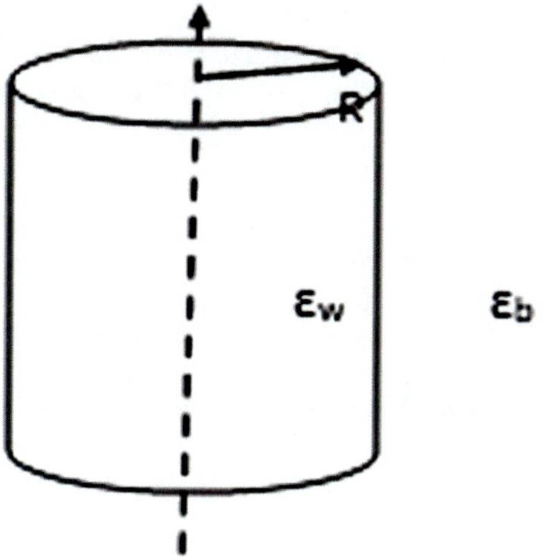

Fig. 4.1 The cylindrical DQWW of a radius R

due to DC effect takes the value $\sim \sqrt{a0_{ex}d}$, while in the direction normal to the DQWW axes the exciton dimension is just $\sim R$ and in the direction parallel to axes the exciton effective radius becomes the value $(a_{0ex}R^2)^{1/3}$.

As follows, in addition to the existing enhancements from the QC effect, the DC effect in turn might lead to the further strong increase of the exciton binding energy and the exciton transition oscillator strength with decreasing of the DQW thickness d or DQWW radius R. The latter creates the more favorable conditions in these structures to provide a substantial enhancement of the light-matter interaction to provide a substantial enhancement of the light-matter interaction with the quantum-confined excitons and thus to support high polariton stability. In this connection in Ref. [11] the contribution of the excitonic transition to the electromagnetic response of the DQW and the possibilities of the propagation of the polariton waves are considered (Fig. 4.1).

The aim of the present paper is to develop an analogous formalism for the analysis of the particular features of the light-matter interaction with the one-dimensional (1D) DC enhanced dipole-allowed excitons in the DQWW with the cylindrical symmetry confinement. Specific azimuthal symmetry here substantially simplifies the convolution of the problem, as together with the 1D wave vector an additional quantum number (azimuthal) is available now.

The outline of presented paper is as follows. In Sect. 4.2, we describe the macroscopic theory of the electromagnetic response of the DQWW to the electromagnetic field with frequency ω and projection of wave vector $\boldsymbol{k}$ on the wire axes ($k_{\|}$). The next section is devoted to obtaining explicitly effective boundary conditions which are considering the dependence of the dielectric properties on the wire radius R. The Sect. 4.4 consists of an application of the obtained effective boundary conditions for the calculation of the light reflection coefficient from the structure under consideration.

4.2 Electromagnetic Linear Response to the Electromagnetic Field in the Macroscopically Homogeneous and Isotropic DQWW

We are considering an infinitely long cylindrical DQWW of a radius R filled by an active material with the dielectric constant ε_w and immersed in a dielectric barrier environment with the dielectric constant ε_b. Let take in the cylindrical polar coordinates (ρ, ϕ, z) , where z axes coincides with the DQWW axes, and the plane of incidence (ρ, ϕ) be normal to the wire axes. Here a strong spatial confinement regime would be assumed presupposing that a quantum wire radius is small compared with the exciton Bohr radius $a_{0ex} = \varepsilon_w \hbar^2 / \mu_{ex}^* e^2$ for bulk samples ($a_{0ex} \gg R$), μ_{ex}^* is the exciton reduced mass. Then, as follows, the distances along the wire axes $|z| \gg R$ would be essential in discussed case and therefore the one-dimensional (1D) long wave region $k_\parallel \ll R^{-1}$ could be appropriate here. For this case, it is quite reasonable to introduce a spatially separated exciton wave function in the form

$$\Psi_{k_\parallel, M_e, M_h} (\boldsymbol{\rho}_e, \boldsymbol{\rho}_h, z_e, z_h) = \phi_l (z_e - z_h)\, \Upsilon_{j,M_e} (\boldsymbol{\rho}_e)\, \Upsilon_{j,M_h} (\boldsymbol{\rho}_h) . \tag{4.1}$$

Here $\phi(z)$ is the $1D$ wave envelope function of the exciton pair relative motion, $l = 1, 2, ...$ numbers the $1D$ exciton sublevels, $\Upsilon (\rho, \phi)$ is the one-particle wave function describing the electron or hole transverse relative to the DQWW axes motion and in the QC model of a square well with infinite walls is characterized by [22]

$$\Upsilon_{j,M} (\boldsymbol{\rho}, \phi) = \frac{1}{s^{1/2}} e^{i|M|\phi} J_{|M|} \left(\lambda_j^{|M|} \frac{\rho}{R}\right) / J_{|M|+1} \left(\lambda_j^{|M|}\right), \tag{4.2}$$

where $M_{e(h)} = 0, 1, 2, \ldots$ is the one-particle angular momenta and $j = 1, 2, ...$ numbers the QWW subbands, $\lambda_j^{|M|}$ are the roots of $J_{|M|} (x) = 0$.

As shown in Ref. [22], for the DQWW satisfying the system of conditions $a_{0ex} \gg R \gg a_{0ex} / (\varepsilon_r \ln \varepsilon_r)^2$, where $\varepsilon_r = \varepsilon_w / \varepsilon_b$, the effective radii of the exciton ground and first excited bound states fall into the range of distances $\varepsilon_r (R/z)^2 \ln (|z|/R) \gg 1$ with $|z| \gg R$, where exciton interaction potential has the form

$$V(z) = -\frac{e^2}{\varepsilon_s R} \sqrt{\frac{\varepsilon_r \ln \varepsilon_r}{2}} \left[1 - \frac{|z|}{R} \sqrt{\frac{2}{\varepsilon_r \ln \varepsilon_r}}\right] \tag{4.3}$$

and for that the 1D wave function of exciton pair relative motion and 1D exciton bound state energy spectrum have the form [25, 26]

$$\phi_l (z) = \frac{N(\mu_l)}{a_{ex}^{1/2}} \Phi \left(\frac{|z|}{a_{ex}} + \mu_l\right), l = 1, 2, 3, \ldots, \text{- even states}, \tag{4.4}$$

$$\phi_{l+1}(z) = \frac{N(\xi_l)}{a_{ex}^{1/2}} \frac{|z|}{z} \Phi\left(\frac{|z|}{a_{ex}} + \xi_{l+1}\right), l = 1, 2, 3, \ldots, \text{- odd states,} \tag{4.5}$$

$$E_l^{even(odd)} = -\frac{e^2}{\varepsilon_s R}\sqrt{\frac{\varepsilon_r \ln \varepsilon_r}{2}}\left[1 - \frac{\mu_l(\xi_{l+1})}{2^{1/3}}\sqrt{\frac{2}{\varepsilon_r \ln \varepsilon_r}\frac{a_{0ex}}{R}}\right], \tag{4.6}$$

where $\Phi(z)$ is the Airy function, μ_l are the solutions of $\Phi'(z) = 0$, ξ_l are the solutions of $\Phi(z) = 0$, $a_{ex} = (a_{0ex} R^2/2)^{1/3}$, N is the normalization constant (for more details see [25, 26]). As follows from Eq. (4.6), the exciton binding energy of the DQWW is determined by the dielectric properties of the surrounding wire medium and increases as $\sim R^{-1}$.

The oscillator strength (per unit length of the DQWW) of the associated optical transition is

$$\frac{f_x}{L} = i\omega \frac{2m_0}{\hbar e^2} \mu_{cv} \left| \int \Psi_M(\rho_e = \rho_h, z_e = z_h)\, d\rho_e dz_e \right|^2, \tag{4.7}$$

where μ_{cv} is the dipole matrix element in the bulk material, L is the wire length, m_0 is the electron free mass. In Eq. (4.7) only the $1D$ excitons with $M = M_e + M_h = 0$ are allowed in the dipole approximation because the others have zero exciton wave function at the origin of the relative coordinates. As follows from Eqs. (4.1), (4.2) and (4.3) the $1D$ exciton effective volume V_{ex} in DQWW decreases as $(R/a_{0ex})^{8/3}$ and since $|\Psi_M(0)|^2 \sim V_{ex}^{-1}$, so the $1D$ exciton line oscillator strength f_{ex} increases as $R^{-8/3}$.

The $1D$ excitonic transition contribution to the electromagnetic linear response of the cylindrical DQWW induces the dielectric polarization connected with the electric field E as

$$P_\alpha(\rho) = \int \chi_{\alpha\beta}\left(k_\parallel, \omega, \rho, \rho'\right) E_{\beta k_\parallel \omega}\left(\rho'\right) d\rho', \tag{4.8}$$

where the nonlocal dielectric Fourier-transformed polarizability $\chi_{\alpha\beta}$ is given by

$$\chi_{\alpha\beta}\left(k_\parallel, \omega, \rho, \rho'\right) = \chi_{\alpha\beta}\left(k_\parallel, \omega\right)\left[\Upsilon_{j,M}(\rho)\right]^2\left[\Upsilon_{j,M}(\rho')\right]^2, \tag{4.9}$$

which will be calculated in accordance with the second order perturbation theory.

Here we assume that the light wave propagate in the plane orthogonal to the DQWW axes and consider only the exciton ground state in the wire which is characterized by $M = 0$ angular momentum. Given the cylindrical symmetry of the exciton ground state, the integral in Eq. (4.8) is nonzero only in the case of cylindrical light waves having zero angular momentum. Thus in the following we limit the discussion to that situation only.

After necessary actions with the excitonic wave functions (4.1), (4.2) and (4.4) for the Fourier-transformed dielectric polarization $P_\alpha\left(k_\parallel, \omega, \rho\right)$ obtain

$$P_{\alpha}\left(k_{\parallel},\omega,\rho\right)=\Lambda_{\alpha\beta\chi}\left(k_{\parallel},\omega;\rho\right)\bar{E}_{\beta}, \quad (4.10)$$

where

$$\chi\left(k_{\parallel},\omega;\rho\right)=\chi\left(k_{\parallel},\omega\right)\left(\frac{J_{|M|}\left(\lambda_j^{|M|}\frac{\rho}{R}\right)}{J_{|M|+1}\left(\lambda_j^{|M|}\right)}\right)^2, \quad (4.11)$$

and $\bar{E}=\int\left[\Upsilon_{j,M}\left(\rho'\right)\right]^2 E_{k_{\parallel}\omega}\left(\rho'\right)d\rho'$.

In the Eq. (4.11)

$$\chi\left(k_{\parallel},\omega\right)=\frac{\left(2\mu_{ex}^{*}e^{8}\hbar^{4}\right)^{1/3}\left[N\left(\mu_{l}\right)\Phi\left(\mu_{l}\right)\right]^{2}}{\pi\varepsilon_{s}^{1/3}R^{8/3}}\frac{\left|V_{cv}\right|^{2}}{\varepsilon_{k_{\parallel}}\left(\varepsilon_{k_{\parallel}}^{2}-(\hbar\omega)^{2}\right)}, \quad (4.12)$$

where V_{cv} is the matrix element of the velocity operator corresponding to the transition from the valence to conduction band, $\varepsilon_{k_{\parallel}}$ is the resonant energy of exciton creation, including its kinetic energy in the state with momentum $\hbar k_{\parallel}$.

In Eq. (4.10) the coefficients $\Lambda_{\alpha\beta}$, where $(\alpha,\beta)=(\rho,\phi,z)$, accounts the polarization structure of the exciton transition, i.e., the symmetry of wave functions of the c and v bands. As a rule, in assumption that the wire and barrier environments are optically isotropic, these coefficients contain just two characteristic terms: $\Lambda_z=\Lambda_{\parallel}$ and $\Lambda_{\rho}=\Lambda_{\perp}$.

Equation (4.11) exhibits clearly the strong increase of the excitonic oscillator strength in DQWW as R decreases. Except that Eqs. (4.10)–(4.11) expresses the nonlocal structure in ***rho***, reflecting, as in the DQW [11–13], the peculiar "firmness" of the exciton transverse to wire axes wave function (4.1). Thereby the exciton creation probability defines by the field strength averaged over the electron and hole transverse to wire axes wave func-tions and the spatial distribution of the induced current itself is proportional to the same wave functions at $\rho_e=\rho_h=\rho'$.

4.3 Effective Boundary Conditions for the Electromagnetic Fields in the Macroscopically Homogeneous and Isotropic DQWW

The distribution of the electromagnetic field in the DQWW in presence of such polarization will be found in analogy with the well established method related the effective boundary conditions in QW systems [1, 11, 27] connected with the field magnitudes in the neighboring media and are followed from the Maxwell's equations. Here we expand this method on the DQWW system case in the first time.

Thereby in accordance with the Eqs. (4.8), (4.9) and (4.10) the corresponding electromagnetic field induction vector is given by

$$D_{\alpha}\left(k_{\parallel},\omega;\rho\right)=\varepsilon_{w}E_{\alpha k_{\parallel}\omega}\left(\rho\right)+4\pi\Lambda_{\alpha\beta}\chi\left(k_{\parallel},\omega,\rho\right)\bar{E}_{\beta}, \tag{4.13}$$

With the aim of to obtain the boundary conditions let now integrate the Maxwell equations and take into consideration only the leading terms in the small quantities $k_{\parallel}R$ and $(\omega/c)R$ together with the Eq. (4.13). In these equations we are dealing with the monochromatic fields $\boldsymbol{E}$, $\boldsymbol{D}$ and $\boldsymbol{H}$ vary in space according to the law

$$\boldsymbol{A}\left(\boldsymbol{r},t\right)=\boldsymbol{A}\left(\boldsymbol{\rho}\right)e^{i\left(k_{\parallel}z-\omega t\right)}=\sum_{|M|}\boldsymbol{A}\left(\boldsymbol{\rho}\right)e^{i|M|\phi}e^{i\left(k_{\parallel}z-\omega t\right)}. \tag{4.14}$$

The resulting solutions will be matched in the barrier and QWW regions and the latter will be occurred only by these boundary conditions. Below, using these boundary conditions, we will analyze the reflection coefficient from the DQWW structure.

Let at first start from the integrating of the Maxwell equation div$\boldsymbol{D}=0$. The latter in the polar planar coordinates (ρ,ϕ) for the case of exciton ground state with the condition $M=0$ takes the form

$$\boldsymbol{e}_{\rho}\int_{0}^{R}\int_{0}^{2\pi}\frac{1}{\rho}\frac{\partial}{\partial\rho}\left(\rho D_{\rho}^{(0)}\left(\rho\right)e^{ik_{\parallel}z}\right)\rho d\rho d\phi+\boldsymbol{e}_{z}\int_{0}^{R}\int_{0}^{2\pi}\frac{\partial}{\partial z}\left(D_{z}^{(0)}\left(\rho\right)e^{ik_{\parallel}z}\right)\rho d\rho d\phi=0, \tag{4.15}$$

which after integrating together with Eq. (4.13) becomes

$$\boldsymbol{E}_{\rho}^{(0)}\left(R\right)=-\frac{iR}{\varepsilon_{w}}\varepsilon_{\parallel}\left(\boldsymbol{k}_{\parallel}\bar{\boldsymbol{E}}_{z}^{(0)}\left(R\right)\right)\boldsymbol{e}_{\rho}, \tag{4.16}$$

where $\bar{E}_{1z}^{(0)}\left(R\right)=\int_{0}^{R}E_{z}^{(0)}\left(\rho\right)\rho d\rho$ and

$$\varepsilon_{\parallel}=\varepsilon_{w}+\frac{4\pi\Lambda_{\parallel}\chi\left(k_{\parallel},\omega\right)}{J_{1}^{2}\left(\lambda_{1}^{0}\right)}, \tag{4.17}$$

In Eq. (4.15) the term associated with an angular variation of the electric field components is obviously omitted. Here Eq. (4.16) links the corresponding components of the electric field.

To receive the complete set of these components let now deal with the equation rot$\boldsymbol{E}=\frac{i\omega}{c}\boldsymbol{H}$ as well. Analogous to the Eq. (4.15) for this case we have

$$-\boldsymbol{e}_{\rho}\int_{0}^{R}\int_{0}^{2\pi}\frac{\partial E_{\phi}^{(0)}}{\partial z}\rho d\rho d\phi+\boldsymbol{e}_{\phi}\int_{0}^{R}\int_{0}^{2\pi}\left(\frac{\partial E_{\rho}^{(0)}}{\partial z}-\frac{\partial E_{z}^{(0)}}{\partial\rho}\right)\rho d\rho d\phi+\boldsymbol{e}_{z}\int_{0}^{R}\int_{0}^{2\pi}\frac{1}{\rho}\frac{\partial(\rho E_{\phi}^{(0)})}{\partial\rho}\rho d\rho d\phi, \tag{4.18}$$

where $\boldsymbol{e}_{\rho}$, $\boldsymbol{e}_{\phi}$ and $\boldsymbol{e}_{z}$ are the polar directional unit vectors.

Let now multiply Eq. (4.18) vectorially by $\boldsymbol{e}_{\rho}$ and make use of the Eq. (4.14) then we receive

$$E_{\phi}^{(0)}\boldsymbol{e}_{\phi}+E_{z}^{(0)}\boldsymbol{e}_{z}=\frac{i\omega R}{c}\left[\bar{\boldsymbol{H}}^{(0)}(R)\,\boldsymbol{e}_{\rho}\right]+\bar{E}_{2z}^{(0)}\boldsymbol{e}_{z}+\frac{ik_{\parallel}R}{2\varepsilon_{w}}\bar{D}_{\rho}^{(0)}\frac{\varepsilon_{w}+8\pi\Lambda_{\perp}\chi\left[J_{1}^{-2}-J_{1}^{-4}\right]}{\varepsilon_{w}+8\pi\Lambda_{\perp}\chi}\boldsymbol{e}_{z},\tag{4.19}$$

where $\bar{\boldsymbol{H}}^{(0)}(R)=\int_{0}^{R}\boldsymbol{H}^{(0)}(\rho)\,\rho d\rho$, $\bar{E}_{2z}^{(0)}(R)=\int_{0}^{R}E_{z}^{(0)}(\rho)\,d\rho$, $\bar{D}_{\rho}^{(0)}(R)=\int_{0}^{R}D_{\rho}^{(0)}(\rho)\,\rho d\rho$.

If combining the Eqs. (4.16) and (4.19) we find the electric field boundary conditions as

$$\boldsymbol{E}(R)=-\frac{iR}{\varepsilon_{w}}\varepsilon_{\parallel}\left(\boldsymbol{k}_{\parallel}\bar{\boldsymbol{E}}_{1z}^{(0)}(R)\right)\boldsymbol{e}_{\rho}-\frac{i\omega R}{c}\left[\bar{\boldsymbol{H}}^{(0)}\boldsymbol{e}_{\rho}\right]+\bar{E}_{2z}^{(0)}\boldsymbol{e}_{z}+\frac{ik_{\parallel}R}{\varepsilon_{s}}\bar{D}_{\rho}\frac{\varepsilon_{w}+8\pi\Lambda_{\perp}\chi\left[J_{1}^{-2}-J_{1}^{-4}\right]}{\varepsilon_{w}+8\pi\Lambda_{\perp}\chi}\boldsymbol{e}_{z},\tag{4.20}$$

From the magnetic field equations $\mathrm{div}\boldsymbol{D}=0$, $\mathrm{rot}\boldsymbol{D}=-\frac{i\omega}{c}\boldsymbol{D}$ the similar manipulations produce finally the corresponding boundary conditions as

$$\boldsymbol{H}(R)=-iR\left(\boldsymbol{k}_{\parallel}\bar{H}_{z}^{(0)}(R)\right)\boldsymbol{e}_{\rho}-\frac{i\omega R}{c}\varepsilon_{\parallel}\left[\bar{\boldsymbol{E}}^{(0)}(R)\,\boldsymbol{e}_{\rho}\right]+\bar{H}_{z}^{(0)}\boldsymbol{e}_{z}+iR\left(\boldsymbol{k}_{\parallel}\boldsymbol{H}_{\rho}^{(0)}\right)\boldsymbol{e}_{z},\tag{4.21}$$

where $\bar{H}_{z}^{(0)}(R)=\int_{0}^{R}H_{z}^{(0)}(\rho)\,\rho d\rho$, $\bar{\boldsymbol{E}}^{(0)}(R)=\int_{0}^{R}\boldsymbol{E}^{(0)}(\rho)\,\rho d\rho$, $\bar{H}_{\rho}^{(0)}(R)=\int_{0}^{R}H_{\rho}^{(0)}(\rho)\,\rho d\rho$.

In Eqs. (4.20) and (4.21) $\boldsymbol{E}(R)$ and $\boldsymbol{H}(R)$ are the boundary values of the electric and magnetic fields in the barrier region, c is the light velocity. Under boundary conditions after Exps. (4.20) and (4.21), a presence of a DQWW, as already emphasized, is taken into account up to the first-order terms in small parameters $\sim k_{\parallel}R$ and $\omega R/c$.

At the same time due to the Eq. (4.17) the first terms of the right hand sides in the Eqs. (4.20) and (4.21), i.e. the terms $\sim\varepsilon_{\parallel}\frac{\omega R}{c}$, would be hold only in the discussions. In turn, the last term in Eq. (4.20) will be taken into account in the narrow frequency range with $\varepsilon_{w}+8\pi\Lambda_{\perp}\chi\ll 1$.

4.4 The Scattering Coefficient of the Light from the Macroscopically Homogeneous and Isotropic DQWW

In this section let calculate the scattering coefficient r_{DQWW} of electromagnetic light wave from the discussed DQWW structure based on the effective boundary conditions after Eqs. (4.20) and (4.21).

As we admitted in Sect. 4.2 the electromagnetic waves of a cylindrical symmetry propagate in the plane orthogonal to the DQWW axes (z axes) and along the latter is directed the electric-field vector (p-polarized light). With that we consider only the exciton ground state in the quantum wire, just characterized by zero angular momentum and are interested in results close to the exciton resonance. We note that in the case with $M=0$ the excitonic polarization lies along the DQWW axes and the mode under consideration has longitudinal nature. Accordingly, as in the [15], we consider the exciton resonant modes appearing as resonances of the Breit-Wigner type in the r_{DQWW} of the barrier waves.

Let establish the scattering coefficient r_{DQWW} as the ratio of the amplitudes of outgoing and incoming cylindrical waves in the DQWW barrier region at $\rho = R$ for which the p-polarized electromagnetic field correspondingly has the forms [15] outgoing and incoming cylindrical waves in the DQWW barrier region at $\rho = R$ for outgoing and incoming cylindrical waves in the DQWW barrier region at $\rho = R$ for which the p-polarized electromagnetic field correspondingly has the forms [15]

$$\begin{cases} E_z^{inc} = H_0^{(1)}(k\rho) = J_0(k\rho) + iN_0(k\rho) \\ H_\phi^{inc} = i\sqrt{\varepsilon_b}\frac{k}{k_0}\left(H_0^{(1)}(k\rho)\right) = i\sqrt{\varepsilon_b}\frac{k}{k_0}\left(J_0'(k\rho) + iN_0'(k\rho)\right) \end{cases}, \tag{4.22}$$

$$\begin{cases} E_z^{out} = H_0^{(2)}(k\rho) = J_0(k\rho) - iN_0(k\rho) \\ H_\phi^{out} = i\sqrt{\varepsilon_b}\frac{k}{k_0}\left(H_0^{(2)}(k\rho)\right) = i\sqrt{\varepsilon_b}\frac{k}{k_0}\left(J_0'(k\rho) - iN_0'(k\rho)\right) \end{cases}, \tag{4.23}$$

where $= \sqrt{k_0^2 - k_\parallel^2}$, $k_0 = \sqrt{\varepsilon_b}\frac{\omega}{c}$, $J_0(x)$ and $N_0(x)$ are the zeroth-order regular and singular Bessel functions, respectively, $J_0'(x) = -J_1(x)$ and $N_0'(x) = -N_1(x)$ are the first derivatives of these functions.

For this type of barrier resonant modes when taking into account the effective boundary conditions after Eqs. (4.20) and (4.21) we have

$$\begin{cases} E_z(k\rho)\,|_{\rho=R^+} = E_{2z}^{(0)}(k\rho)\,|_{\rho=R^-} \\ H_\phi(k\rho)\,|_{\rho=R^+} = -\frac{i\omega R}{c}\varepsilon_\parallel\gamma E_{2z}^{(0)}(k\rho)\,|_{\rho=R^-} \end{cases}, \tag{4.24}$$

where

$$E_z(k\rho)\,|_{\rho=R^+} = E_z^{inc}(k\rho)\,|_{\rho=R^+} + r_{DQWW}E_z^{out}(k\rho)\,|_{\rho=R^+}, \tag{4.25}$$

$$H_\phi(k\rho)\,|_{\rho=R^+} = H_\phi^{inc}(k\rho)\,|_{\rho=R^+} + r_{DQWW}H_\phi^{out}(k\rho)\,|_{\rho=R^+} \tag{4.26}$$

and $\gamma = \bar{E}_{1z}^{(0)}(R)\,/\bar{E}_{2z}^{(0)}(R) \sim 1$.

After substituting the Eqs. (4.22), (4.23), (4.25) and (4.26) in Eq. (4.24) we get for the scattering coefficient r_{DQWW} as

$$r_{DQWW} = \frac{[\Lambda N_0(kR) - N_1(kR)] - i\,[\Lambda J_0(kR) - J_1(kR)]}{[\Lambda N_0(kR) - N_1(kR)] + i\,[\Lambda J_0(kR) - J_1(kR)]}, \tag{4.27}$$

where $\Lambda = \frac{\omega R}{c\sqrt{\varepsilon_b}}\varepsilon_\parallel\gamma$ and $\varepsilon_\parallel$ is defined after Eq. (4.17).

Equation (4.27) presents the scattering coefficient of p-polarized light from the DQWW structure at normal to the wire axes incidence near $\omega = \omega_{ex} = \varepsilon_{k_\parallel} \hbar^{-1}$ and strongly depends from the DQWW material parameters such as: (a) quantum wire radius R and dielectric constant ε_w, (b) barrier dielectric constant ε_b.

Since here we are dealing with an accuracy of small parameters $\sim k_\parallel R$ and $\omega R/c$ then Eq. (4.27) will take the following limiting form

$$r_{DQWW} = \frac{(2/\pi)\left[\Lambda \ln(kR) + (kR)^{-1}\right] - i\Lambda}{(2/\pi)\left[\Lambda \ln(kR) + (kR)^{-1}\right] + i\Lambda}, \tag{4.28}$$

where the asymptotic small parameter limits of the Bessel functions are taken into account. If now substitute corresponding expressions for Λ, $\varepsilon_\parallel$ and χ parameters in Eq. (4.28) then after simple transformations we receive as in Ref. [15] a Breit-Wigner type final form for the scattering coefficient of the p-polarized light

$$r_{DQWW} = \frac{\omega - \widetilde{\omega}_p\left(k_\parallel, \omega_{ex}\right) - i\Gamma\left(k_\parallel, \omega_{ex}\right)}{\omega - \widetilde{\omega}_p\left(k_\parallel, \omega_{ex}\right) + i\Gamma\left(k_\parallel, \omega_{ex}\right)}, \tag{4.29}$$

with

$$\widetilde{\omega}_p\left(k_\parallel, \omega_{ex}\right) = \omega_{ex} + \frac{\alpha(k_0 R)^2 \ln(kR)}{\varepsilon_b + \alpha\varepsilon_w (k_0 R)^2 \ln(kR)} \Omega_{ex}, \tag{4.30}$$

where

$$\Omega_{ex} = \frac{(2\mu_{ex}^* e^8)^{1/3} \left[N(\mu_l)\, \Phi(\mu_l)\right]^2 |V_{cv}|^2}{\pi \varepsilon_w^{1/3} \hbar^5 \omega_{ex}^6 R^{8/3}}. \tag{4.31}$$

Here the dispersion law for the DQWW structure exciton-polaritons and their ra-diative broadening would be found from the following expressions

$$\omega = \widetilde{\omega}_p\left(k_\parallel, \omega_{ex}\right) \tag{4.32}$$

and

$$\Gamma\left(k_\parallel, \omega_{ex}\right) = \frac{\alpha\varepsilon_b (k_0 R)^2 \ln(kR)}{(\varepsilon_b + \alpha\varepsilon_w (k_0 R)^2 \ln(kR))(\varepsilon_b + \alpha(k_0 R)^2 \ln(kR))} \Omega_{ex}. \tag{4.33}$$

The Eqs. (4.30)–(4.33) made it possible to obtain the dispersion spectrum and life-time for the $1D$ exciton-polaritons of the DQWW near the exciton resonance with the opportunities of the valuable manipulations along with the magnitudes of the DQWW material parameters.

In conclusion, following the typical procedure for the optics of $2D$ system [1, 2, 27], we have developed a theory to solve Maxwell's equations together with the non-local $1D$ excitonic response to the dielectric polarization of the DQWW in the presence of the dielectric mismatch effect. This permits to calculate the dispersion spectrum of 1D exciton-polaritons of the DQWW near the excitonic resonance.

Acknowledgements The authors acknowledge funding by the State Committee of Science of RA for financial support under the grant 21T-1C275.

References

1. Agranovich, V.M., Mills, D.L.: Surface polaritons. North-Holland, Electromagnetic waves at surfaces and interfaces (1982)
2. Agranovich, V.M., Bassani, G.F.: Thin Films and Nanostructures: Electronic Excitations in Organic Based Nanostructures. Elsevier Academic Press (2003)
3. Cottam, M.G., Tilley, D.R.: Introduction to Surface and Superlattice Excitations. Cambridge University Press (1989)
4. Weisbuch, C., Nishioka, M., Ishikava, A., Arakawa, Y.: Observation of the coupled exciton-photon mode splitting in a semiconductor quantum microcavity. Phys. Rev. Lett. **69**, 3314 (1992)
5. Tignon, J., Voisin, P., De La Lande, C., Voos, M., Houdre, R., Osterle, U., Stanley, R.P.: From Fermi's golden rule to the vacuum rabi splitting: magnetopolaritons in a semiconductor optical microcavity. Phys. Rev. Lett. **74**, 3967 (1995)
6. Tassone, F., Bassani, F.: Reflectivity of the quantum grating and the grating polariton. Phys. Rev. B **51**, 16973 (1995)
7. Ivchenko, E.L., Kavokin, A.V.: Two-dimensional excitonic polaritons in a microcavity with quantum wires. JEETP Lett. **62**, 710 (1995)
8. Kaliteevski, M.A., Brand, S., Abram, R.A.: Exciton polaritons in a cylindrical Microcavitity with an embedded quantum wire. Phys. Rev. B **61**, 13791 (2000)
9. Kavokin, A.V., Baumberg, J.J., Malpuech, G., Laussy, F.P.: Microcavities. Oxford University Press (2007)
10. Trichet, A., Sun, L., Pavlovic, G., Gippius, N.A., Malpuech, G., Xie, W., Chen, Z., Richard, M., Dang, L.S.: One-dimensional ZnO exciton polaritons with negligible thermal broadening at room temperature. Phys. Rev. B **83**, 041302(R) (2011)
11. Keldish, L.V.: Polaritons in thin semiconducting films. JEETP Lett. **30**, 224 (1979)
12. Keldish, L.V.: Excitons and polaritons in semiconductor/insulator quantum wells and superlattices. Superlattices Microstruct. **4**, 637 (1988)
13. Keldish, L.V.: Excitons in semiconductor-dielectric nanostructures. Phys. Status Solidi (a) **164**, 3 (1997)
14. Tassone, F., Bassani, F., Andreani, L.C.: Resonant and surface polaritons in quantum wells. Nuovo Cimento **D12**, 1673 (1990)
15. Tassone, F., Bassani, F.: Quantum wire polaritons. Nuovo Cimento **D14**, 124 (1992)
16. Miller, R.C., Kleinman, D.A., Tsang, W.T., Gossard, A.C.: Observation of the excited level of excitons in GaAs quantum wells. Phys. Rev. B **24**, 1134 (1981)
17. Zhukov, E.A., Muljarov, E.A., Romanov, S., Masumoto, Y.: Pump-probe studies of photoluminescence of InP quantum wires embedded in dielectric matrix. Solid St. Commun. **112**, 575 (1999)
18. Shinada, M., Sugano, S.: Interband optical transitions in extremely anisotropic semiconductors. I. Bound and unbound exciton absorption. J. Phys. Soc. Jpn. **21**, 1936 (1966)
19. Kazaryan, E.M., Enfiadjyan, R.L.: On the theory of the light absorption in thin semiconductor films in presence of the quantum size effect. Semiconductors **5**, 2002 (1971)
20. Rytova, N.S.: Coulomb interaction of the electrons in thin film. Soviet Phyics-Doklady **10**, 754 (1965)
21. Rytova, N.S.: Screened potential of the point charge in thin film. Vestnik Mosk. Univ. Fizika, Astronomia 30 (1967)
22. Babichenko, V.S., Keldysh, L.V., Silin, A.P.: Coulomb interaction in thin semiconductor and semimetal wires. Phys. Solid State **22**, 1238 (1980)

23. Chaplik, A.V., Entin, M.V.: Charged impurities in very thin layers. JETP **61**, 2469 (1971)
24. Keldysh, L.V.: Coulomb interaction in thin semiconductor and semimetal films. JEETP Lett. **29**, 658 (1979)
25. Aharonyan, K.H., Kazaryan, E.M.: Charged impurities in thin quantized wires. Semiconductors **16**, 122 (1982)
26. Aharonyan, K.H., Kazaryan, E.M.: Exciton states in thin semiconductor wires and their influence on the coefficient of the interband absorption. Semiconductors **17**, 1140 (1983)
27. Aharonyan, K.H., Tilley, D.R.: Propagating electromagnetic modes in thin semiconductor films. J. Phys. Condens. Matter **1**, 5391 (1989)

Chapter 5
Broadband Infrared Absorption Due to Low Q-factor Dipole Modes of Cr Strips

H. A. Parsamyan, D. S. Hambaryan, and H. S. Haroyan

Abstract The absorption properties of a metamaterial absorber based on the chromium rectangular strips is studied in the long-wavelength infrared spectrum. Theoretical analysis in the framework of electrostatic approximation is performed for an individual elongated oblate ellipsoid, and the resonant condition ensuring strong absorption and negligible scattering is established. The analysis shows that the absorption cross-section of such a particle exceeds the geometrical cross-section by about 15 times with the absorption line width of the order of the resonant wavelength. The results were applied to an equivalent rectangular strip to design a broadband absorber. The analysis reveals that the absorption bandwidth of a periodic system of strips distributed on a metal-grounded dielectric surface can reach 10.7 μm in the spectrum from 8.88 to 19.58 μm.

5.1 Introduction

During the past two decades due to advances in nanotechnology investigations of the interaction of subwavelength metallic configurations with electromagnetic radiation and their resonant characteristics have been grown enormously. Metallic configurations with subwavelength sizes, such as spheres, rods, disks, cubes etc. show resonant behaviour under the incident electromagnetic waves due to so-called localized surface plasmon resonance (LSPR) [1, 2]. The LSPR of noble metal particles (e.g., Au and Ag), conditioned by the collective oscillations of the conduction electrons of metals with electromagnetic waves mostly lies in the visible and near-infrared (NIR) spectrum. However, resonant properties of noble metal configurations disappear when one aims to shift the working spectrum to the longer wavelength range due to the rapid increase in the imaginary part and the absolute value of the real

H. A. Parsamyan (✉) · D. S. Hambaryan · H. S. Haroyan
Department of Radiophysics, Yerevan State University, A. Manoogian 1, Yerevan 0025, Armenia
e-mail: hparsamyan@ysu.am

H. A. Parsamyan
Center for Nanoscience and Technology, Institute of Chemical Physics, NAS RA, P. Sevak 5/2, Yerevan 0014, Armenia

D. Blaschke et al. (eds.), *Optics and Its Applications*, Springer Proceedings in Physics 281, https://doi.org/10.1007/978-3-031-11287-4_5

part of dielectric permittivities of these metals. This is mainly conditioned by the extremely high carrier concentrations of the order of $\sim 10^{-22} - 10^{23}\ \text{cm}^{-3}$ such that the plasma frequency of these metals lies in the ultraviolet [3]. Because of the LSPR resulting in the strong absorption of the noble metal particles in the visible and NIR, various optical devices for label-free biosensing, photodetection, electromagnetic wave energy confinement and absorption have been proposed [4].

The plasma frequency can be effectively manipulated, for instance, by considering semiconductor configurations with various doping levels allowing one to shift the LSPR to the mid- and long-wave infrared [5]. This opens up new benefits for controlling absorption and scattering at longer wavelengths by subwavelength configurations and particularly to increase the absorption of the electromagnetic waves [1]. Metamaterial- and metasurface-based configurations have been widely used for absorption and filtering of electromagnetic waves [6]. Such systems are mainly composed of orderly arranged subwavelength periodic units [7]. The common method to achieve wideband or multiband absorption is based on combining several resonant configurations in a single unit cell of the metastructure and thus merging the absorption spectra of each element, which also makes systems more complex and difficult to analyze [8–10]. Near-perfect metamaterial absorbers were proposed across the electromagnetic spectrum from the visible to microwaves [11–13]. Although in the visible and NIR perfect absorption is associated with the plasmonic resonances and the absorber structures are straightforward, at longer wavelengths due to the absence of any plasmonic effects in the common configurations (spheres, cubes, rods etc.), the researchers modify the configurations in a certain way to reduce the effective dielectric permittivity of the whole system. A cure to this can be so-called poor metals, such as chromium (Cr), nickel (Ni), titanium (Ti) and tungsten (W) whose real and imaginary parts of dielectric permittivities are of the same order in magnitude [14–16].

In this article, we suggest to exploiting a broadband mid- and long-wave infrared absorber composed of rectangular strips made of Cr. First, absorption and scattering properties of an individual oblate ellipsoidal particle are theoretically investigated by applying electric-dipole approximation. A functional dependence between geometrical and material parameters of the strip is derived ensuring strong absorption and negligible scattering. Numerical analysis based on the finite element method (FEM) is carried out for an equivalent rectangular strip-like configuration to derive the absorption and scattering cross-sections (ACS and SCS, respectively). The simulation and theoretical results were compared showing good agreement. Using the results obtained for an individual strip, a 3D periodic structure was designed composed of a number of such strips and a dielectric layer. Efficient ($>90\%$) broadband and polarization-sensitive absorption is achieved in such a system from $\sim$8.88 to 19.58 μm corresponding to the absolute bandwidth (BW) of 10.7 μm and the relative bandwidth (RBW) which is the ratio of the BW to the center wavelength of the absorption band in precents is around 75%.

5.2 Results and Discussion

5.2.1 Theory

Let's assume an oblate ellipsoidal configuration with semi-axes $a_x < a_y \ll a_z$ and complex dielectric permittivity of ε_1 embedded in a medium with a real dielectric constant of ε_3 is illuminated by electromagnetic waves polarized along the long semi-axis. By applying electric-dipole approximation, i.e., considering that $a_z \ll \lambda$, where λ is the incident wavelength in the surrounding, one can derive the absorption and scattering cross-section of the prolate ellipsoid as follows [17, 18]:

$$\sigma_{abs} = \frac{k}{\varepsilon_0}\mathrm{Im}\left(\alpha\right), \tag{5.1}$$

$$\sigma_{scat} = \frac{k^4}{6\pi\varepsilon_0}\left|\alpha\right|^2, \tag{5.2}$$

where

$$\alpha \approx \frac{4\pi\varepsilon_0}{3}\cdot\frac{a_z^3}{0.4-\ln\eta}\cdot\frac{n^{(z)}}{\varepsilon_2/(\varepsilon_1-\varepsilon_2)+n^{(z)}}, \tag{5.3}$$

is the polarizability of the particle, $\eta = a_y/a_z$, $\gamma = a_x/a_z$ and $n^{(z)} = \eta\gamma\ln\left(1.5/\eta\right)$ is the so-called depolarization factor which depends only on the shape of a particle. The wavelength-dependent dielectric permittivity of a metallic particle can be presented by the form $\varepsilon_1\left(\lambda\right) = \varepsilon_{1r}\left(\lambda\right) + i\cdot\varepsilon_{2i}\left(\lambda\right)$, where $\varepsilon_{1r} < 0$. The absorption resonance of a particle occurs when the following condition between the dielectric permittivities of the particle and surrounding (the right-hand side) and the depolarization factor related to the geometrical parameters of the particle is met:

$$\frac{a_x a_y}{a_z^2}\left(0.4-\ln\frac{a_y}{a_z}\right) = \frac{\left|\varepsilon_{1r}\right|\varepsilon_2}{\left|\varepsilon_1\right|^2}. \tag{5.4}$$

Thus, the absorption of a particle with given metallic material can be maximized by varying the sizes of a particle so that condition (5.4) is satisfied. Throughout the paper, we will use ACS and SCS normalized by the longitudinal geometrical cross-section of a particle ($\sigma_{geom} = \pi a_y a_z$) known as absorption and scattering efficiencies [12]:

$$Q_{abs} = \frac{\sigma_{abs}}{\sigma_{geom}} \text{ and } Q_{scat} = \frac{\sigma_{scat}}{\sigma_{geom}}. \tag{5.5}$$

Investigating the optical response of an ellipsoid made of poor metals (e.g., Cr or Ti) in the long-wavelength infrared spectrum, where the real and imaginary parts of the dielectric permittivities of these metals are of the same order in magnitude, one obtains that

$$Q_{abs} \sim 8\pi a_x |\varepsilon_1|^2/(3\lambda\varepsilon_2\varepsilon_{1i}), \tag{5.6}$$

which is noticeably larger than the unity, whereas the scattering efficiency

$$Q_{scat} \sim (2\pi)^2 a_x^2 a_y a_z |\varepsilon_1|^2/\lambda^4. \tag{5.7}$$

is negligibly small.

Figure 5.1 left axis plots the real (solid black) and imaginary (dotted black) parts of the dielectric permittivity of Cr [19].

5.2.2 Numerical Analysis

Full-wave numerical analysis based on the FEM was conducted to reveal the dependence of the absorption and scattering properties of the geometrical parameters. To simplify the simulated geometry, the oblate spheroid with semi-axes a_i $(i = x, y, z)$ was replaced by an equivalent rectangular strip having width $d_x = 2a_x$, depth $d_y = 2a_y$ and height $d_z = 2a_z$, schematically illustrated in Fig. 5.2a. The strip is illuminated by a z-polarized plane wave propagating along the x-axis.

We choose Cr as the material of the configurations since the real and imaginary parts of the dielectric permittivity of this metal in the infrared are of the same order and for given geometrical parameters condition (5.4) can be satisfied ensuring strong absorption and low scattering. Theoretical values of absorption (Q_{abs}) and scattering (Q_{scat}) efficiencies of an oblate ellipsoid as a function of the wavelength obtained

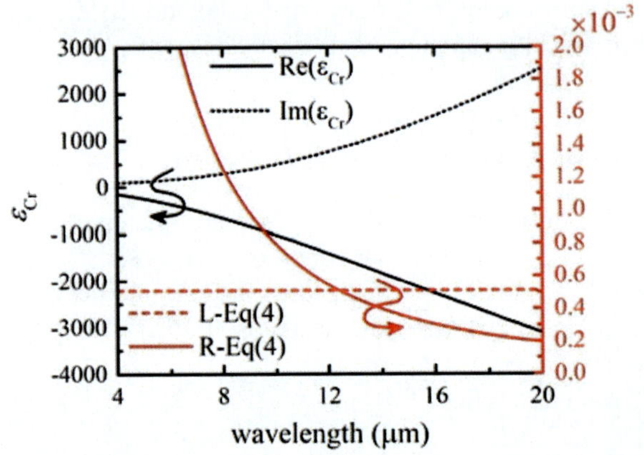

Fig. 5.1 Real (solid black) and imaginary (dotted black) parts of the dielectric permittivity of Cr according to Ref. [19] (left axis). Graphical representation of condition (5.4) of the absorption resonance (right axis). Dashed and solid red lines stand for the left-hand [L-Eq. (5.4)] and right-hand [R-Eq. (5.4)] sides of condition (5.4). The geometrical parameters are: $a_x = 5$ nm, $a_y = 25$ nm and $a_z = 1$ μm. As a metal and surrounding Cr and air ($\varepsilon_2 = 1$) were used, respectively

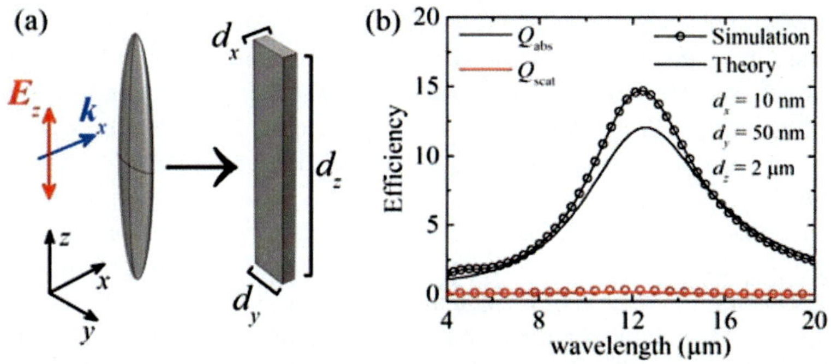

Fig. 5.2 **a** Schematic sketch of the oblate elongated ellipsoid and the equivalent rectangular strip and **b** absorption and scattering efficiencies of the oblate ellipsoidal Cr particle obtained according to Eqs. (5.1) and (5.2); represented by black and red solid lines, respectively. The same for a rectangular strip represented by black and red rings. The sizes of the strip are: $d_x = 10$ nm, $d_y = 50$ nm and $d_z = 2$ μm. The surrounding medium is air with $\varepsilon_2 = 1$

according to Eqs. (5.1) and (5.2) are presented in Fig. 5.2b by solid black and red lines, respectively. Simulated Q_{abs} and Q_{scat} spectra of a strip with sizes of $d_x = 10$ nm, $d_y = 50$ nm and $d_z = 2$ μm are depicted by black and red rings, correspondingly. One sees that the theoretical and simulated results are in good correspondence. Particularly, the resonant wavelength in both cases is around 12.5 μm and the slight difference between the Q_{abs} values ($Q_{abs}^{th} \sim 12$ and $Q_{abs}^{sim} \sim 14.7$) can be associated with the strong concentrations of charges at the sharp corners of the strip. In the theoretical part, an oblate ellipsoid is considered. The approximation of the oblate elliptical configuration to a rectangular strip is valid since condition $a_z \gg a_x, a_y$ is applied to semi-axes of the ellipsoid. Note that longitudinal cross-section areas of an ellipsoid and equivalent strip are close enough ($S_{el}/S_{st} \sim 0.8$). The validity of the electrostatic approximation is worsened as a_y and a_x increase. On the other hand, although S_{st} is slightly larger than S_{el}, at the resonance $Q_{st} > Q_{el}$ meaning that the absorption properties are improved with a strip. The scattering efficiency is negligibly small compared to the Q_{abs} so that the absorption-to-scattering ratio of the suggested configuration is about 50, the full width at half maximum (FWHM) of the absorption resonance is ~5.6 μm, which is of the same order as the resonant wavelength. The complex dielectric permittivity of Cr is $\varepsilon_{\text{Cr}} = -1523.8 + 844.26i$ (see Fig. 5.1) at the absorption resonance $\lambda \sim 12.5$ μm [19].

The spectral dependencies of the absorption and scattering efficiencies on the strip d_x and d_y sizes are plotted in Fig. 5.3a–b and c–d, respectively. The absorption resonance in both cases blueshifts by increasing d_x and d_y. Increasing d_x from 10 to 50 nm results in a rapid increase of the Q_{abs} from 14 to 39, nevertheless, Q_{scat} also increases drastically from 0.3 to about 7.5. The absorption-to-scattering ratio, in this case, decreases from 50 to 4.9. However, the scattering efficiency is quite small compared with the absorption for d_x values up to around 35 nm. One sees from Eqs. (5.6) and (5.7) that Q_{abs} linearly depends on d_x, whereas Q_{scat} is a quadratic function of d_x and such dependencies are conformed by numerical simulations. On the contrary, both the absorption and scattering efficiencies slightly depend on d_y. The peak value of Q_{abs} practically remains unchanged as d_y increased from 50 up to 100 nm and only FWHM undergoes a minor decrease. Note that the absorption-to-scattering ratio is more tolerant to d_y variations in contrast to d_x changes thus fabrication imperfections in the d_y will not noticeably affect the absorption and scattering performance of the configuration.The blueshift of the absorption resonance as d_x and d_y are increased directly follows from condition (5.4). The graphical representation of condition (5.4) is illustrated in Figure 5.1 right axis. Here the right-hand side of Eq. (5.4) depending on the metal and surrounding dielectric permittivities is a monotonically decreasing function from wavelength [R-Eq. (5.4)]. The left-hand side of Eq. (5.4) depending only on the geometrical parameters is depicted by the red dashed line [L-Eq. (5.4)]. The intersection point corresponds to the absorption peak wavelength ($\lambda = 12.5\ \mu m$ in the considered case), which means that increasing d_x or d_y will lead condition (5.4) to be satisfied at lower wavelengths as the intersection point moves to left. Note that the natural logarithm in the left-hand side is a slowly varying function of $d_x/d_z \ll 1$. Recall that Cr is used since the real and imaginary parts of its dielectric permittivity are of the same order in magnitude. Comparing the absorption and scattering properties with Au strip, one will see that the Q_{abs} and Q_{scat} are of the same order. It is due to the significant contrast between the real and imaginary parts of the dielectric permittivity of Au.

In this case, for a strip with the same sizes, condition (5.4) is satisfied at $\lambda \sim 6\ \mu\text{m}$ leading to Q_{abs} and Q_{scat} being of the same order since the resonant wavelength approaches to the d_z height of the strip. Dielectric permittivity of Au at $\lambda = 6\ \mu\text{m}$ is $\varepsilon_{\text{Au}} = -1818.4 + 428.48i$ [20].

Until this point, we investigated the interaction of the incident plane waves with a rectangular strip-like particle by means of electrostatic approximation using scattering and absorption cross-sections. However, for practical applications, absorptance of the system should be introduced. To use such strips for efficient absorption of electromagnetic waves one needs to follow the conventional design path [21] and create a periodic system composed of a bottom reflector, dielectric spacer and rectangular strips. The lateral (along the incident plane) and azimuthal (the surface of the dielectric) cross-sections of the unit cell are illustrated in Fig. 5.4a. Here the unit cell is composed of a Cr reflector, a dielectric spacer and a Cr strip. The unit cell is periodic along y- and z-axes ("PC" notations) with periodicities of p_y and p_z. t_m, t_d and d_x are the thickness of metal ground, dielectric spacer and strip, respectively. The unit cell is illuminated by a z-polarized plane wave. First, to explore the absolute absorptance of the strip only, a system whose unit cell consists of a dielectric substrate and a strip (i.e., the bottom reflector is absent) is considered and its performance is compared with that of the substrate coated by a uniform Cr layer. The results are shown in Fig. 5.4b, where the solid line stands for the case of strip and the dashed line for the uniform metal layer.

The absorptance in such a system is calculated by general expression $A = 1 - R - T$, where R is the reflection and T is the transmission coefficient. One sees that the absorptance of the uniform metal film is approximately constant in the studied

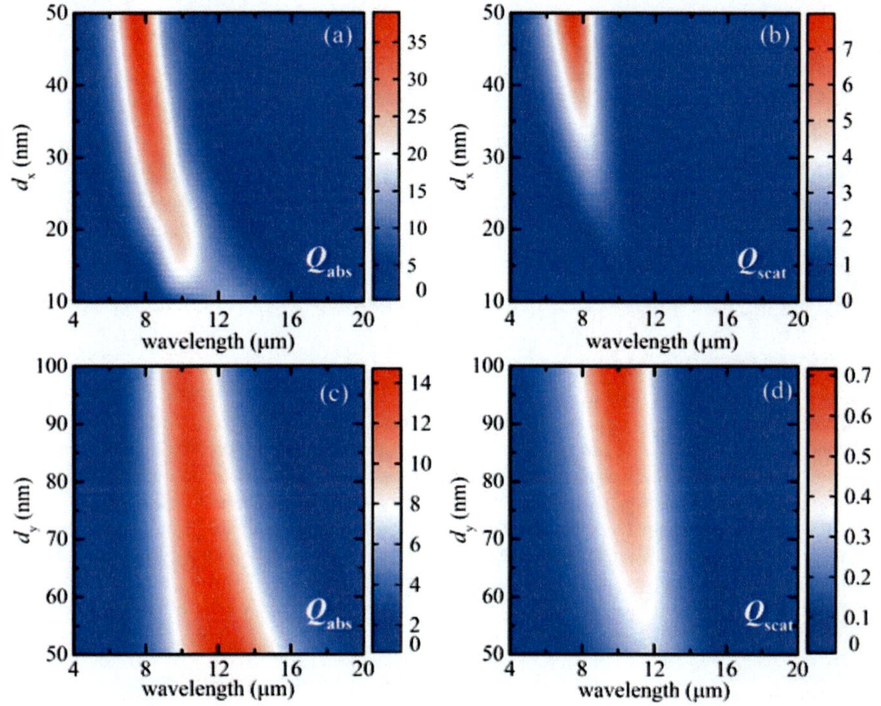

Fig. 5.3 Evaluations of the absorption (the first column) and scattering (the second column) efficiency as a function of the wavelength and $(a-b)d_x$ and $(c-d)d_y$. The height $d_z = 2$ μm is set in all cases. The surrounding medium is air

spectrum and is of the order of 0.18. On the contrary the absorption spectrum of a system with a strip has resonant nature with an absorptance peak value ∼0.42 at $\lambda \sim 13$ μm.

Introducing a metal ground layer of a thickness $t_m = 100$ nm reasonably improves the absorptance of the system. Here again comparison between a system composed of a strip with the one consisting of a uniform metal layer of the same thickness is performed and the results are shown in Fig. 5.4c. A narrow absorption resonance with a peak value of ∼0.93 at $\lambda = 4.08$ μm is observed of a system with a uniform Cr layer which is associated to the asymmetric Fabry-Perot mode of the structure. In contrast, the ground reflector increases the absorptance as the wave interacts twice with the strip (through the forward wave and the wave reflected from the ground metal), and also broadens the bandwidth. Since the ground plane prevents light transmission in the forward direction and $T = 0$, the absorptance of such a system is calculated by a simplified expression $A = 1 - R$. Two absorption peaks are observed with the absorption rates of ∼0.7 and 0.81 at resonant wavelengths $\lambda_1 = 5.7$ μm and $\lambda_2 = 15.3$ μm, respectively. The physical origin of broadband absorption of the structure is associated with low-Q-factor dipole modes of the rectangular strip, as

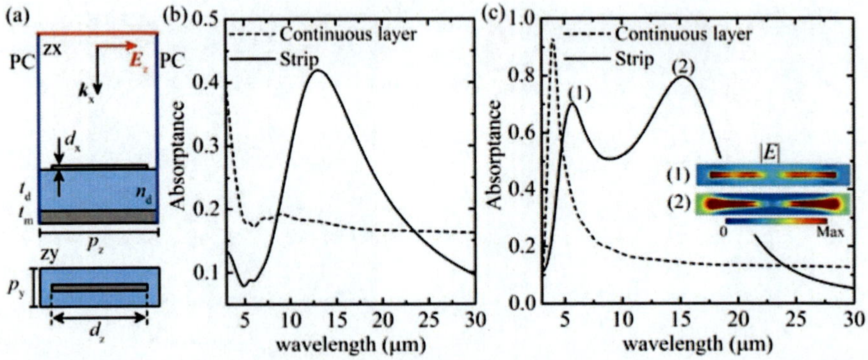

Fig. 5.4 Absorption performance of the periodic structure composed of a Cr strip, dielectric spacer and Cr reflector. **a** zx- and zy-plane view of the unit cell of the periodic structure. p_y and p_z are periodicities, t_m, t_d and d_x are the thickness of the bottom reflector, the dielectric spacer and the strip. **b** Comparison of the absorptance of a system composed of a dielectric substrate and orderly arranged rectangular Cr strips (without a bottom reflector) with a continuous Cr layer. **c** The same as in (**b**) but with a bottom Cr reflector. The inset images show the distributions of the electric field magnitude at (1) and (2) resonances. During the analysis the following parameters were used: $t_d = 1$ μm, $p_z = 2.5$ μm, $p_y = 350$ nm, $d_x = 10$ nm, $dz = 2$ μm, $n_d = 1.5$. Both the strip and continuous layer have the same thickness d_x. $t_m = 100$ nm is used for the metal ground thickness throughout the analysis

follows from Fig. 5.4c insets showing electric field distributions along the strip at two absorption maxima (5.7 μm and 15.3 μm). The low-Q-factor modes are due to the large loss tangent of Cr in the considered part of the infrared spectrum. For instance, at $\lambda_2 = 15.3$ μm the real and the imaginary parts of the dielectric permittivity of Cr are $\varepsilon_r = \sim 2120.7$ and $\varepsilon_{im} = 1372.1$, respectively, yielding $\tan\delta \sim 0.647$. At the same time, loss tangent of Au is $\tan\delta \sim 0.55$.

This basic structure of an absorber can be further optimized to achieve higher absorptance and wider bandwidth. Particularly, Fig. 5.5a shows the absorption spectrum at the dielectric spacer thickness of 1, 1.5 and 2 μm. One sees that the absorptance level, as well as the absorption bandwidth, are changed noticeably by increasing the dielectric thickness. Here >90% absorption from about 8.88 up to 19.58 mm with the absorption $BW \sim 10.7$ μm is achieved for a metamaterial structure with unit cell sizes as follows: $p_z = 2.5$ μm, $p_y = 350$ nm and $t_d = 2$ μm. The sizes of the strip are $d_z = 2$ μm, $d_y = 50$ nm and $d_x = 10$ nm. The relative bandwidth of the absorber is calculated according to the expression $RBW = BW/\lambda_c \times 100\%$, where λ_c is the central wavelength of the efficient absorption band. Considering that at the optimized thickness of the dielectric spacer $\lambda_c \sim 14.23$ μm and $BW \sim 10.7$ μm, RBW of the absorber is calculated to be around 75%.

The absorption spectrum at p_y values of 300 nm (dots), 350 nm (dashed line) and 400 nm (solid line) are depicted in Fig. 5.5b. The dielectric layer thickness and z-periodicity are $t_d = 2$ μm and $p_z = 2.5$ μm. It is seen that the absorption bandwidth slightly decreases by increasing the periodicity in the y-direction. However,

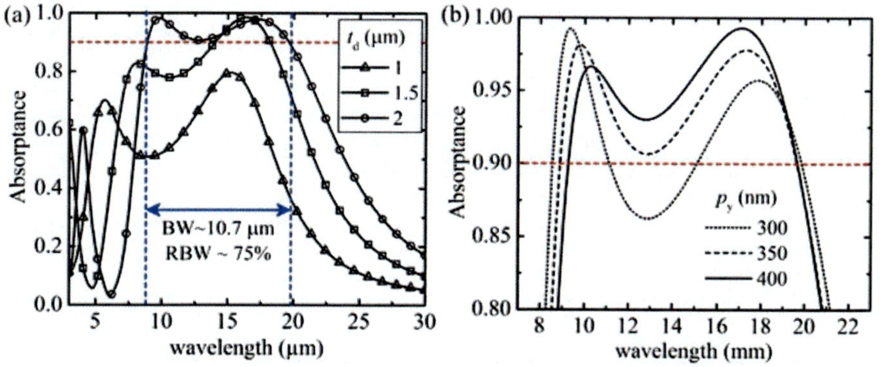

Fig. 5.5 Absorption spectrum at different values of **a** dielectric spacer thickness t_d of 1 μm (triangles), 1.5 μm (squares) and 2 μm (rings) and **b** periodicity p_y of 300 nm (dots), 350 nm (dashed line) and 400 nm (solid line). The following geometrical parameters are used if otherwise is not stated: $p_z = 2.5$ μm, $p_y = 350$ nm, $t_d = 2$ μm, $d_z = 2$ μm, $d_y = 50$ nm and $d_x = 10$ nm

the absorptance is improved. For instance, at $p_y = 300$ nm the absorptance breaks $A = 0.9$ level with an absorption dip of ~0.86 in the ~$11.15 - 15$μm wavelength range. As p_y increases, the average absorptance in the spectrum corresponding to also increases from 94.3% at $p_y = 350$ nm to ~95.2% at $p_y = 400$ nm. The total thickness of the absorber optimized unit cell is 2.11 μm, which is ~6.75 times smaller than the central wavelength of the efficient absorption band.

5.3 Conclusion

Summarizing, we investigated absorption and scattering features of an elongated oblate ellipsoid and equivalent rectangular strip in the framework of electrostatic approximation. The analysis reveals that dominant-over-scattering absorption can be ensured by careful choice of the geometrical and material parameters leading to the absorption cross-section being 15 times larger than the lateral geometrical cross-section and 50 times larger than that of the scattering. The performance of such configurations was studied as unit cell elements of a metamaterial structure. Broadband and efficient ($A \geq 90\%$) absorption in the spectrum from 8.88 μm up to 19.58 μm with the bandwidth of 10.7 μm is achieved by considering a metal-dielectric-metal system, where the top metal is a Cr strip. Such favourable properties are achieved with a structure of a total thickness of 6.75 times smaller than the central wavelength of the efficient absorption band.

Acknowledgements This work was supported by scientific research grants through the Science Committee of MESCS of Armenia (20DP-1C05, 21AG-1C061 and 21AA-1C011), by a faculty research funding program 2021 implemented by Enterprise Incubator Foundation (EIF) with the support of PMI science and PhD support program 2021 by EIF and PMI science.

References

1. Kasani, S., Curtin, K., Wu, N.: A review of 2D and 3D plasmonic nanostructure array patterns: fabrication, light management and sensing applications. Nanophotonics **8**, 2065–2089 (2019)
2. Yao, Y., Zhou, J., Liu, Z., Liu, X., Fu, G., Liu, G.: Refractory materials and plasmonics based perfect absorbers. Nanotechnology **32**, 132002 (2021)
3. Maier, S.A.: Plasmonics: Fundamentals and Applications. Springer, US (2007)
4. Hedayati, M., Faupel, F., Elbahri, M.: Review of plasmonic nanocomposite metamaterial absorber. Materials (Basel). **7**, 1221–1248 (2014)
5. Kapetanovic, V., Bicket, I.C., Lazar, S., Lagos, M.J., Botton, G.A.: Tunable infrared plasmon response of lithographic Sn-doped indium oxide nanostructures. Adv. Opt. Mater. **8**, 2001024 (2020)
6. Ogawa, S., Kimata, M.: Metal-insulator-metal-based plasmonic metamaterial absorbers at visible and infrared wavelengths: a review, Materials (Basel). **11**, 458 (2018)
7. Kim, J., Han, K., Hahn, J.W.: Selective dual-band metamaterial perfect absorber for infrared stealth technology. Sci. Rep. **7**, 6740 (2017)
8. Liu, J., Ma, W.-Z., Chen, W., Yu, G.-X., Chen, Y.-S., Deng, X.-C., Yang, C.-F.: Numerical analysis of an ultra-wideband metamaterial absorber with high absorptivity from visible light to near-infrared. Opt. Express **28**, 23748 (2020)
9. Ma, W., Wen, Y., Yu, X.: Broadband metamaterial absorber at mid-infrared using multiplexed cross resonators. Opt. Express **21**, 30724 (2013)
10. Liang, C., Yi, Z., Chen, X., Tang, Y., Yi, Y., Zhou, Z., Wu, X., Huang, Z., Yi, Y., Zhang, G.: Dual-band infrared perfect absorber based on a Ag-Dielectric-Ag multilayer films with Nanoring Grooves Arrays. Plasmonics **15**, 93–100 (2020)
11. Parsamyan, H., Haroyan, H., Nerkararyan, K.: Broadband microwave absorption based on the configuration resonance of wires. Appl. Phys. A **126**, 773 (2020)
12. Parsamyan, H.A., Nerkararyan, K.V., Bozhevolnyi, S.I.: Efficient broadband infrared absorbers based on core-shell nanostructures. J. Opt. Soc. Am. B **36**, 2643 (2019)
13. Seren, H.R., Zhang, J., Keiser, G.R., Maddox, S.J., Zhao, X., Fan, K., Bank, S.R., Zhang, X., Averitt, R.D.: Nonlinear terahertz devices utilizing semiconducting plasmonic metamaterials. Light Sci. Appl. **5**, e16078–e16078 (2016)
14. Yu, P., Yang, H., Chen, X., Yi, Z., Yao, W., Chen, J., Yi, Y., Wu, P.: Ultra-wideband solar absorber based on refractory titanium metal. Renew. Energy **158**, 227–235 (2020)
15. Parsamyan, H.: Near-perfect broadband infrared metamaterial absorber utilizing nickel. Appl. Opt. **59**, 7504 (2020)
16. Rana, A.S., Mehmood, M.Q., Jeong, H., Kim, I., Rho, J.: Tungsten-based ultrathin absorber for visible regime. Sci. Rep. **8**, 2443 (2018)
17. Landau, L.D., Pitaevskii, L.P., Lifshitz, E. M.: Electrodynamics of Continuous Media, 2nd ed. Elsevier Science & Technology (1984)
18. Carlos, E.: Solivérez. The Depolarization Tensor Method (Free Scientific Information, Electrostatics and Magnetostatics of Polarized Ellipsoidal Bodies (2016)
19. Rakić, A.D., Djurišić, A.B., Elazar, J.M., Majewski, M.L.: Optical properties of metallic films for vertical-cavity optoelectronic devices. Appl. Opt. **37**, 5271 (1998)
20. Ordal, M.A., Bell, R.J., Alexander, R.W., Long, L.L., Querry, M.R.: Optical properties of Au, Ni, and Pb at submillimeter wavelengths. Appl. Opt. **26**, 744 (1987)
21. Yu, P., Besteiro, L.V., Huang, Y., Wu, J., Fu, L., Tan, H.H., Jagadish, C., Wiederrecht, G.P., Govorov, A.O., Wang, Z.: Broadband metamaterial absorbers. Adv. Opt. Mater. **7**, 1800995 (2019)

Chapter 6
Microwave and Joule Heating Visualization by a Thermo-Elastic Microscope for Carbon Composite Materials

Sh. Arakelyan, A. Babajanyan, G. Berthiau, B. Friedman, and K. Lee

Abstract Microwave heating visualization for the carbon fiber/polyether ether ketone composite material was implemented using the thermo-elastic optical indicator microscopy (TEOIM) system. Losses of an anisotropic composite material were characterized due to the microwave radiation influence. TEOIM visualization showed that the near-field distribution stretches along the composite material high conductivity axis and takes different forms when some defects in the composite material were analyzed.The visualized microwave heat was measured to be 0.2 K when the probe feeding power was around 15 dBm.The Obtained microwave heat visualization data was in good agreement with the simulated result. Additionally, noninvasive Joule heating (around 1 K) made by 0.3 A DC flux was observed in order to characterize mechanical faults in a composite material.

6.1 Introduction

Reinforced carbon fiber composites have wide range of applications in medicine (surgical and dental pharmaceutical applications), aircraft construction, engineering, industry, etc. [1–4]. The carbon fibers are stable under high mechanical stress, have refined elastic properties and can resist high mechanical deformation by changing shape whereas other materials such as metals or ceramics would be destroyed. In some applications, carbon fibers composite materials can successfully replace metals [5]. Depending on the required properties carbon fibers with various structural com-

Sh. Arakelyan · K. Lee
Department of Physics, Sogang University, Seoul 121-742, Korea

A. Babajanyan (✉)
Department of Radiophysics, Yerevan State University, Yerevan 0025, Armenia
e-mail: barsen@ysu.am

G. Berthiau
IREENA, Boulevard de L'iniversite 37, BP-406, 44602, Saint-Nazaire Cedex, France

B. Friedman
Department of Physics, Sam Houston State University, Huntsville, TX 77341, USA

D. Blaschke et al. (eds.), *Optics and Its Applications*, Springer Proceedings in Physics 281, https://doi.org/10.1007/978-3-031-11287-4_6

binations are used. An example of such a fiber configuration is a unidirectional carbon fiber/polyether ether ketone (PEEK) composite material, the mechanical properties of which are well studied [6–8].

Due to its nature the electrical and mechanical properties of the carbon fiber/PEEK composite material are spatially anisotropic. Such anisotropic properties are attractive from the point of view material science particularly, when investigating the interaction between the electromagnetic field and a material. Recent examples of such studies include: microwave absorption properties investigation [9, 10] and its enhancement for radar absorption purposes [11, 12], carbon-based composites mechanical properties enhancement by microwave curing [13, 14], characterization of anisotropic electrical properties of a carbon fiber/PEEK composite [15], testing and inspection of reinforced carbon composites by a microwave microscopy [16] etc. The heating process due to the microwave absorption is important in the above studies and thus we have undertaken a detailed investigation. In the heating of high electrical conductivity materials induction currents are responsible whereas for low conductivity materials the heat generally is created due to displacement currents [17, 18]. From this point of view, the investigation of composite material anisotropy properties presents a wide area of interest, since the anisotropic electrical conductivity of the unidirectional carbon fiber/PEEK composite material is drastically changed depending on the fibers orientation. However, to characterize the microwave heating mechanisms and losses in the composite material one must use temperature detection systems with a high sensitivity. An example of such a system is the thermo-elastic optical indicator microscopy (TEOIM) technique developed by H. Lee and co-authors for the temperature and microwave near-field visualization [19]. The TEOIM technique with slide glass indicators visualizes the thermal distribution without a scanning process and is more sensitive (up to 4 mK) compared to other thermal detectors such as infrared devices [20, 21].

In this paper, we present microwave energy losses characterization in carbon fibers/PEEK composite material by the TEOIM visualization. We visualize microwave heating of the composite material sheet and observe its changes when artificial defects in the form of hole have been made in a composite material sheet. We also provide a COMSOL Multiphysics simulation for the best understanding of the anisotropic material heating phenomenon and comparable analysis of obtained experimental results. Taking into account the thermal distributions due to the conductive and dielectric losses it will be possible to realize a qualitative estimation of the electromagnetic field distribution. In addition, to characterize mechanical defects in a composite material we perform Joule heat visualization caused by a direct current (DC) flux. The proposed defect measurement technique may become a supplement to the mechanical property testing used for carbon fiber reinforced composites. Such investigation may also be a useful tool for the study of electromagnetic shielding properties in an anisotropic material.

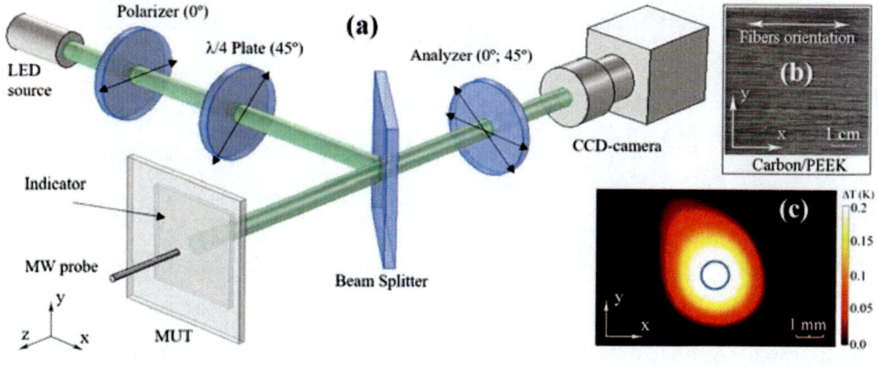

Fig. 6.1 **a** The experimental setup of the thermo-elastic imaging system, **b** the fibers orientation in the composite material, and **c** the visualized thermal pattern caused by microwave field at 5 GHz from end of MW probe at 1 cm distance

6.2 Materials and Methods

The temperature distribution in a material with electrical and thermal anisotropy was measured by the TEOIM technique [19]; a schematic diagram is shown in Fig. 6.1a.

The composite material with 6 cm × 6 cm surface area used in our experiments consists of PEEK and unidirectional aligned carbon fibers with diameters of 1 μm Fig. 6.1b. The volume fraction of fibers in the composite material is 3 : 5 (or 60%). The electrical and thermal conductivities of the composite sheet are diagonal tensors with elements of $\sigma_{xx} \cong 3.9 \times 10^4 S/m$, $\sigma_{yy} = \sigma_{zz} \cong 7.7 S/m$ [22] and $k_{xx} \cong 4.5 W/(m \cdot K)$, $k_{yy} = k_{zz} \cong 0.67 W/(m \cdot K)$ [23] respectively. In all experiments, composite material fibers were aligned along the x-axis. As an indicator, we have used a borosilicate glass substrate (Corning Eagle XG) with a 200 nm Pt deposited layer. Indicator has 4 cm × 4 cm sizes and placed at distance of 5 mm from sample. An Agilent E5071B network analyzer in the continuous mode was used as a power source for feeding of a microwave (MW) probe with diameter of 1 mm. All measurements were performed by a microwave signal with 15 dBm power and 5 GHz frequency at 1 cm distance from sample. The visualized thermal distribution (without sample) caused by microwaves at 5 GHz and at 1 cm distance from MW probe end is presented in Fig. 6.1c. For Joule heating measurements, an HP E3631A programmable DC power supply has replaced the network analyzer. A reliable contact on the carbon fiber/PEEK sheet has been provided by using a silver paste on the composite material edges, alight emitted diode ($\lambda = 530$ nm) matrix was used as a light source, a polarizer and a quarter-wave plate provide the circular polarized light incidence on the indicator film as shown in the Fig. 6.1a configuration. By a beam splitter the circular polarized light is being directed to the indicator, then during the reflection from the indicator the light polarization state is being changed to elliptical due to the mechanical stresses in the indicator and recorded by CCD camera with resolution of 1024 × 640.

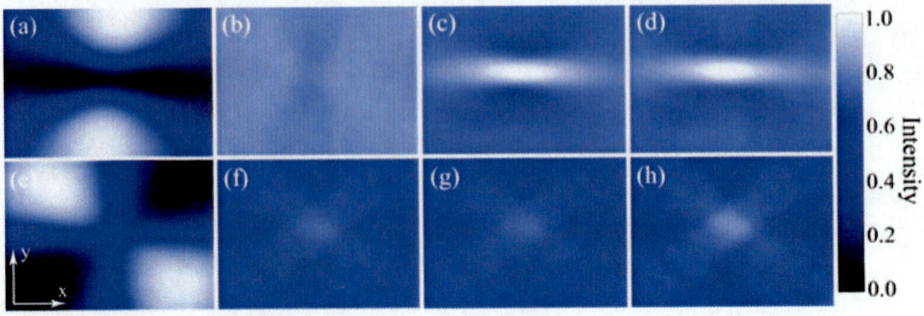

Fig. 6.2 Image-processing steps in LabVIEW: **a** and **e** correspond to the measured β_1 and β_2 images, respectively; **b** and **c** are $d^2\beta_1/dx^2$ and $-d^2\beta_1/dy^2$, respectively; **f** and **g** are $d^2\beta_2/dxdy$ and $d^2\beta_2/dydx$, respectively. **d** and **h** represent **(b)**+**(c)** and **(f)**+**(g)**, respectively. The final obtained image is the sum of **(d)** and **(h)**

During the measurements the analyzer orientation has been changed to be 0° or 45° with respect to the xy-plane and by choosing the corresponding orientation of the analyzer the charge-coupled device (CCD)-camera detects mechanical stress distribution images (β_1 and β_2). For the temperature distribution calculation, we solved the inverse problem of mechanical stress formation caused by the heat in a thin indicator film. The solution of that inverse problem i.e. the relation between the heat source density and mechanical stress distributions has the following form

$$q = C\left(\frac{\partial^2\beta_1}{\partial x^2} - \frac{\partial^2\beta_2}{\partial y^2} + 2\frac{\partial^2\beta_2}{\partial x\partial y}\right), \tag{6.1}$$

where q is the heat source density, β_1 and β_2 are linear birefringent distribution images corresponding to 0° and 45° of the analyzer orientation, and C is the constant related to the indicator parameters and the wavelength of the probing light.

Figure 6.2 illustrates image processing steps corresponding to the measured mechanical stress distributions. As Eq. (6.1) shows the obtained final image will be (b) + (c) + (f) + (g) which will represent the temperature difference between two states i.e. in the presence and absence of the heat source. More details about the visualization technique used and image processing are described in [19].

6.3 Results and Discussion

One should take into consideration the microwave loss mechanisms in a material to understand the origin of microwave heating. Since the carbon/PEEK does not have magnetic losses the losses due to the imaginary part of the relative permeability can be neglected. Thus in a composite material heating, the conductive and dielectric losses are responsible [18]. In our experiment, at the microwave probe apex the

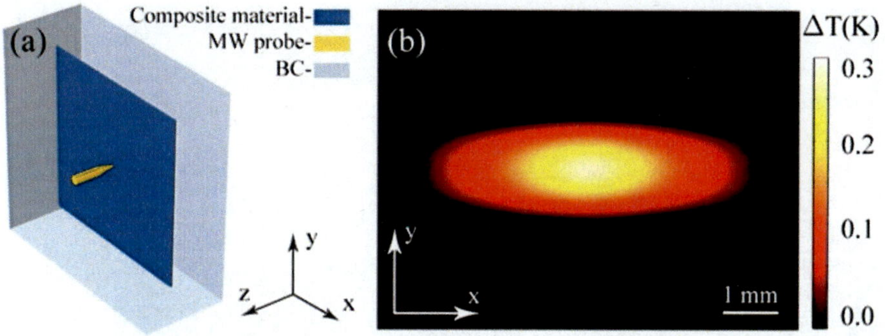

Fig. 6.3 COMSOL Multiphysics simulation: **a** geometry of the model with indications of MW probe and boundary condition (BC), **b** simulated microwave heat distribution in the carbon/PEEK

electromagnetic near-field symmetry following to the symmetry of its geometry; the electric field is concentrated around the probe head area whereas the magnetic field intensity is zero in that area due to eddy behavior. When the composite sheet is placed close to the microwave probe, the field symmetry is stretched along the high conductivity axis and the field distribution becomes elliptical. The induced current in the composite material leads heating of the sample.

We modeled the carbon fiber/PEEK microwave heating in COMSOL Multiphysics interface to compare obtained experimental data with the theoretical model. The geometry of the simulated microwave heating is shown in Fig. 6.3a. The model was built by the multiphysical coupling of two modules: the "Electromagnetic Wave" and "Heat Transfer in Solids". In the model, we used "Coaxial Lumped Port" with power of 15 dBm as a microwave source for the probe feeding, we set "Scattering" and "Temperature" boundary conditions for the "Electromagnetic Wave" and "Heat Transfer in Solids" modules, respectively, and configured the "Frequency Domain" and "Stationary" solvers for them. Thermal and electrical conductivities of the carbon/PEEK were set corresponding to the data mentioned in section "Material and Methods", 1320 kg/m^3 and 6 J(kg · K) were chosen as a density and heat capacity respectively, air was taken as an ambient environment, and the microwave probe was set to be perfect electric conductor. The simulation result is shown in Fig. 6.3b where the elliptic distribution is caused by the anisotropy of thermal and electrical conductivities, since the induced current flows easily through the high electrical conductivity direction and the largest portion of its generated heat transfers along the same direction.

Figure 6.4a shows the microwave heat distribution in the carbon fiber/PEEK visualized by the TEOIM technique. Here the orientation of the ellipse foci coincides with the high electrical conductivity direction as it was theoretically assumed and simulated above. Note that the 0.1 K of difference of the simulated and measured heat magnitude may be caused by choosing the microwave probe as a perfect conductor or some material properties mismatch. By varying simulation parameters one can perfectly match the experimental and simulated results, however we skip that since

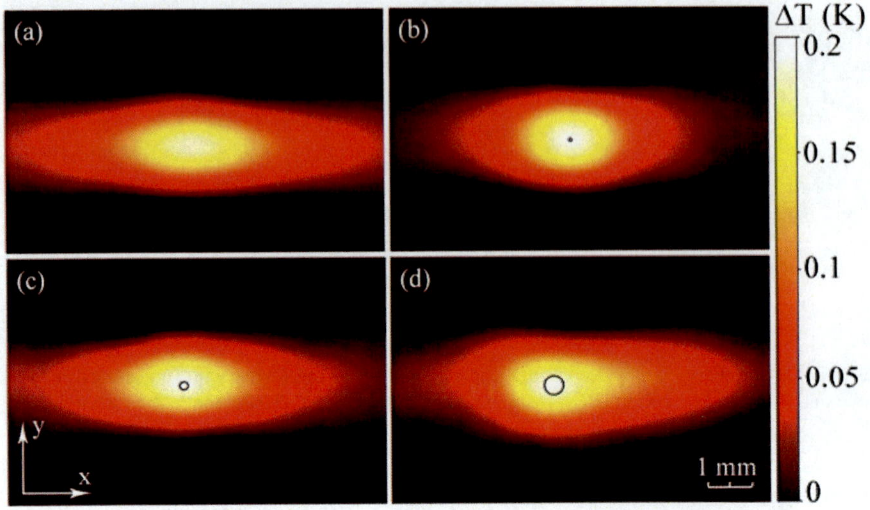

Fig. 6.4 Measured images for microwave heating distribution in the carbon/PEEK: **a** defect-free sample and sample with defect diameters of **b** 0.1 mm, **c** 0.2 mm, and **d** 0.4 mm

the simulated model initially has been built for the qualitative understanding of the microwave heat origin in the carbon fibers/PEEK. We made some artificial defects in the composite material in order to observe the heat distribution changes caused by defects. We varied the diameter of circle shaped hole type defects in the range of 0.1–0.4 mm.

As shown in Fig. 6.4b–d the hole in the center of the sample changes the heat distribution. One can see that even for the smallest defect size (0.1 mm) the shape of the visualized heat distribution has shrunken along the high conductivity direction. The reason for such shrinking phenomenon is that the defect interrupts the carbon fibers continuity and leads to the changes of the induced current distribution around the defect location. Depending on the growth of the defect size the changes in the heat distribution become more important. The image asymmetry in Fig. 6.4d comes from the position of the microwave probe with respect to the hole center. The symmetry in the resulting image was drastically changed only when the defect size was more than the microwave probe size (0.1 mm). Summarizing the microwave heat visualizations, we claim that fabrication faults and defects in a composite material sheet can be discovered by the microwave heating visualization and comparison of defect-free and custom/testing samples.

Figure 6.5 shows the Joule heat visualization for the defective carbon/PEEK composite material when the DC flux has been 0.3 A. Note that along the high electrical conductivity direction the Young's modulus (a measure of the stiffness) also is high, that is to say the mechanical defect can be discovered by the Joule heating visualization. In Fig. 6.5b high intensities correspond to the areas where the high electrical conductivity was interrupted by the defects. Those areas represent themselves the

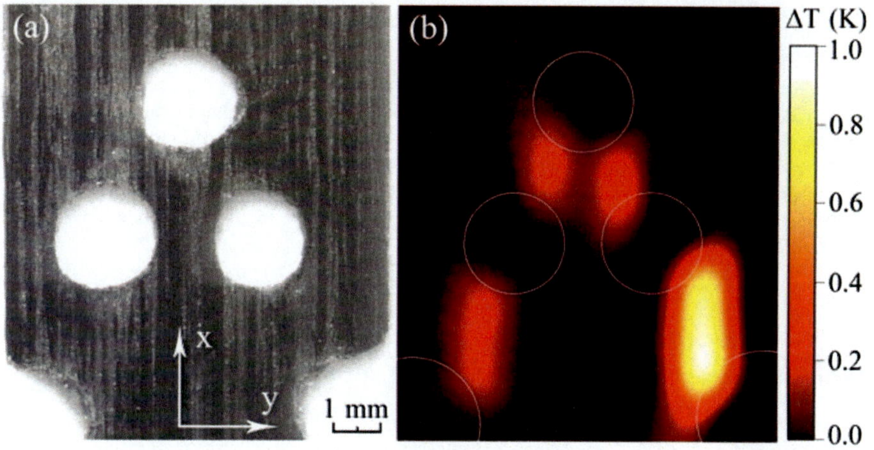

Fig. 6.5 **a** Optical image and **b** heat distribution visualization of the defective carbon/PEEK sample under the 0.3 A DC flux

mechanical defect areas. In this experiment the diameters of defective holes were around 2 mm which simply could be observed by the CCD camera without applying any heat visualization technique. However, in many cases optical observation is impossible e.g. when the defect is located inside of the material or when it is too small for the CCD camera observation. Note that the spatial resolution of the TEOIM technique is around 200 nm [19] which makes possible even single fiber defect detection. Since the system works as polarization microscope, the TEOIM sensitivity and spatial resolution caused by ability of high accuracy polarization detection and CCD camera pixel density, as well as data processing but not by optical light intensity and wavelength. We used linear polarized green light (530 nm) but similar measurement accuracy provided using a blue light (400 nm) and even white light sources. We suggest that this technique can be applied successfully for non-destructive testing and examination purposes.

6.4 Conclusion

The loss properties investigation for the carbon/PEEK composite material was performed by microwave heat visualization. The described loss mechanisms for microwave power give information about the heat mapping which can be used for material defect characterization and discovery. We succeeded in visualizing microwave and Joule heat in a carbon fiber/PEEK composite material.

Taking into account that this visualization technique is appropriate for investigations of DC and microwave influence on the composite material, we assert that this technique can be a promise tool for nondestructive testing and analysis during design and fabrication processes.

Acknowledgements This work was supported under the framework of international cooperation program managed by National Research Foundation of Korea (NRF-2020K2A9A2A08000165, FY2021) and funded by the Korea government (MSIT) (NRF-2021R1A2C1007334), by a scientific research grant through the Science Committee of MESCS of Armenia (21AG-1C061) and by a faculty research funding program 2021 implemented by Enterprise Incubator Foundation with the support of PMI Science.

References

1. Harris, B.: Engineering Composite Materials. IOM (1999)
2. Gruner, G.: Carbon nanotube films for transparent and plastic electronics. J. Mater. Chem. **16**, 3533 (2006)
3. Saiello, S., Kenny, J., Nicolais, L.: Interface morphology of carbon fibre/PEEK composites. J. Mater. Sci. **25**, 3493–3496 (1990)
4. Lonjon, A., Demont, P., Dantras, E., et al.: Electrical conductivity improvement of aeronautical carbon fiber reinforced polyepoxy composites by insertion of carbon nanotubes. J. Non Cryst. Solids **358**, 1859–1862 (2012)
5. Kadirvelu, K., Thamaraiselvi, K., Namasivayam, C.: Removal of heavy metals from industrial wastewaters by adsorption onto activated carbon prepared from an agricultural solid waste. Bioresour. Technol. **76**, 63–65 (2001)
6. Wang, H., Vu-Khanh, T.: Damage extension in carbon Fiber/PEEK Crossply laminates under low velocity impact. J. Compos. Mater. **28**, 684–707 (1994)
7. Qiang Yuan, Q., Bateman, S.A., Friedrich, K.: Thermal and mechanical properties of PAN- and Pitch-based carbon fiber reinforced PEEK composites. J. Thermoplast. Compos. Mater. **21**, 323–336 (2008)
8. Curson, A.D., Leach, D.C., Moore, D.R.: Impact failure mechanisms in carbon Fiber/PEEK Composites. J. Thermoplast. Compos. Mater. **3**, 24–31 (1990)
9. Liu, L., Zhou, K., He, P., et al.: Synthesis and microwave absorption properties of carbon coil-carbon fiber hybrid materials. Mater. Lett. **110**, 76–79 (2013)
10. Choi, J., Jung, H.-T.: A new triple-layered composite for high-performance broadband microwave absorption. Compos. Struct. **122**, 166–171 (2015)
11. Neo, C.P., Varadan, V.K.: Optimization of carbon fiber composite for microwave absorber. IEEE Trans. Electromagn. Compat. **46**, 102–106 (2004)
12. Zang, Y., Xia, S., Li, L., et al.: Microwave absorption enhancement of rectangular activated carbon fibers screen composites. Compos. Part B Eng. **77**, 371–378 (2015)
13. Xu, X., Wang, X., Cai, Q., et al.: Improvement of the compressive strength of carbon fiber/epoxy composites via microwave curing. J. Mater. Sci. Technol. **32**, 226–232 (2016)
14. Lee, H.S., Kim, S., Noh, Y.J., et al.: Design of microwave plasma and enhanced mechanical properties of thermoplastic composites reinforced with microwave plasma-treated carbon fiber fabric. Compos. Part B Eng. **60**, 621–626 (2014)
15. Lee, H., Galstyan, O., Babajanyan, A., et al.: Characterization of anisotropic electrical conductivity of carbon fiber composite materials by a microwave probe pumping technique. J. Compos. Mater. **50**, 1999–2004 (2015)
16. Rufail, L., Laurin, J.-J., Moupfouma, F.: On the use of microwave microscopy for detecting defects in the lightning strike protection mesh for carbon fiber composite aircraft. In: 2016 17th International Symposium on Antenna Technology and Applied Electromagnetics (ANTEM). IEEE, pp. 1–2

17. Bailleul, M.: Shielding of the electromagnetic field of a coplanar waveguide by a metal film: Implications for broadband ferromagnetic resonance measurements. Appl. Phys. Lett. **103**, 192405 (2013)
18. Mekhitarian, V.M.: The Faraday law of induction for an arbitrarily moving charge. J. Contemp. Phys. (Armenian Acad. Sci.) **51**, 108–126 (2016)
19. Lee, H., Arakelyan, S., Friedman, B., et al.: Temperature and microwave near field imaging by thermo-elastic optical indicator microscopy. Sci. Rep. **6**, 39696 (2016)
20. Pris, A.D., Utturkar, Y., Surman, C., et al.: Towards high-speed imaging of infrared photons with bio-inspired nanoarchitectures. Nat Photonics **6**, 564–564 (2012)
21. LeMieux, M.C., McConney, M.E., Lin, Y.-H., et al.: Polymeric nanolayers as actuators for ultrasensitive thermal bimorphs. Nano Lett. **6**, 730–734 (2006)
22. Bui, H.K., Wasselynck, G., Trichet, D., et al.: 3-D modeling of thermo inductive non destructive testing method applied to multilayer composite. IEEE Trans. Magn. **49**, 1949–1952 (2013)
23. Pan, C.T., Hocheng, H.: Evaluation of anisotropic thermal conductivity for unidirectional FRP in laser machining. Compos. Part A Appl. Sci. Manuf. **32**, 1657–1667 (2001)

Chapter 7
Acceptor-Assisted Intraband Photoconductivity in $GaAs/AlGaAs$ Quantum Wells

Maxim Vinnichenko, Ivan Makhov, Vadim Panevin, Ratmir Ustimenko, Grigorii Melentev, Sergey Sorokin, Irina Sedova, David Hayrapetyan, and Dmitry Firsov

Abstract The present work is devoted to the experimental investigation of the far-, mid- and near-infrared photoconductivity related to the optical hole transitions involving acceptor states in $GaAs/AlGaAs$ quantum wells. Photoconductivity spectra are studied at low lattice temperatures. It is shown that the main contribution to the far- and mid-infrared photoconductivity is associated with the optical hole transitions from the ground acceptor state to the delocalized states of the valence subbands, delocalized states above the quantum well and to the excited states of the acceptors. The relaxation times of impurity-assisted photocurrent in quantum wells were also studied. The ionization energies of the acceptor impurity obtained by various experimental methods are in a good agreement with theoretical calculations.

7.1 Introduction

Technological progress in opto- and nanoelectronics allows to develop devices operating in the near-, mid- and far-infrared (IR) spectral ranges for a wide range of applications. Far-IR radiation can pass through a large number of different non-conducting materials. This feature of radiation gives a promising list of applications for devices based on it, for example, spectroscopy, astronomy, physical research of materials, medicine and security systems, etc. [1]. The mid-infrared spectral range as well as far-infrared one contains a large number of absorption lines of various gases

M. Vinnichenko (✉) · V. Panevin · R. Ustimenko · G. Melentev · D. Firsov
Peter the Great St. Petersburg Polytechnic University, St. Petersburg 195251, Russia
e-mail: mvin@spbstu.ru

I. Makhov
National Research University Higher School of Economics, St. Petersburg 194100, Russia

S. Sorokin · I. Sedova
Ioffe Institute, St. Petersburg 194021, Russia

D. Hayrapetyan
Russian-Armenian University, Yerevan 0051, Armenia

D. Blaschke et al. (eds.), *Optics and Its Applications*, Springer Proceedings in Physics 281, https://doi.org/10.1007/978-3-031-11287-4_7

and molecules that are important for gas and bio sensing technologies. In accordance with the above, the development of sources and detectors operating in far- and mid-infrared spectral ranges is a promising task nowadays.

To date, the most efficient and compact source of far- and mid-infrared radiation is a quantum cascade laser [2, 3]. It has high enough output power, a fairly high efficiency and, most importantly, it has a small physical size. Despite the fact that quantum cascade lasers have several serious disadvantages, such as the complexity of the technological manufacturing and, accordingly, the high cost, they remain the most promising far- and mid-IR radiation sources.

The most well-proven detectors of far- and mid-IR radiation realized on semiconductors are based on intersubband optical transitions of charge carriers in 2D nanostructures with quantum wells (so-called quantum well infrared photodetectors (QWIP) [4, 5]) and interband optical transitions in bulk materials. The main difficulty with using QWIPs is associated with the need to fulfill the conditions of the selection rules for intersubband optical transitions according to which intersubband optical transitions of charge carriers can occur only for the light polarized along the growth axis of the structure. As a result, it leads to a complication of the photodetector structure design. Mid- and far-infrared photodetectors based on interband transitions are realized on semiconductors with a very narrow energy gap, such as $In_xAs_{1-x}Sb$ [6] and $Hg_xCd_{1-x}Te$ [7] alloys. Having a high sensitivity such alloys are difficult to grow as well as difficult to achieve the required doping profiles.

On the other hand, detection of infrared radiation can be realized with the use of charge carrier transitions involving impurity states. For example, the effective detection of far- and mid-infrared radiation is realized with bulk $Ge : Ga$ and $Ge : Hg$ photoresistors, respectively. The properties of impurities in bulk materials are comprehensively investigated to date. However, the energy spectrum of a certain impurity in a bulk semiconductor remains unchanged without external influences. In quantum well (QW) structures binding energy of an impurity center can be significantly varied by changing the QW parameters. The position of the impurity center in QW and the width of QW affect the energy spectrum of the impurity center [8]. Thus, doped QWs could expand the operating spectral range of devices based on impurity-related optical transitions. There is a significant amount of works related to the investigation of optical and photoelectrical properties of heterostructures with QWs doped with donor impurity centers. For example, far-infrared absorption [9, 10], photoconductivity [11, 12] and photoluminescence [12–15] are studied in $GaAs/AlGaAs$ quantum wells doped with shallow donors.

The optical properties of acceptors in QWs have not been studied in detail due to the non-trivial energy spectrum of acceptor and valence subband states. However, this feature allows to significantly expand the spectral range in which various acceptor-related optical phenomena can be observed. Moreover, p-type QWs have several advantages over n-type QWs. Firstly, acceptors have higher binding energy, it can even exceed the energy of optical phonons in $GaAs$. This can significantly increase the lifetime of photoexcited charge carriers, which is important for the development of mid- and far-infrared photodetectors. Secondly, according to the selection rules, intersubband optical transitions in p-type QWs are possible for radiation with a polarization vector lying in the plane of the QW layers. This makes devices based on such structures promising for detectors and modulators of infrared radiation [16].

The present work is devoted to the experimental investigation of the infrared photoconductivity related to the optical hole transitions involving acceptor states in $GaAs/AlGaAs$ quantum wells.

7.2 Samples and Methods

200 $GaAs/Al_{0.4}Ga_{0.6}As$ quantum wells were MBE grown on a semi-insulating GaAs substrate. Each QW was 3 nm wide; the barriers between QWs were 7 nm wide. The central part of each QW ($\sim$ 0.7 nm) was doped with beryllium (acceptor) with a surface concentration of about 10^{11} cm^{-2}. For photoconductivity studies, the ohmic indium contacts were deposited on the sample surface. The contacts were annealed in a dry nitrogen atmosphere with smooth heating of the sample to 450 °C for 5 minutes, then this temperature was maintained for 5 min, and then sample was cooled to the room temperature for 5 min. The distance between the contacts was 5.1 mm, the length of the contacts was 5.8 mm.

The results of numerical calculations of the energy spectrum of acceptor states and valence subbands of the structure under study were previously presented by us in [17]. The system of differential equations for bulk semiconductors can be solved analytically, but for the case under study with QW, the eigenvalues of the Hamiltonian can be obtained only numerically, for example, by the finite difference method [18] or using the transfer matrix formalism [19]. In [17], well-proven methods were used for determining the energy of impurity levels by solving the Schrödinger equation with the Luttinger-Kohn Hamiltonian with the addition of the confining potential of the quantum well and the energy of the Coulomb interaction with a charged acceptor [20]. The calculated energy spectrum of valence subbands and acceptor states for a 3 nm wide $GaAs/Al_{0.4}Ga_{0.6}As$ QW is shown in Fig. 7.1. In the calculations, it was assumed that the impurity is located in the center of the QW. In [19] the momentum-dependent decomposition coefficients were used to calculate the optical matrix elements of the hole transitions from localized acceptor states to higher energy states in QW in dipole approximation. Calculated spectral and polarization dependencies of the infrared optical absorption due to photoionization of acceptors were verified by experimentally measured impurity-assisted absorption for two polarizations of light [17]. In this work, we continue the investigations of impurity-assisted optical transitions in p-doped $GaAs/Al_{0.4}Ga_{0.6}As$ quantum wells started in [17].

Experimental studies were performed using the Bruker Vertex 80v vacuum Fourier transform spectrometer operating in a rapid- or step-scan mode (see scheme of the experiment presented in Fig. 7.2) with spectral resolution of about 1 meV. The spectrometer was equipped with a Mylar, KBr or CaF_2 beam splitters (BMS) for the far-infrared, mid-infrared and near-infrared studies, respectively. A globar was used as a source of broadband infrared radiation. For good heat dissipation, the sample

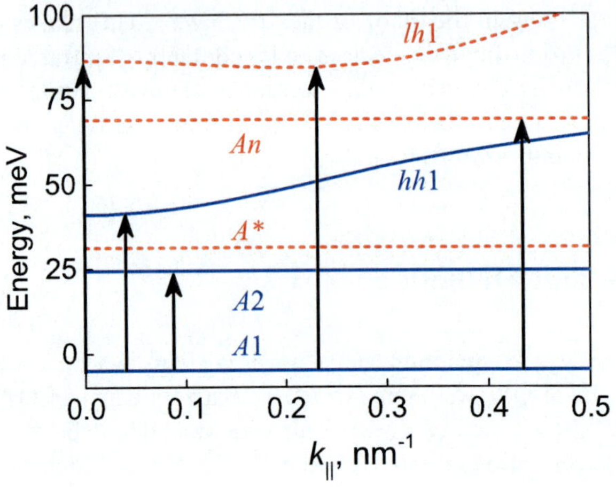

Fig. 7.1 Calculated energy spectrum of acceptor states and valence band subbands for a $GaAs/Al_{0.4}Ga_{0.6}As$ QW of 3 nm wide. Blue curve corresponds to the first hole subband (*hh*—heavy holes), and red one corresponds to the second hole subband (*lh*—light holes). The extension of impurity levels is shown schematically with solid and dashed horizontal lines without considering the degree of localization of impurity states in the *k*-space

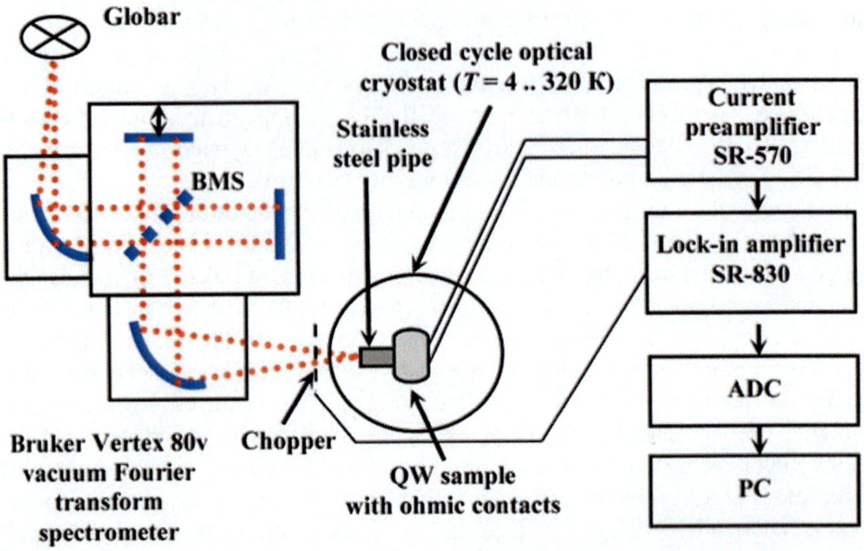

Fig. 7.2 Experimental setup scheme for investigation of the photoconductivity spectra

was indium-soldered to a copper holder of a Janis PTCM-4-7 closed-cycle optical cryostat with an operating temperature range of 4 – 320 K. The entrance window of the cryostat was made of KBr or TPX for mid-infrared and far-infrared studies, respectively. The sample in the cryostat was shielded all around from external background radiation to prevent undesirable photoionization of impurities.

Photoconductivity measurements were carried out under 5 V bias voltage applied to the sample using a SR-570 current preamplifier. The light from the globar passed through the interferometer of the Fourier spectrometer and then was modulated with an optical chopper. Infrared radiation was directed to the sample through a cold polished stainless steel pipe of a 5 mm diameter. A chopper was synchronized with lock-in amplifier SR-830. The photocurrent from the sample was registered by the above-mentioned SR-570 current preamplifier, and then measured by lock-in amplifier SR-830. This signal after analog-digital converter (ADC) was accumulated in personal computer (PC), where obtained interferogram was converted to spectrum by Fourier transform in OPUS software.

7.3 Experimental Results

7.3.1 Photoconductivity Spectra

The interband photoconductivity spectra of the $GaAs/Al_{0.4}Ga_{0.6}As$ quantum well structure in the near-IR spectral range at various cryogenic temperatures are shown in Fig. 7.3. The photocurrent increases at the photon energy close to the $GaAs$ energy gap value, which corresponds to the interband transitions of charge carriers in $GaAs$ (marked with arrow E_g in Fig. 7.3). It is associated with the contribution to the photocurrent from the $GaAs$ substrate and the $p - GaAs$ cap layer of the nanostructure. We observed in the photoconductivity spectra two dips corresponding to the photon energies of optical transitions associated with formation of heavy hX and light lX free excitons in QW (the corresponding photon energies are shown with arrows in Fig. 7.3). Photon energy values for these optical transitions were obtained using free heavy and light exciton binding energies as 10 meV and 13 meV, respectively, calculated in [21]. Spectral position of heavy free exciton line was also verified by low-temperature near-infrared photoluminescence studies (see inset to Fig. 7.3). These excitons are formed by charge carriers located in the ground QW subbands. The total photocurrent of the structure decreases at the hX and lX photon energies at rather low crystal lattice temperatures because excitons with such high binding energies cannot contribute to the photocurrent. The positive contribution of excitons to the photocurrent may be observed at higher lattice temperatures, when the energy of thermal motion of charge carriers is comparable to the binding energy of the exciton. In this case, the excitons formed under the action of radiation can contribute to the photocurrent due to thermal dissociation into a nonequilibrium electrons and holes in the subbands of the conduction and valence bands. The luminescence band related to the optical transitions of electrons from the first electron subband to the ground acceptor state was also observed in the near-infrared photoluminescence spectrum (see inset to Fig. 7.3). The arrow $e1 - A1$ in inset to Fig. 7.3 shows the

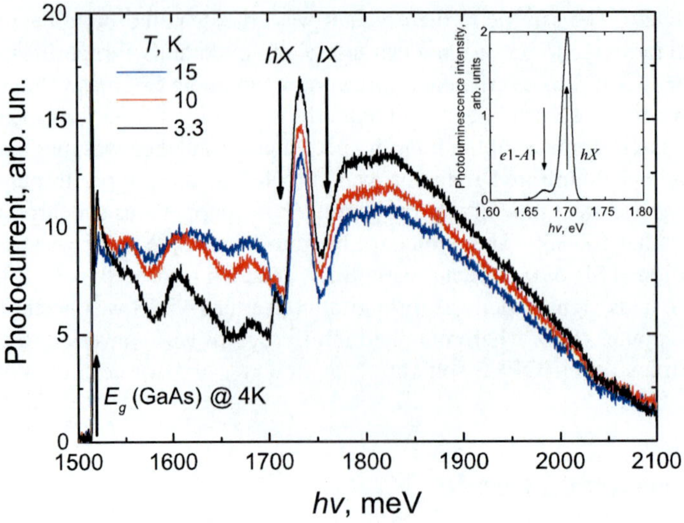

Fig. 7.3 Spectra of interband near-infrared photoconductivity at different temperatures. The inset shows near-infrared photoluminescence spectrum measured at $T = 4$ K

calculated position of the energy of optical electron transition from the first electron subband to the ground state of acceptor impurity. It should be noted that contribution of the optical transitions of charge carriers from the acceptor states to the first electron subbands was not revealed in the near-infrared photoconductivity spectra because in equilibrium the acceptor states are unoccupied by electrons at low crystal lattice temperatures.

The long-wavelength part of the photoconductivity spectra (photon energy below 1700 meV) is related to the charge carriers generated in the near-surface heavily doped (near the critical acceptor concentration for Mott transition) $GaAs$ cap layer as well as in semi-insulating compensated $GaAs$ substrate, which contains substantial concentration of ionized donors. At low temperatures, ionized impurity scattering is the main scattering mechanism indoped $GaAs$ [22]. With increasing temperature, the intensity of the impurity scattering decreases, which leads to an increase in the mobility of photoexcited charge carriers and to an increase in photoconductivity with temperature. The inverse temperature dependence of photoconductivity at high photon energies (photon energies above 1750 meV) can be explained by the fact that layers with quantum wells make a significant contribution to photoconductivity in this spectral region. At low temperatures, the acceptors in quantum wells are non-ionized and the mobility of charge carriers is determined by phonon scattering, the intensity of which increases with temperature.

The impurity-assisted time-resolved photoconductivity spectra were measured in step-scan mode of Fourier-spectrometer in mid-infrared spectral range. The time-resolved photoconductivity spectra at $T = 4$ K are shown in Fig. 7.4. Photoconductivity spectra at the liquid helium temperature were also measured in rapid scan mode

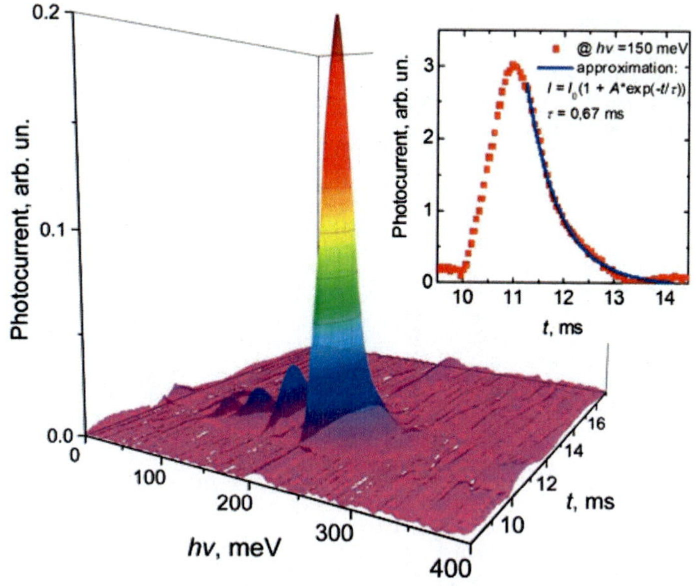

Fig. 7.4 Time-resolved spectra of impurity-assisted photoconductivity in the mid-IR range at $T = 7$ K. Inset: photocurrent dynamic at the photon energy of 150 meV

of the Fourier-transform spectrometer for different values of scanning velocity f of the interferometer. The results are shown in the Fig. 7.5. It should be noted that current preamplifier bandwidth was of about 1 MHz. The wide peak observed near the photon energy of 205 meV is associated with the optical transitions of holes from the ground acceptor state $A1$ to the continuum states above the QW. The small shoulder in the photocurrent at a photon energy of about 240 meV could be associated with the contribution of deep impurity centers in the layers of the nanostructure to the photocurrent [23] or with some features of joint density of states for hole transitions. We associate the photocurrent peak at the photon energy of about 150 meV with optical transitions of holes from the ground impurity state $A1$ to the bottom of the third hole subband (or the second subband of heavy holes $hh2$, which is not shown in Fig. 7.1). It is important to note that in the calculation presented in Fig. 7.1, we used the expansion of the hole wave function in the states of the first two valence subbands only. So, in order to obtain third hole subband $hh2$ we used the standard solution of the Schrödinger equation in the framework of the parabolic model. The photocurrent peak at the photon energy of about 90 meV (see Fig. 7.5) is associated with the optical transitions of holes from the ground state of acceptor impurity $A1$ to states near the bottom of the second hole subband $lh1$ near $k_{\parallel} = 0$ (arrow $A1 - lh1$ in Fig. 7.1). In general, from Fig. 7.5 it is clearly seen that with an increase in the scanning speed of the interferometer, the intensity of the photocurrent decreases. It is connected with finite relaxation time of holes from the valence subbands and excited impurity states to the ground impurity states. This time is estimated as a few mil-

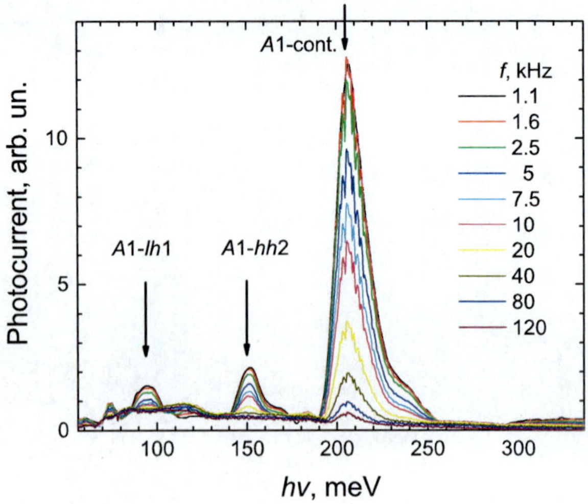

Fig. 7.5 Impurity-assisted photoconductivity spectra in the mid-IR range at $T = 7$ K for different scanning velocity f of interferometer

liseconds. Also, this time can be obtained from the time-resolved spectrum (see Fig. 7.4). We approximated the photocurrent dynamics (see inset to Fig. 7.4 for photon energy of about 150 meV) with an expression derived from the rate equation:

$$I = I_0 \left(1 + A \cdot \exp\left(t/\tau\right)\right), \tag{7.1}$$

where I—photocurrent intensity; I_0 and A—fitting parameters; τ—relaxation time. Using this approximation method, we got relaxation times for three observed optical transitions in the range of 0.58–0.67 ms. The obtained relaxation time for the acceptor-related optical transitions is quite large. It can be related to the large binding energy of acceptors in our 3 nm wide quantum wells. The calculated value of acceptor binding energy is about 40 meV, that exceeds the $GaAs$ optical phonon energy resulting in a significant suppression of the carrier capture to acceptor with optical photon scattering. It also should be noted that in semi-insulating $GaAs$, the relaxation time for optical transitions of electrons from the conduction band to the ground acceptor state at low crystal lattice temperatures is about tens microseconds or more [24].

We also measured the spectra of impurity-assisted photoconductivity in the mid-IR spectral range for different cryogenic temperatures from 4 K with a step of 2 K. All photocurrent peaks gradually decrease and at a temperature of about 30 K the photocurrent disappears due to impurity ionization. The reference samples without QWs did not exhibit any features in the photocurrent spectra. This confirms that all the features observed in the mid-infrared photoconductivity spectra of QWs are associated with impurity transitions in QWs and agrees well with theoretical calculations.

The far-infrared impurity-assisted photoconductivity spectra are shown in the Fig. 7.6 for different lattice temperatures. The wide long-wavelength peak near a photon

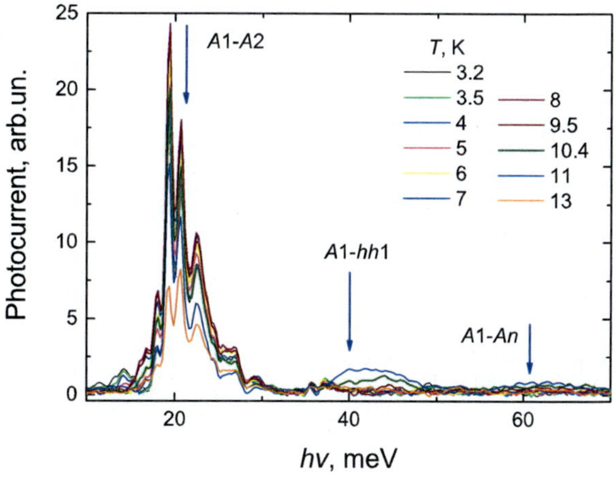

Fig. 7.6 Impurity-assisted photoconductivity spectra in the far-infrared spectral range at different temperatures

energy of about 20 meV is associated with the optical hole transitions from the ground acceptor state $A1$ to the first excited state of the acceptor $A2$ (arrow $A1 - A2$ in Fig. 7.1). These excited carriers in $A2$ state can contribute to the photocurrent due to thermal ejection into the first subband of the valence band $hh1$. The peak in photoconductivity spectra located near the photon energy of about 40 meV could be associated with the hole transitions from the ground state of the acceptor impurity $A1$ to the first subband of heavy holes $hh1$, which also agrees well with the calculated value of the acceptor binding energy (arrow $A1 - hh1$ in Fig. 7.1). The photocurrent peak at the photon energy of about 60 meV is associated with intracenter optical transitions of holes between the ground state of the impurity $A1$ and the impurity states located below the second hole subband $lh1$ (arrow $A1 - An$ in Fig. 7.1). All photocurrent peaks gradually decrease with temperature and at the temperature of about 15 K the photoconductivity disappear due to impurity ionization.

7.3.2 Temperature Dependence of Electroconductivity

It was shown above that investigations of photoconductivity have confirmed our calculation of the energy spectrum of impurity states and the band diagram of QWs. Additionally, we have determined the binding energy of acceptors in the QW from the analysis of the temperature dependence of the electroconductivity σ (see red dots in Fig. 7.7). Firstly, we assumed that electroconductivity is proportional to the concentration of ionized holes in the valence band. This is correct if the hole mobility is assumed to be temperature independent. Secondly, it was important to consider that hole gas is non-degenerate. To prove this, let's find the temperature dependence of the concentration. In the impurity conductivity range, two-dimensional concentration of ionized holes in the valence band n_s can be written as [25]

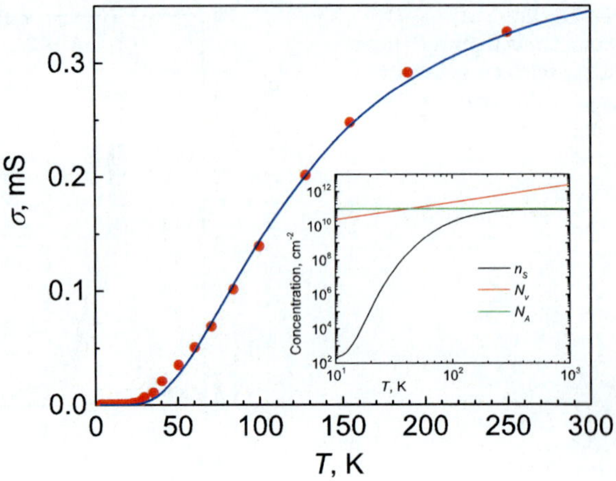

Fig. 7.7 Temperature dependence of electroconductivity. Red dots—experimental data, blue line—theoretical calculations according to Eq. (7.6). Inset: calculated temperature dependence of two-dimensional concentration of ionized holes in the valence band n_s, N_V—effective density of states in valence subband, N_A—concentration of acceptors

$$n_s = N_A - N_h, \tag{7.2}$$

where N_A—concentration of acceptors, N_h—concentration of unionized holes at the impurity levels. The hole concentration at acceptor levels according to the Fermi statistics is

$$N_h = \frac{N_A}{1 + \beta \exp\left(\frac{-E_A - \mu}{kT}\right)}, \tag{7.3}$$

where β—impurity degeneracy factor of the ground acceptor state, equal to 4 for holes in the QW; E_A—ionization energy of an acceptor impurity; μ—Fermi energy. Substituting expression (7.3) into (7.2) one can obtain the Fermi energy

$$\mu = -E_A - kT \ln\left(\frac{n_s}{N_h \beta}\right), \tag{7.4}$$

It is also known that for the two-dimensional Fermi-Dirac statistics, Fermi energy can be written as

$$\mu = kT \ln\left(\exp\left[\frac{n_s}{N_V}\right] - 1\right), \tag{7.5}$$

where $N_V = mkT/\left(\pi \hbar^2\right)$ is the two-dimensional effective density of states in valence subband, m—effective mass of holes. Equating μ from (7.4) and (7.5), we obtain the relationship between concentration of ionized holes in the valence band n_s and temperature with E_A as parameter

$$\exp\left(\frac{E_A}{kT}\right) = \frac{n_s}{(N_A - n_s)\beta}\left(\exp\left[\frac{n_s}{N_V}\right] - 1\right), \tag{7.6}$$

Using Eq. (7.6) and assuming that $\sigma(T) \sim n_s(T)$ we plotted dependence $\sigma(T)$ (see blue line in Fig. 7.7) with fitting parameter $E_A = 40$ meV. It gives the best fit between the experimental and calculated data. Also, from Eq. (7.6) we received dependence $n_s(T)$ (see black line in the inset to Fig. 7.7). One can see that concentration of free holes n_s is much less than the effective density of states N_V in the hole subband. This confirms that the hole gas is non-degenerate.

The value of the acceptor ionization energy obtained from $\sigma(T)$ is compatible with calculated one (see Fig. 7.1) and with values obtained from near-infrared photoluminescence (see inset to Fig. 7.3) and far-infrared photoconductivity studies (see Fig. 7.6). Also, the obtained ionization energy of acceptors in the QW is in a good agreement with the results of other authors [26, 27].

7.4 Conclusion

The results of studies of low-temperature acceptor-assisted photoconductivity of $p - GaAs/AlGaAs$ QW were presented in the near-, mid- and far-infrared spectral ranges. The mid- and far-infrared photoconductivity spectra contain the bands associated with the hole transitions from the ground acceptor state to the delocalized states of the valence subbands, delocalized states above the QW and to the excited states of the acceptor. The features of near-infrared photoluminescence and photoconductivity spectra are related to the interband optical transitions with assistance of impurities as well as with exciton transitions. Time-resolved photoconductivity spectra allowed us to find the relaxation time of impurity-assisted transitions. The ionization energies of the acceptor impurity determined by various experimental methods are in a good agreement with theoretical calculations.

Acknowledgements The work was partially funded by the Ministry of Science and Higher Education of the Russian Federation as part of thematic work "Activities to support of efficiency of Russian-Armenian and Belorussian-Russian Universities" (supplement agreement contract No. 075-03-2021-050/5 dated 08.07.21). Also, work was partially supported by the Russian Science Foundation under grant No. 18-72-00034 and the Ministry of Science and Higher Education of the Russian Federation (state assignment No. 075-03-2021-050). IM acknowledges support from the Basic Research Program of the National Research University Higher School of Economics.

References

1. Tonouchi, M.: Cutting-edge terahertz technology. Nat. Photon. **1**, 97–105 (2007)
2. Bosco, L., et al.: Thermoelectrically cooled THz quantum cascade laser operating up to 210 K. Appl. Phys. Lett. **115**, 010601 (2019)
3. Razeghi, M., et al.: Quantum cascade lasers: from tool to product. Opt. Express **23**(7), 8462–8475 (2015)
4. Gueriaux, V., et al.: Quantum well infrared photodetectors: present and future. Opt. Eng. **50**(6), 061013 (2011)

5. Liu, H.C., Song, C.Y., SpringThorpe, A.J., Cao, J.C.: Terahertz quantum-well photodetector. Appl. Phys. Lett. **84**(20), 4068–4070 (2004)
6. Rogalski, A., et.al.: InAsSb-based infrared photodetectors: Thirty years later on. Sensors (Switzerland) **20**(24), 7047, 1–74 (2020)
7. Rogalski, A.: HgCdTe infrared detector material: history, status and outlook. Rep. Prog. Phys. **68**(10), 2267–2336 (2005)
8. Masselink, W.T., et al.: Shallow impurity levels in AlGaAs/GaAs semiconductor quantum wells. Solid-State Electron. **29**(2), 205–214 (1986)
9. Jiang, P., Kok, W.C.: Far-infrared absorption spectra of doped quantum well structures. J. Appl. Phys. **90**(3), 1271–1274 (2001)
10. Helm, M., et al.: Far-infrared spectroscopy of minibands and confined donors in GaAs/AlxGa1-xAs superlattices. Phys. Rev. B **43**(17), 13983–13991 (1991)
11. Mercy, J.M., Jarosik, N.C., McCombe, B.D. Photoconductivity of confined donors in Al0.3Ga0.7As/GaAs quantum wells. J. Vacuum Sci. Technol. B Microelectron. Process. Phenom. **4**, 1011–1013 (1986)
12. Firsov, D.A., et al.: Terahertz emission and photoconductivity in n-type GaAs/AlGaAs quantum wells: the role of resonant impurity states. Semiconductors **44**(11), 1394–1397 (2010)
13. Makhov, I.S., et al.: Terahertz photoluminescence of the donor doped GaAs/AlGaAs quantum wells controlled by the near-infrared stimulated emission. J. Lumin. **210**, 352–357 (2019)
14. Makhov, I.S., et al.: Impurity-assisted terahertz photoluminescence in compensated quantum wells. J. Appl. Phys. **126**(17), 175702 (2019)
15. Makhov, I.S., et al.: The effect of stimulated interband emission on the impurity-assisted far-infrared photoluminescence in GaAs/AlGaAs quantum wells. Superlattices Microstruct. **112**, 79–85 (2017)
16. Chen, H.H., Wang, Y.-H., Houng, M.-P.: Near 10 μm intervalence sub-band optical transitions in p-type In0.49Ga0.51P-GaAs quantum well structures. IEEE J. Quant. Electron. **32**(3), 471–477 (1996)
17. Vinnichenko, M.Ya. et. al.: Acceptor-related infrared optical absorption in GaAs/AlGaAs quantum wells. Phys. E: Low-dimensional Syst. Nanostruct. **124**, 114601 (2020)
18. Ikonić, Z.: Milanović, V: Hole-bound-state calculation for semiconductor quantum wells. Phys. Rev. B **45**(15), 8760–8762 (1992)
19. Tadić, M., Ikonić, Z.: Bound-free intersubband absorption in p-type doped semiconductor quantum wells. Phys. Rev. B **52**(11), 8266–8275 (1995)
20. Brown, G.J. et. al.: Intersubband hole absorption in GaAs-GaInP quantum wells grown by gas source molecular beam epitaxy. Appl. Phys. Lett. **65**(9), 1130–1132 (1994)
21. Belov, P.A.: Energy spectrum of excitons in square quantum wells. Physica E **112**, 96–108 (2019)
22. Blakemore, J.S.: Semiconducting and other major properties of gallium arsenide. J. Appl. Phys. **53**, R123–R181 (1982)
23. Bourgoin, J.C., Von Bardeleben, H.J., Stievenard, D.: Native defects in gallium arsenide. J. Appl. Phys. **64**(9), R65–R92 (1988)
24. Kurdyubov, A.S. et. al.: Impurity-induced modulation of terahertz waves in optically excited GaAs. AIP Adv. **7**(11), 115222 (2017)
25. Blakemore, J.S.: Semiconductor Statistics. Courier Corporation (2002)
26. Masselink, W.T., Chang, Y.C., Morkoc, H.: Acceptor spectra of AlxGa1-xAs-GaAs quantum wells in external fields: electric, magnetic, and uniaxial stress. Phys. Rev. B **32**(8), 5190–5201 (1985)
27. Einevoll, G.T., Chang, Y.C.: Effective bond-orbital model for shallow acceptors in $GaAs - Al_x Ga_{1-x} As$ quantum wells and superlattices. Phys. Rev. B **41**(3), 1447–1460 (1990)

Chapter 8
Charge Carriers' States and Optical Transitions in $CdS/HgS/CdS$ Core/Shell/Shell Cylindrical Nanostructure in the Presence of Strong Uniform Electrostatic Field

V. A. Harutyunyan

Abstract The single-particle states of charge carriers and optical transitions in a cylindrical layered β-CdS/β-HgS/β-CdS core/shell/shell nanoheterostructure (quantum nanotube) in the presence of a strong lateral uniform electrostatic field are considered in this report. The consideration is carried out for the case, when the strong size quantization regime is realized for charge carriers in the HgS layer of the structure under consideration. It is shown that a strong external field radically changes the nature of the movement of charge carriers along the angular variable. Expressions for the envelope wave functions and energy spectrum of charge carriers in β-HgS in the presence of an external field are obtained in an explicit analytical form. The threshold frequency of interband transitions under the action of an electrostatic field is shifted towards low frequencies. The external field also leads to an explicit dependence of the intensity of interband transitions on the effective masses of charge carriers.

8.1 Introduction

Nanoparticles based on metacinnabar (β-HgS) in the form of planar quantum layers and films, homogeneous and inhomogeneous quantum dots, various layered structures such as core/shell/shell, etc. (see e.g. Refs. [1–10] and literature therein) have been intensively researched in recent decades. These nanoparticles are currently widely used. They are used in various electronic and optoelectronic devices with low energy consumption [11–14], solar and photo-electrochemical batteries [6, 15–17], as a promising material for the implementation of new quantum states—excitonic insulators [6, 13, 18], in areas of biomedicine [19, 20], ecology [20, 21], etc.

Among the aforementioned quantum nanostructures based on β-HgS, a separate class is made up of core/shell/shell structures with cylindrical symmetry. In particu-

V. A. Harutyunyan (✉)
Russian-Armenian University, 0051 Yerevan, Armenia
e-mail: volodya.harutyunyan@rau.am

D. Blaschke et al. (eds.), *Optics and Its Applications*, Springer Proceedings in Physics 281, https://doi.org/10.1007/978-3-031-11287-4_8

lar, layered core/shell/shell cylindrical nanoheterostructures β-CdS/β-HgS/β-CdS (β-HgS nanotubes), the study of which is devoted to a large number of works (see e.g. Refs. [22–28] and references in them). The interest of researchers to these heterophase structures is primarily due to the fact that they combine the properties of both quantum films and quantum wires and, from an applied point of view, are more multifunctional than individual quantum films and quantum wires. In Refs. [22–24], the energy spectrum of charge carriers in a layered cylindrical β-CdS/β-HgS/β-CdS heterostructure was calculated, and the specificity of the conduction mechanism in such structure was considered. In Ref. [25], Raman scattering of electrons in a cylindrical layer of a β-CdS/β-HgS quantum dot was considered, and Ref. [26] was devoted to the study of the influence of hydrostatic pressure and temperature factor on the diamagnetic properties of the electron subsystem in a β-CdS/β-HgS cylindrical layered structure.

One of the powerful factors of modulating effect on semiconductor materials, as it is well known, is an external static electric field. At the same time, one of the most productive methods for studying of the band structure of semiconductors is the optical method. In a number of previous works by the author and coauthors, optical transitions and exciton states in the cylindrical structure of β-CdS/β-HgS/β-CdS have been investigated both in the absence and in the presence of electrostatic fields of various intensities [27–32].

In this work, the effect of a strong lateral uniform electrostatic field on the energy spectrum of charge carriers and on the optical absorption spectrum in the cylindrical core/shell/shell structure β-CdS/β-HgS/β-CdS is considered theoretically, when strong quantization mode is realized in β-HgS layer.

8.2 Model Approximations

Further calculations will be carried out within the framework of the two-zone model in the approximation of an isotropic effective mass. Table 8.1 shows the physical characteristics of the bulk crystals of β-CdS and β-HgS required in the following.

It is clear from the data presented that, due to the large value of the energy offset at the interface of contacting materials, in the radial direction the HgS layer plays

Table 8.1 Some characteristics of bulk semiconductors β-CdS and β-HgS (Data taken from Refs. [8, 27, 32, 33])

Material	Latt. const a_0 (nm)	Static dielect. const ε	Electron effect. mass μ_e/μ_0	Hole effect. mass μ_h/μ_0	Band gap E_g (eV)	c-band energy offset ΔU_c (eV)	v-band energy offset ΔU_v (eV)
CdS	0.5818	9.1	0.2	0.7	2.5	–	–
HgS	0.5851	18.2	0.036	0.044	0.5	1.2	0.8

Here: m_0—is the mass of free electron

Table 8.2 Excitonic characteristics of the structural components of the β-CdS/β-HgS/β-CdS structure (Data taken from Refs. [8, 27, 32])

Material	3D Exciton binding energy E_{ex}^{3D} (meV)	3D Exciton Bohr radius a_{ex}^{3D} (nm)	Electron 3D Bohr radius a_e^{3D} (nm)	Hole 3D Bohr radius a_h^{3D} (nm)
CdS	27	3	2.4	0.69
HgS	0.81	50	27	22

the role of a two-dimensional quantum well, and the core and the outer layer-shell of CdS play the role of two-dimensional barriers. Let us present now the exciton characteristics of the components of the structure under consideration obtained using Table 8.2.

It is easy to conclude from this table that for the thickness of the mercury sulfide layer $L = 5 \div 15$ nm, when the condition

$$L < a_{ex}^{3D}, a_e^{3D}, a_h^{3D} \tag{8.1}$$

within the β-HgS layer, the strong quantization mode will be realized with sufficient accuracy for charge carriers. At the same time, we assume that the following condition is also fulfilled in the considered composition:

$$L^2 \ll R_1^2, R_2^2. \tag{8.2}$$

Here R_1, $R_2 = R_1 + L$ are the inner and outer radii of β-HgS layer, respectively. For specific calculations, in what follows, we will take $R_1 = 30$ nm.

From Table 8.1 for the size quantization energy of charge carriers at $L = 5 \div 15$ nm, we have the following estimates:

$$E_{rad}^e \sim \frac{\hbar^2}{2\mu_e L^2} = 42,4 \div 4,7\text{meV} \ll \Delta U_c = 1,2\text{eV—for electrons} \tag{8.3}$$

$$E_{rad}^h \sim \frac{\hbar^2}{2\mu_h L^2} = 34,7 \div 3.8\text{meV} \ll \Delta U_v = 0,8\text{eV—for holes} \tag{8.4}$$

It has been shown in a number of works (see, for example, the review [32]) that under conditions (8.1)–(8.4) in the corresponding Schrödinger equation in cylindrical coordinates (r, ϕ, z), it is possible to separate the variables, and the β-HgS quantum well can be approximated by a rectangular infinitely deep potential well in radial direction. As a result, for the wave functions and energy of single-particle states in the β-HgS layer in the absence of an external field, we arrive to the following expressions:

$$\Psi(r, \phi, z) = \Phi_n(r) f_m(\phi) u(z) = \sqrt{\frac{2}{L}} \frac{\sin \frac{\pi n}{L}(r - R_1)}{\sqrt{r}} \frac{e^{i|m|\phi}}{\sqrt{2\pi}} \frac{e^{ikz}}{\sqrt{d}} \tag{8.5}$$

$$E = E_{rad}^{e,h} + E_{rot}^{e,h} + E_{long}^{e,h} = \frac{\pi^2 \hbar^2 n_{e,h}^2}{2\mu_{e,h} L^2} + \frac{\hbar^2 m_{e,h}^2}{2\mu R_0^2} + \frac{\hbar^2 k_{e,h}^2}{2\mu_{e,h}}; \tag{8.6}$$

$$n_{e,h} = 1, 2, \ldots; m_{e,h} = 0, \pm 1, \pm 2, \ldots; \tag{8.7}$$

$$R_0^{-2} = 0.5 \left(R_1^{-2} + R_2 {-} 2 \right); \mathbf{k} = \mathbf{k}\,(0, 0, k) \tag{8.8}$$

Here d is the normalizing length of the structure along the symmetry axis (z). Using wave functions (8.5), the intensity and selection rules will be determined, and using the energy spectrum (8.6), it is possible to calculate the values of the threshold frequencies and the frequency dependence of interband and intersubband transitions.

Let us now turn, within the framework of the proposed model, to consider single-particle states in the β-HgS layer of the structure under consideration in the presence of a strong transverse electrostatic field.

8.3 Single-Particle States in a Layer in the Presence of a Strong Transverse Field

An external field of strength F is assumed to be directed along the x axis: $\boldsymbol{F} = \boldsymbol{F}\,(F, 0, 0)$. For the potential energy of a particle with a charge q within the layer, in this case we have [32]:

$$V\,(r, \phi) = qF \left(Br + \frac{C}{r} \right) \cos\phi = V\,(r) \cos\phi;\; B = \frac{6R_2^2}{9R_2^2 - R_1^2};\; C = \frac{BR_1^2}{3}. \tag{8.9}$$

In the Schrödinger equation, the motion along the variable z, as in the case of absence of a field, is separated (see Exps. (8.5)–(8.6)). Correspondingly for the transverse motion of a particle we will have the following equation:

$$-\frac{\hbar^2}{2\mu} \left(\frac{1}{r} \frac{\partial}{\partial r} r \frac{\partial}{\partial r} + \frac{1}{r^2} \frac{\partial^2}{\partial \phi^2} \right) \Psi\,(r, \phi) + qF \left(Br + \frac{C}{r} \right) \cos\phi \Psi\,(r, \phi) = E_{tr} \Psi\,(r, \phi) \tag{8.10}$$

Here E_{tr}^F is the total transversal energy of the particle in β-HgS layer in the presence of external field. Let us make the following substitution in Eq. (8.10):

$$\Psi\,(r, \phi) = \frac{\chi\,(r, \phi)}{\sqrt{r}} \tag{8.11}$$

Instead of Eq. (8.10) we now get:

$$-L^2\frac{\partial^2\chi(r,\phi)}{\partial r^2}-\frac{\hbar^2}{2\mu r^2}\frac{2\mu L^2}{\hbar^2}\chi(r,\phi)+\frac{V(R)}{\hbar^2/2\mu L^2}\cos\phi\chi(r,\phi)$$
$$-\frac{\hbar^2}{2\mu r^2}\frac{2\mu L^2}{\hbar^2}\frac{\partial^2\chi(r,\phi)}{\partial\phi^2}=\frac{E_{tr}^F\chi(r,\phi)}{\hbar^2/2\mu L^2} \quad (8.12)$$

Let us now turn to the strong field factor. Within the framework of this problem, the field can be considered strong if the energy received by the particle within the layer from the external field is much greater than its dimensional quantization energy, i.e. if the following condition is met:

$$\frac{|V(r)|}{\hbar^2/2\mu L^2}\gg 1 \quad (8.13)$$

In this case, the function can be expanded in a series and limited to the first terms of the expansion:

$$\cos\phi\sim 1-\phi^2/2 \quad (8.14)$$

Taking into account conditions (8.2) and (8.13), the adiabatic approximation can be used to solve Eq. (8.12). Equation (8.12) in this case splits into the following two equations:

$$-\frac{\hbar^2}{2\mu}\frac{d^2G(\rho)}{d\rho^2}+\beta\rho G(\rho)=E_{rad}^F G(\rho)\,;\,\rho=R_2-r;\,\rho[0;L]\,;\,\beta=\frac{1}{2}|e|F; \quad (8.15)$$

$$-\frac{\hbar^2}{2\mu}\frac{d^2g(\phi)}{d\phi^2}+\beta\left(Br+\frac{C}{r}\right)\phi^2 g(\phi)=E_{ang}^F g(\phi)\,; \quad (8.16)$$

Here $\chi(\rho,\phi)=G(\rho)g(\phi)$, $E_{rad}^F+E_{ang}^F=E_{tr}^F$.

In the presence of size quantization, the effect is most pronounced for the lowest energy states. Therefore, when solving Eq. (8.15), we restrict ourselves to calculating the ground state. When condition (8.13) is fulfilled in the lowest states, the particle is "captured" by the external field, and the motion in this case will actually occur already in the triangular potential well. The ground electronic state can be described with sufficient accuracy by the variational method. We choose the trial function in the following form:

$$G(\rho)=A\rho\exp-\frac{\alpha\rho^2}{2}. \quad (8.17)$$

Here A, α are the variation parameters. Carrying out standard calculations, for the trial wave function and the ground state energy of Eq. (8.15), we obtain:

$$G_e(\xi) = \sqrt{\frac{2}{L}}\sqrt{\frac{4a_e}{3\pi}}\xi \exp - \left(\frac{2}{3}\frac{a_e^{2/3}}{\pi^{1/3}}\frac{\xi^2}{2}\right)^{-2/3}; \xi = \frac{\rho}{L}; a_e = \frac{\beta L}{\hbar^2/2\mu_e L^2} \quad (8.18)$$

$$\frac{\left(E^F_{rad}\right)_{e,min}}{\hbar^2/2\mu_e L^2} \approx 2.346\left(\frac{\beta L}{\hbar^2/2\mu_e L^2}\right)^{2/3} \quad (8.19)$$

As you can see, with an increase of the strength of the external field, the main energy level of the radial motion of the particle in the layer-well rises, which is quite natural.

When solving equation (8.16), firstly, taking into account condition (8.2) and result (8.6), the value $\hbar^2/2\mu_e r^2$ of centrifugal energy within the layer can be replaced by the value $\hbar^2/2\mu_e R_0^2$. Secondly, due to the fulfillment of condition (8.13), the quantity $\beta(Br + C/r)$ without distorting the physical essence of the problem can be replaced simply by the average value of the potential energy of a particle within the β-HgS layer:

$$\langle\beta(Br + C/r)\rangle = \frac{|e|F}{2}\left[(BR_2 + C/R_2) - (BR_1 + C/R_1)\right] \approx \frac{\beta L}{2} \quad (8.20)$$

Then we come to the following equation:

$$\frac{d^2 g(\phi)}{d\phi^2} + \left(\gamma^2 - \lambda^2\phi^2\right) g(\phi) = 0; \gamma^2 = \frac{2\mu_e R_0^2 E^F_{ang}}{\hbar^2}; \lambda^2 = \frac{\beta L \mu_e R_0^2}{2\hbar^2} \quad (8.21)$$

The solutions of this equation are the functions of the harmonic oscillator. For the envelope wave functions and the energy of motion along the angular variable, we obtain, respectively:

$$g_{s_e}(\phi) = \left(\frac{1}{2^{s_e} s_e!}\sqrt{\frac{\lambda}{\pi}}\right)^{1/2} H_{s_e}\left(\sqrt{\lambda}\phi\right) \exp\left[-\frac{\lambda\phi^2}{2}\right] \quad (8.22)$$

$$\frac{\left(E^F_{ang}\right)_{s_e}}{\hbar^2/2\mu_e L^2} = \left(\frac{\beta L}{\hbar^2/2\mu_e L^2}\right)^{1/2} \frac{L}{R_0}\left(s_e + \frac{1}{2}\right); s_e = 0, 1, 2, \ldots. \quad (8.23)$$

As we see, in the presence of a strong external field, the motion of an electron along the angular coordinate is reduced to oscillations in a narrow angular segment in the vicinity of $\phi = 0$. The solution of the equations for a particle with a positive charge shows that under the action of a strong external field, holes will oscillate in a narrow segment in the vicinity of $\phi = \pi$.

8.4 Discussion of Results

The results obtained show that in the presence of a strong external field, the energy of charge carriers in the β-HgS layer, where the strong size quantization regime is realized, increases significantly. According to expression (8.19), the position of the main level of radial motion in the presence of a field is determined by the expression

$$\left(E_{rad}^{F}\right)_{e,min} = 2,345\left(\frac{\beta L}{E_{e}^{rad}}\right) \cdot E_{e}^{rad}. \tag{8.24}$$

So, even with not too strong fields, for the ground level in the presence of a field we have:

$$\left(E_{rad}^{F}\right)_{e,min} \sim 11E_{rad}^{e} > \frac{\pi^2\hbar^2}{2\mu_e L^2}. \tag{8.25}$$

As for the motion along the angular variable, a strong external field in this case radically changes the structure of the energy spectrum: the rotational motion of carriers around the circumference of the layer (expressions (8.5)–(8.6)) transforms into oscillations in a narrow angular segment in the vicinity of the opposite ends of the layer diameter: in the vicinity angle $\phi = 0$ for electrons, and $\phi = \pi$ for holes, respectively. Moreover, the energy of these angular oscillations is determined not only by the magnitude of the external field, but also by the relationship between the thickness and the average radius of the layer:

$$\left(E_{ang}^{F}\right)_{min} = \frac{1}{2}\left(\frac{\beta L}{\hbar^2/2\mu_e L^2}\right)^{1/2}\frac{L}{R_0}\frac{\hbar^2}{2\mu_e L^2} > E_{rot}^{e} \sim \frac{\hbar^2}{2\mu_e R_0^2}. \tag{8.26}$$

A change in the energy spectrum of carriers is also reflected in the parameters of optical absorption in the sample. In the absence of a field, the threshold frequency of interband transitions was determined by the size quantization energy of the transverse motion of charge carriers in the β-HgS layer:

$$\hbar\omega_{c,v}^{(0)} = E_g^L + \left(E_{rad}^{e}\right)_{min} + \left(E_{h}^{F}\right)_{min} + \left(E_{rot}^{e}\right)_{min} + \left(E_{rot}^{h}\right)_{min}. \tag{8.27}$$

In the presence of an external field instead Exp. (8.27) we get:

$$\hbar\omega_{c,v}^{(F)} = E_g^L + \left(E_{rad}^{F}\right)_{e,min} + \left(E_{rad}^{F}\right)_{h,min} + \left(E_{ang}^{F}\right)_{e,min} + \left(E_{ang}^{F}\right)_{h,min} - \beta L. \tag{8.28}$$

Here E_g^L is the band gap of bulk β-HgS, and the energy count is carried out from the bottom of the conduction band. As can be seen, the external field shifts the threshold of interband absorption in a layer of β-HgS to the region of low frequencies.

As for the frequency dependence of interband absorption, in this case it will be determined by the multiplier $\left(\hbar\omega - \hbar\omega_{c,v}^{F}\right)^{-1/2}$ which is characteristic for quasi-one-dimensional systems.

The external field also leads to the dependence of the intensity of interband transitions $I_{c,v}$ from the effective mass of charge carriers. In particular, for the intensity of transitions on the threshold frequency, we have:

$$I_{c,v} \sim \frac{(\mu_e \mu_h)^{3/2}}{(\mu_e + \mu_h)\left(\mu_e^{2/3} + \mu_h^{2/3}\right)^3}. \tag{8.29}$$

8.5 Conclusions

The proposed model allows us to get the desired quantitative description of the problem under consideration and obtain results explicitly analytically. The results of the work show that combining changes in the geometric sizes of the system with changes in the intensity of the external field, you can obtain the desired interval of frequency of transmission and absorption for incident on the sample light.

References

1. Siemsen, K.J., Riccius, H.D.: Preparation and optical properties of evaporated $\beta - HgS$ films. Phys. Stat. Sol. B **37**, 445–451 (1970)
2. Mohammed, N.J., Dagher, H.F.: Synthesis and characterization of mercuric sulfide nanoparticles thin films by Pulsed Laser Ablation (PLA) in Distilled Water (DW). Iran. J. Mater. Sci. & Eng. **17**(3), 11–16 (2020)
3. Haus, J.W., Zhou, H.S., Honma, I., Komiyama, H.: Quantum confinement in semiconductor heterostructure nanometer-size particles. Phys. Rev. B **47**(3), 1359–1365 (1993)
4. Schoss, D., Mews, A., Eychmüller, A., Weller, H.: Quantum-dot quantum well CdS/HgS/CdS: theory and experiment. Phys. Rev. B **49**(24), 17072–17078 (1994)
5. Rogach, A.: Semiconductor Nanocrystal Quantum Dots: Synthesis, Assembly, Spectroscopy and Applications, p. 372. Springer, Berlin (2008)
6. Xu, X., Carraway, E.R.: Sonication-assisted synthesis of β-mercuric sulphide nanoparticles. Nanomater. Nanotechnol. **2**, 1–6 (2012). Art. 17:2012
7. Liu, K., Gao, G., Sun, T.: $\beta - HgS$ quantum dots: preparation, properties and applications. Prog. Chem. **29**(7), 776–784 (2017)
8. Benhaddou, F., Zorkani, I., Jorio, A.: The confinement effect in spherical inhomogeneous quantum dots and stability of excitons. AIP Adv. **7**, 065112-1-6 (2017)
9. Sun, Ch.Q.: Electron and Phonon Spectrometrics, 511p. Springer Nature (2020)
10. Akinlami, J.O., Onyeanu, F.C.: Electronic and optical properties of $\beta - HgS$. SPQEO **22**(2), 150–155 (2019)
11. Chakraborty, I., Mitra, D., Moulik, S.P.: Spectroscopic studies on nanodispersions of CdS, HgS, their core-shells and composites prepared in micellar medium. J. Nanopart. Res. **7**, 227–236 (2005)
12. Kershaw, S.V., Harrison, M., Rogach, A.L., Kornowski, A.: Development of IR-emitting colloidal II-VI quantum-dot materials. IEEE J. Sel. Top. Quantum Electron. **6**, 534–543 (2000)

13. Wichiansee, W., Nordin, M.N., Green, M., Curry, R.J.: Synthesis and optical characterization of infra-red emitting mercury sulfide (HgS) quantum dots. J. Mater. Chem. **21**, 7331–7336 (2011)
14. Kim, J., Yoon, B., Kim, J., Choi, Y., Kwon, Y.-W., Park, S.K., Jeong, K.S.: High electron mobility of $\beta - HgS$ colloidal quantum dots with doubly occupied quantum states. RSC Adv. **7**(61), 38166–70 (2017)
15. Roberts, G.G., Lind, E.L., Davis, E.A.: Photoelectronic properties of synthetic mercury sulphide crystals. J. Phys. Chem. Solids **30**, 833–844 (1969)
16. Galain, I., Aguiar, I., Eugenia, M., Barthaburu, P., Fornaro, L.: First steps to use $\beta - HgS$ nanostructures in solution as electron acceptor in organic-inorganic solar cells. In: Proceedings of XIV Brazil MRS Meeting - Rio de Janeiro, from 27 September to 01 October 2015, p. 99 (2015)
17. Barthaburu, M., Galain, I., Pereira, A.H.B., Bethencourt, L., Miranda, P.B., Sampaio, M.F.B., Fornaro, L.: Hybrid $\beta - HgS$ nanoparticles and P3HT layers for solar cells applications. Nano-Struct. & Nano-Objects **10**, 15–21 (2017)
18. Virot, F., Hayn, R., Richter, M., van den Brink, J.: Metacinnabar ($\beta - HgS$): a strong 3D topological insulator with highly anisotropic surface states. Phys. Rev. Lett. **106**, 236806-1–4 (2011)
19. Rajendran, S.: Metal, Metal Oxides and Metal Sulphides for Biomedical Applications, 360p. Springer Nature (2020)
20. Yang, F., Gao, G., Wang, J., Chen, R., Zhu, W., Wang, L., Ma, Z., Luo, Z., Sun, T.: Chiral $\beta - HgS$ quantum dots: aqueous synthesis, optical properties and cytocompatibility. J. Colloid Interface Sci. **573**, 422–430 (2019)
21. Joy, K.M.F., Jaya, N.V.: Electrical resistivity of Nano-Hgs under high pressure and temperature. J. Chem. Pharm. Sci. **9**(4), 2387–2390 (2016)
22. Pronishin, I.V., Pazyuk, V.V., Tkach, N.V.: Electronic spectrum in a quantum double $CdS/\beta - HgS/H_2O$ cylindrical wire. Russ. Phys. J. **40**(5), 475–480 (1997)
23. Tkach, M., Pronyshyn, I., Makhanetz, O., Zharkoy, V.: Electron and hole spectra in cylindrical quant wire superlattice. In: Yen W.M. (eds.) Proceedings of the Third International Conference on Excitonic Processes in Condensed Matter, EXCON '98, The Electrochemical Society, pp. 320–327 (1998)
24. Tkach, M.V., Makhanets, O.M., Seti, Ju.O., Dovganiuk, M.M., Voitsekhivska, O.M.: Electronic conductivity in open cylindrical two-barrier symmetric resonance tunnel structure. Acta Phys. Polon. A **117**(6), 965–970 (2010)
25. Zhong, Q., Liu, C.: Studies of electronic Raman scattering in CdS/HgS cylindrical quantum dot quantum well structures. Thin Sol. Films **516**(10), 3405–3410 (2008)
26. Edrissi, S.J., et al.: CdS/HgS Cylindrical Core/Shell: the simultaneous effect of the hydrostatic pressure and temperature on the diamagnetic susceptibility of hydrogenic donor impurity. J. Phys.: Conf. Ser. **1292**, 012004-1–5 (2019)
27. Harutyunyan, V.A., Hayrapetyan, D.B., Kazaryan, E.M.: Optical transitions and photoluminescence in cylindrical core/layer/shell $\beta - CdS/\beta - HgS/\beta - CdS$ heterostruture. Phys. Solid State **62**(8), 1305–1316 (2020)
28. Harutyunyan, V.A., Hayrapetyan, D.B., Kazaryan, E.M.: Excitonic states and optical transitions in CdS/HgS/CdS nanocylindrical core/shell/shell heterostructure (Abstract Book). In: 8th International Conference on Nanostructures (ICNS8), 18–20 November 2020, Tehran, Iran, p. 188 (2020)
29. Harutyunyan, V.A., Hayrapetyan, D.B., Kazaryan, E.M.: Interband absorption and photoluminescence in the cylindrical layered CdS/HgS/CdS heterostructure. J. Contemp. Phys. **53**(1), 48–57 (2018)
30. Harutyunyan, V.A., Harutyunyan, S.L., Demirjian, G.H., Petrosyan, H.S.: Optical transitions in a quantized cylindrical layer under the presence of homogeneous electric field. Semiconductors **39**(7), 839–843 (2005)
31. Harutyunyan, V.A., Demirjian, G.H., Gasparyan, N.H.: Semiconductor nanotube in strong radial field: spectrum of carriers and intraband transitions. Phys. E **43**(2), 614–619 (2010)

32. Harutyunyan, V.A.: Effect of Static Electric Fields on The Electronic And Optical Properties of Layered Semiconductor Nanostructures: PART I: Effect of Static Electric Fields on The Electronic Properties of Layered Semiconductor Nanostructures, 243p. Bentham Science Publishers (2015)
33. Harutyunyan, V.A., Aramyan, K.S., Petrosyan, H.S.: Confinement Stark effect and electroabsorption in semiconductor cylindrical layer. Phys. E **21**(1), 111–116 (2004)

Chapter 9
Quasi-conical Quantum Dot Helium

K. S. Khachatryan and M. A. Mkrtchyan

Abstract By the analogy with the real helium atom theory, a model of a two-electron system, localized in a quasi-conical quantum dot, (quasi-conical helium atom) is constructed in the framework of the perturbation theory. The presence of the electron spin is considered in the Russell–Saunders approximation and the wave function is considered as the multiplication of the coordinate and spin parts. The Coulomb and exchange corrections to the energy states of the system are determined. The behavior of the Coulomb and exchange energy on the geometric parameters of the quantum dot is discussed. The exchange time of states between the electrons is calculated.

9.1 Introduction

The investigation of the few- and multi-particle complexes in the semiconductor quantum dots (QD) continue to be an actual problem in the physics of quantum nanostructures [1–8]. Such problems are of great practical importance and can reveal the characteristic features of physical data. In particular, in [5] the authors have considered a system, consisting of two electrons in a QD with a three-dimensional potential of harmonics limitation under the action of a magnetic field. Two different confinement conditions were considered: isotropic three-dimensional and anisotropic quasi-two-dimensional conditions. The properties of singlet and triplet lowest states, such as energy and exchange coupling, are analyzed. In [8], the authors studied one- and two-particle states in electrostatically induced QDs in gapped bilayer graphene. Based on these single-particle states, they obtained the orbital, spin, and valley states of two interacting electrons in a QD. The interaction strength of two-particle were also found.

The simplest few particle system is a two-electron system, localized inside of QD. Such systems are very often called as "artificial" helium atoms [9, 10]. An essential advantage of these helium-like systems in comparison with the real helium

K. S. Khachatryan (✉) · M. A. Mkrtchyan
Russian-Armenian University, Yerevan, Armenia
e-mail: khachik.khachatryan@rau.am

D. Blaschke et al. (eds.), *Optics and Its Applications*, Springer Proceedings in Physics 281, https://doi.org/10.1007/978-3-031-11287-4_9

atom is the possibility of the flexible manipulation of a two-electron artificial helium atom by changing the geometric sizes and shape of a QD. If the localization of electrons in real helium atom is realized due to the attractive nuclear forces, then the localization of electrons in an "artificial" Helium atom occurs due to the influence of the QD's confining potential. It is clear the spectrum of two-electron system in QD can be theoretically controlled by changing the confining potential. From the mathematical description point of view, the closest systems to a real helium atom are spherical QDs. The symmetry in the case of single-particle states is spherical. For such systems one can successfully adapt the description methods such as perturbation theory and the Hartree Fogh method, as well as give a successful implementation of two-particle model of the Mashinsky atom [11–14]. On the other hand, modern methods for the growth of QDs make it possible to implement zero-dimensional structures of various geometric shapes such as cylindrical, ellipsoidal, pyramidal, lens-shaped, etc. [15–24]. It is obvious that the investigation of two-electronic states in the above mentioned systems can lead to the revealing of the interesting features of the studied energy spectrum. The above is all the more relevant, because the above mentioned QDs are not systems with spherical symmetry and the classification of states of helium-like systems in them requires a separate consideration. Two-electron states were considered in QDs with different geometries [25–27]. Relatively recently, conical QDs have been implemented, in which states close in character to quantum wires (strongly prolate conical QDs) and quantum wells (strongly oblate conical QDs) can be realized [28–30]. Such structures are considered as successful candidates for the role of an element base for LEDs based on QDs. In particular, in [29] the prospect of "core/shell" QDs has been described, which may further to contribute to the development of highly efficient and stable sensitized QDs of solar cells. In the theoretical work [31] the possibility of the implementation of the active medium for LEDs, using ensembles of conical QDs of various sizes, was theoretically demonstrated.

The theoretical description of the single-particle states in conical QDs requires independent consideration; in the general case this problem requires numerical modeling. In the case of a strongly prolate or strongly oblate QD geometry it is possible to successfully apply the adiabatic description [32], which allows obtaining a number of analytical results. On the other hand, a conical QD can be approximately modeled as a spherical segment. The convexity of the con's bottom in several cases may have an insignificant effect on the nature of the charge carrier energy. However, the one-particle Schrodinger equation for such geometry can be solved exact analytically. The optical properties of such system were investigated in the work [33] in which, the selection rules for interband transitions for models of simple zones are determined. It is clear that this model of a conical QD makes it possible to describe relatively simple two-electron states in the framework of perturbation theory, since the analytical form of the wave functions of one-electron states is known. In this case, a significant difference between the considered QD and the spherical one is that the orbital quantum number ceases to be integer. In this paper, a quasi-conical helium atom is investigated in the framework of perturbation theory and the Russell–Saunders approximation.

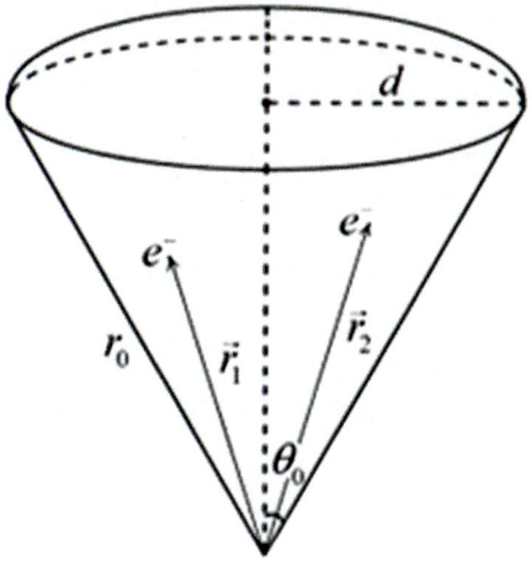

Fig. 9.1 Quasi-conical QD

9.2 Theory

Let's discuss two electron states quasi-conical QD made from $InAs$ (Fig. 9.1). The confining potential in the radial and polar directions has the following forms:

$$V_{conf}^{rad}(r) = \begin{cases} 0, r < r_0 \\ \infty, r \geq r_0 \end{cases}, \tag{9.1}$$

$$V_{conf}^{pol}(\theta) = \begin{cases} 0, \theta < \theta_0 \\ \infty, \theta \geq \theta_0 \end{cases}, \tag{9.2}$$

where d is the quasi-con's base radius, θ_0 is the half of the opening angle, r_0 is the side edge of the quasi-con, $\boldsymbol{r}$ is the electrons' radius vector.

The Hamiltonian of the system can be written as:

$$\hat{H} = \hat{H}_1 + \hat{H}_2 + V_{e-e}, \tag{9.3}$$

where

$$\hat{H}_i = -\frac{\hbar^2}{2\mu}\nabla_i^2 + V_{conf}(\boldsymbol{r}_i), \tag{9.4}$$

is the one-particle Hamiltonian ($i = 1, 2$), μ is the electron effective mass (for $InAs$ $\mu = 0.023 \cdot m$, where m is electron mass), $V_{conf}(\boldsymbol{r}_i) = V_{conf}^{rad}(r) + V_{conf}^{pol}(\theta)$ and V_{e-e} is the Coulomb interaction term between electrons:

$$V_{e-e} = \frac{e^2}{\varepsilon_d |\boldsymbol{r}_i - \boldsymbol{r}_j|}, \tag{9.5}$$

where $\varepsilon_d = 15.5$ is the dielectric constant of the material and e is electron charge.

Let's proceed to the discussion of the one-particle Schrodinger equation, which in spherical coordinates can be written as:

$$-\frac{\hbar^2}{2\mu}\nabla^2_{r,\theta,\phi}\psi(r,\theta,\phi) + \left(V^{rad}_{conf}(r) + V^{pol}_{conf}(\theta)\right)\psi(r,\theta,\phi) = E\psi(r,\theta,\phi). \tag{9.6}$$

The wave function for (9.6) has the form [33]:

$$\psi(r,\theta,\phi) = R(r)P(\theta)e^{im\phi}, \tag{9.7}$$

where $R(r)$ and $P(\theta)$ can be determined by solving the following equations [33]:

$$\frac{d^2P(\theta)}{d\theta^2} + \cot\theta\frac{dP(\theta)}{d\theta} + \left(l(l+1) - \frac{m^2}{\sin^2\theta}\right)P(\theta) = 0, \tag{9.8}$$

$$\frac{d^2R(r)}{dr^2} + \frac{2}{r}\frac{dR(r)}{dr} + \left(k^2 - \frac{l(l+1)}{r^2}\right)R(r) = 0. \tag{9.9}$$

Here $m = 0; \pm1; \pm2; \ldots$ is the magnetic quantum number, $k = \sqrt{\frac{2\mu E}{\hbar^2}}$ and l are radial and orbital quantum numbers, respectively. It is worth to note, that l is not integer in this case. The solution of (9.8) is a linear combination of two functions, namely:

$$\begin{cases} m \geq 0,\ P(\theta) = C\sin^m\frac{\theta}{2}\cos^m\frac{\theta}{2}\,{}_2F_1\left(m+l+1, m-l, m+1, \sin^2\frac{\theta}{2}\right) \\ m < 0,\ P(\theta) = C\sin^{-m}\frac{\theta}{2}\cos^m\frac{\theta}{2}\,{}_2F_1\left(l+1, -l, 1-m, \sin^2\frac{\theta}{2}\right) \end{cases}, \tag{9.10}$$

To continue, solution of the radial part of wave function (9.9) has the form:

$$R(r) = D\frac{1}{\sqrt{r}}J_{l+\frac{1}{2}}\left(\sqrt{\frac{2\mu E}{\hbar^2}}r\right), \tag{9.11}$$

where ${}_2F_1$ is the hypergeometric function and $J_{l+\frac{1}{2}}$ is the Bessel function.

The values of the quantum numbers and as well as energy can be found from the fulfillment of the boundary conditions:

$$R(r_0) = 0,\ P(\theta_0) = 0. \tag{9.12}$$

It should be highlighted, that in the further calculations we will be focused on the first two $|klm\rangle$ states of the one-particle problem, namely, the ground and the first excited states, which will be denoted by $|110\rangle$ and $|210\rangle$ by the analogy with work [32].

The total wave function of the problem in the scope of Russell–Saunders approximation is considered as a product of two parts, namely, coordinate and spin parts.

Within the framework of the theory of the helium atom, we have investigated two electronic states, considering the Coulomb interaction between electrons as a perturbation. By analogy with this theory, the wave function for the singlet state can be written as follows:

$$\psi_{singlet} = \frac{1}{\sqrt{2}}\left[\psi_1(r_1,\theta_1,\phi_1)\,\psi_2(r_2,\theta_2,\phi_2) + \psi_2(r_1,\theta_1,\phi_1)\,\psi_1(r_2,\theta_2,\phi_2)\right] \chi_a(\sigma_1,\sigma_2), \tag{9.13}$$

where $\psi_1(r_1,\theta_1,\phi_1)$, $\psi_2(r_2,\theta_2,\phi_2)$ are ground and first exited states of (9.7), $\chi_a(\sigma_1,\sigma_2)$ is the spin wave function, which is asymmetric to the permutation and has well-known form:

$$\chi_a(\sigma_1,\sigma_2) = \frac{1}{\sqrt{2}}\left[\alpha_1\beta_2 - \alpha_2\beta_1\right], \alpha = \frac{\hbar}{2}\begin{pmatrix}1\\0\end{pmatrix}, \beta = \frac{\hbar}{2}\begin{pmatrix}0\\1\end{pmatrix} \tag{9.14}$$

Here α describes the state of spin up and β—state of spin down.

The wave function for the triplet state is as follows:

$$\psi_{triplet} = \frac{1}{\sqrt{2}}\left[\psi_1(r_1,\theta_1,\phi_1)\,\psi_2(r_2,\theta_2,\phi_2) - \psi_2(r_1,\theta_1,\phi_1)\,\psi_1(r_2,\theta_2,\phi_2)\right] \chi_s(\sigma_1,\sigma_2), \tag{9.15}$$

where $\chi_s' = \alpha_1\alpha_2$, $\chi_s'' = \beta_1\beta_2$, $\chi_s''' = \frac{1}{\sqrt{2}}\left[\alpha_1\beta_2 + \alpha_2\beta_1\right]$.

The singlet and triplet states' energies have the following forms, respectively [33]:

$$E_{singlet} = E_1 + E_2 + K + P, \tag{9.16}$$

$$E_{triplet} = E_1 + E_2 + K - P, \tag{9.17}$$

where E_1 and E_2 are the single particle energies for the first and the second states, respectively, K is the energy correction of Coulomb interaction and P is the exchange interaction energy.

For V_{e-e} one can write:

$$V_{e-e} = \frac{e^2}{\varepsilon_d\sqrt{r_1^2 + r_2^2 - 2r_1r_2\left(\cos\theta_1\cos\theta_2 + \sin\theta_1\sin\theta_2\cos(\phi_1-\phi_2)\right)}}. \tag{9.18}$$

It is worth to mention, that the relative angle θ between the r_1 and r_2 is taken into account and for θ we have the expression: $\cos\theta = \cos\theta_1\cos\theta_2 + \sin\theta_1\sin\theta_2 \cos(\phi_1-\phi_2)$. The schematic plot for θ is presented in Fig. 9.2.

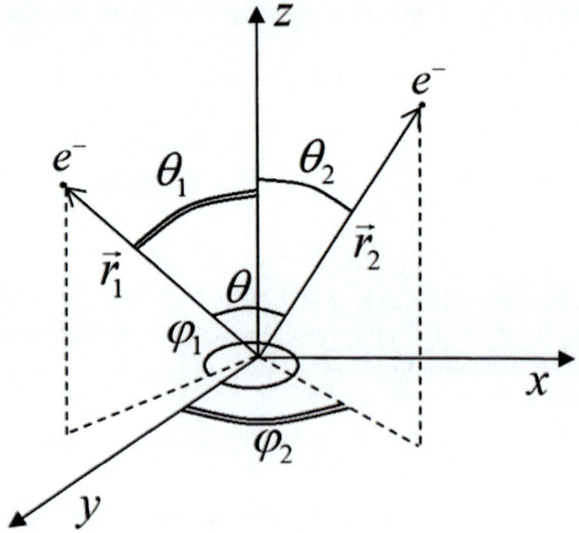

Fig. 9.2 Schematic plot of the relative angle between two electrons

For K and P we have the following integrals:

$$K = \int_0^R \int_0^{2\pi} \int_0^{\theta_0} \int_0^R \int_0^{2\pi} \int_0^{\theta_0} \frac{\Psi_{01}^* \Psi_{01} \cdot \sin\theta_1 d\theta_1 d\phi_1 r_1^2 dr_1 \cdot \sin\theta_2 d\theta_2 d\phi_2 r_2^2 dr_2}{\sqrt{r_1^2 + r_2^2 - 2r_1 r_2 \left(\cos\theta_1 \cos\theta_2 + \sin\theta_1 \sin\theta_2 \cos\left(\phi_1 - \phi_2\right)\right)}}, \quad (9.19)$$

$$P = \int_0^R \int_0^{2\pi} \int_0^{\theta_0} \int_0^R \int_0^{2\pi} \int_0^{\theta_0} \frac{\Psi_{01}^* \Psi_{02} \cdot \sin\theta_1 d\theta_1 d\phi_1 r_1^2 dr_1 \cdot \sin\theta_2 d\theta_2 d\phi_2 r_2^2 dr_2}{\sqrt{r_1^2 + r_2^2 - 2r_1 r_2 \left(\cos\theta_1 \cos\theta_2 + \sin\theta_1 \sin\theta_2 \cos\left(\phi_1 - \phi_2\right)\right)}}, \quad (9.20)$$

where

$$\Psi_{01} = \psi_1 \left(r_1, \theta_1, \phi_1\right) \psi_2 \left(r_2, \theta_2, \phi_2\right), \quad (9.21)$$

$$\Psi_{02} = \psi_1 \left(r_2, \theta_2, \phi_2\right) \psi_2 \left(r_1, \theta_1, \phi_1\right), \quad (9.22)$$

where $\psi_1 \left(r_2, \theta_2, \phi_2\right)$ is the wave function of the first particle in the second state, $\psi_2 \left(r_1, \theta_1, \phi_1\right)$ is the wave function of the second particle in the first state, respectively. The calculations of these integrals will be carried out numerically.

9.3 Results

Based on the theory, described in Sect. 9.2, let's proceed to the discussion of the obtained results. Figure 9.3 shows the dependences of the Coulomb energies on the geometrical parameters of the considered QD. In particular, Fig. 9.3a shows the

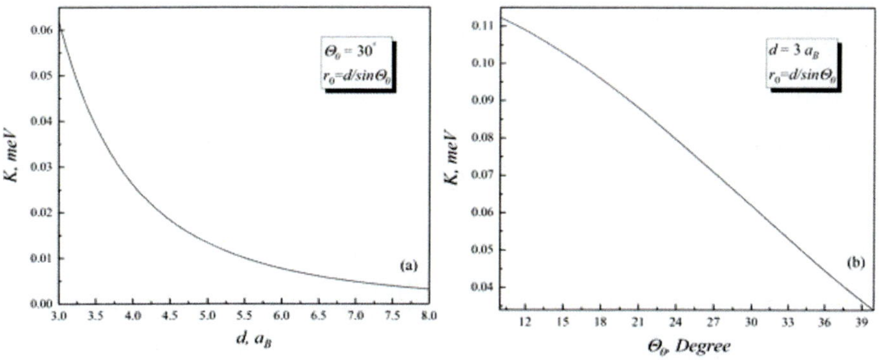

Fig. 9.3 Dependence of the Coulomb interaction energy on the quasi-conical QD's base radius (**a**) and on the opening angle (**b**)

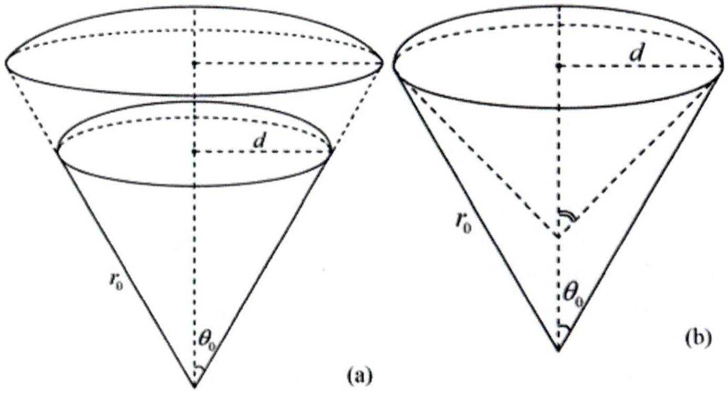

Fig. 9.4 The schematic plot of changing the QD's base radius d and the angle θ_0. **a**—the angle θ_0 is fixed and the base radius d changes, **b**—the base radius d is fixed and the angle θ_0 changes

dependence K on the quasi-con's base radius d, when the θ_0 is fixed, and Fig. 9.3b—the dependence of K on θ_0 for the fixed value of d.

It is obvious from the Fig. 9.3a, that the Coulomb interaction between electrons decreases with the increase of the QD's base radius. This behavior can be explained by the fact, that with the increase of the quasi-con's geometrical parameter the electrons' localization region increases, and, as a sequence, they are located at a greater distance from each other, due to the repulsion. The same behavior for the Coulomb interaction occurs in the Fig. 9.3b, where θ_0 is changed. It is worth to note, that the interaction energy between electrons is more sensitive to the decrease of θ_0 angle. The changes of the QD's geometric parameters are shown in the Fig. 9.4.

The result of the state exchange between electrons for a certain time is interpreted as exchange energy between electrons [34]. Adhering to this interpretation, let us consider two electrons' case in the quasi-conical QD.

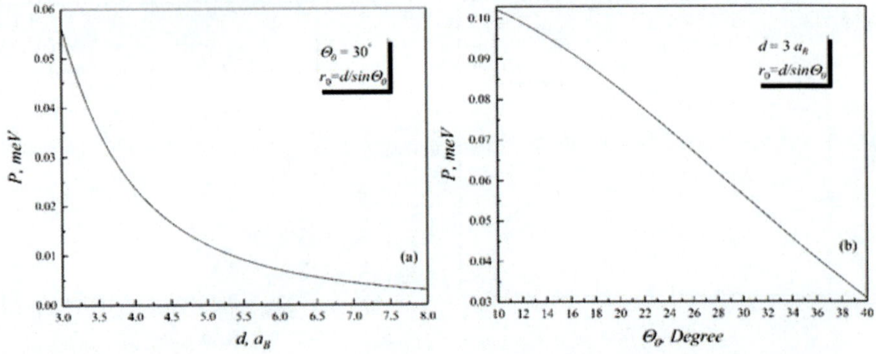

Fig. 9.5 Dependence of the exchange energy P on the quasi-conical QD's base radius (**a**) and on the opening angle (**b**)

In order to calculate time, needed for the transition of the system from Ψ_{01} state to the Ψ_{02} state, we have used the following formula [34]:

$$\tau = \frac{\pi \hbar}{2P}. \tag{9.23}$$

The dependence of the exchange energy on the quasi-conical QD's radius d and angle θ_0 is shown in Fig. 9.5.

It follows from the Fig. 9.5, that the dependences both in (a) and (b) have the similar behaviors as in the case of the Coulomb interaction. The exchange interaction is due to the degree of the electrons' wave functions' overlapping, localized in the QD. It is also clear, that the wave functions' overlapping decreases with the increase of the QD's geometrical parameters, since the distance between electrons increases, which, in its turn, leads to the decrease of the exchange energy. Note, that ground and first exited states are again considered here.

Finally, Fig. 9.6 shows the dependences of τ on the quasi-conical QD's geometrical parameters. In particular, Fig. 9.6a shows the dependence on the base radius, while Fig. 9.6b shows the dependence of the exchange time on the opening angle.

It can be seen from the figure, that the dependences for the exchange time have the opposite behavior for the exchange energy, i.e. the exchange time increases with the increase of the quasi-cone's base radius (Fig. 9.6a). It is also obvious, that the discussion for Fig. 9.6a is correct for Fig. 9.6b. Note, that in the case of the spherical QD [35], when the electrons are located diametrically at a large distance, the exchange time τ is about 10^{-10} s, and in the case of a spherical sector, due to the wave functions' overlap, the exchange time decreases and has order of 10^{-11} s.

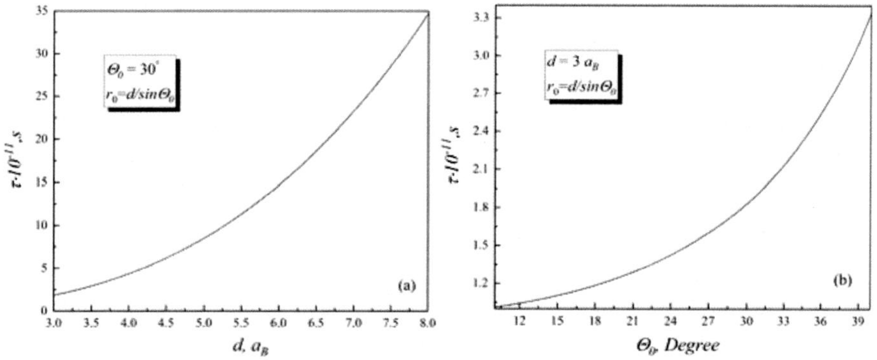

Fig. 9.6 Dependences of the electrons' exchange time on the base radius (**a**) and on the opening angle (**b**)

9.4 Conclusion

Two electronic states in the quasi-conical QD are theoretically investigated. The differences between Coulomb and exchange interactions have been discussed. The increase of the quasi-conical QD's base radius results in the decrease of the energy of Coulomb and exchange interactions. It has been shown, that exchange energy as well as the state exchange time can be controlled by changing the geometrical parameters of the QD.

Acknowledgements This work was supported by the RA MESCS State Committee of Science, in the frames of the basic research project No. 10-2/I-5.

References

1. Merkt, U., Huser, J., Wagner, M.: Energy spectra of two electrons in a harmonic quantum dot. Phys. Rev. B **43**(9), 7320 (1991)
2. Zhu, J.L., Yu, J.Z., Li, Z.Q., Kawazoe, Y.: Exact solutions of two electrons in a quantum dot. J. Phys.: Condens. Matter **8**(42), 7857 (1996)
3. Dybalski, W., Hawrylak, P.: Two electrons in a strongly coupled double quantum dot: from an artificial helium atom to a hydrogen molecule. Phys. Rev. B **72**(20), 205432 (2005)
4. He, L., Bester, G., Zunger, A.: Singlet-triplet splitting, correlation, and entanglement of two electrons in quantum dot molecules. Phys. Rev. B **72**(19), 195307 (2005)
5. Maniero, A.M., de Carvalho, C.R., Prudente, F.V., Jalbert, G.: Oscillating properties of a two-electron quantum dot in the presence of a magnetic field. J. Phys. B: At. Mol. Opt. Phys. **53**(18), 185001 (2020)
6. Dineykhan, M., Nazmitdinov, R.G.: Two-electron quantum dot in a magnetic field: analytical results. Phys. Rev. B **55**(20), 13707 (1997)
7. Besombes, L., Kheng, K., Marsal, L., Mariette, H.: Few-particle effects in single CdTe quantum dots. Phys. Rev. B **65**(12), 121314 (2002)

8. Knothe, A., Fal'ko, V.: Quartet states in two-electron quantum dots in bilayer graphene. Phys. Rev. B **101**(23), 235423 (2020)
9. Pfannkuche, D., Gudmundsson, V., Maksym, P.A.: Comparison of a Hartree, a Hartree-Fock, and an exact treatment of quantum-dot helium. Phys. Rev. B **47**(4), 2244 (1993)
10. Pfannkuche, D., Gerhardts, R.R., Maksym, P.A., Gudmundsson, V.: Theory of quantum dot helium. Phys. B **189**(1–4), 6–15 (1993)
11. Abolfath, R.M., Hawrylak, P.: Real space Hartree-Fock configuration interaction method for complex lateral quantum dot molecules. J. Chem. Phys. **125**(3), 034707 (2006)
12. Reusch, B., Grabert, H.: Unrestricted Hartree-Fock for quantum dots. Phys. Rev. B **68**(4), 045309 (2003)
13. Hayrapetyan, D.B., Kazaryan, E.M., Mkrtchyan, M.A., Sarkisyan, H.A.: Long-wave absorption of Few-Hole gas in prolate ellipsoidal Ge/Si quantum dot: implementation of analytically solvable Moshinsky model. Nanomaterials **10**(10), 1896 (2020)
14. Moshinsky, M.: How good is the Hartree-Fock approximation. Am. J. Phys. **36**(1), 52–53 (1968)
15. Atoyan, M.S., Kazaryan, E.M., Sarkisyan, H.A.: Direct interband light absorption in a cylindrical quantum dot in quantizing magnetic field. Phys. E **22**(4), 860–866 (2004)
16. Sargsian, T.A., Mkrtchyan, M.A., Sarkisyan, H.A., Hayrapetyan, D.B.: Effects of external electric and magnetic fields on the linear and nonlinear optical properties of InAs cylindrical quantum dot with modified Pöschl-Teller and Morse confinement potentials. Phys. E **126**, 114440 (2021)
17. Hayrapetyan, D.B., Bleyan, Y.Y., Baghdasaryan, D.A., Sarkisyan, H.A., Baskoutas, S., Kazaryan, E.M.: Biexciton, negative and positive trions in strongly oblate ellipsoidal quantum dot. Phys. E **105**, 47–55 (2019)
18. Hayrapetyan, D.B., Kazaryan, E.M., Sarkisyan, H.A.: Implementation of Kohn's theorem for the ellipsoidal quantum dot in the presence of external magnetic field. Phys. E **75**, 353–357 (2016)
19. Osorio, J.A., Caicedo-Paredes, D., Vinasco, J.A., Morales, A.L., Radu, A., Restrepo, R.L., Martínez-Orozco, J.C., Tiutiunnyk, A., Laroze, D., Hieu, N.N., Phuc, H.V., Mora-Ramos, M.E., Duque, C.A.: Pyramidal core-shell quantum dot under applied electric and magnetic fields. Sci. Rep. **10**(1), 1–14 (2020)
20. Nenashev, A.V., Dvurechenskii, A.V.: Variational method of energy level calculation in pyramidal quantum dots. J. Appl. Phys. **127**(15), 154301 (2020)
21. Restrepo, R.L., Ospina-Muñoz, W.A., Feddi, E., Mora-Ramos, M.E., Vinasco, J.A., Morales, A.L., Duque, C.A.: Excitons in spherical quantum dots revisited: analysis of colloidal nanocrystals. Eur. Phys. J. B **93**(6), 1–9 (2020)
22. Ungan, F., Martínez-Orozco, J.C., Restrepo, R.L., Mora-Ramos, M.E., Kasapoglu, E., Duque, C.A.: Nonlinear optical rectification and second-harmonic generation in a semi-parabolic quantum well under intense laser field: effects of electric and magnetic fields. Superlattices Microstruct. **81**, 26–33 (2015)
23. Zamani, A., Azargoshasb, T., Niknam, E., Mohammadhosseini, E.: Absorption coefficient and refractive index changes of a lens-shaped quantum dot: Rashba and Dresselhaus spin-orbit interactions under external fields. Optik **142**, 273–281 (2017)
24. Aderras, L., Feddi, E., Bah, A., Dujardin, F., Duque, C.A.: On the electronic states in lens-shaped quantum dots. Phys. Status Solidi (b) **254**(10), 1700144 (2017)
25. Kandemir, B.S.: Variational study of two-electron quantum dots. Phys. Rev. B **72**(16), 165350 (2005)
26. Bachau, H., Nikolopoulos, L.A.: Direct and sequential two-photon double ionization of two-electron quantum dots. J. Phys. Chem. A **122**(6), 1574–1583 (2018)
27. Garagiola, M., Pont, F.M., Osenda, O.: Binding of two-electron metastable states in semiconductor quantum dots under a magnetic field. J. Phys. B: At. Mol. Opt. Phys. **51**(7), 075504 (2018)
28. Dong, Y., Wang, Y.K., Yuan, F., Johnston, A., Liu, Y., Ma, D., Choi, M.J., Chen, B., Chekini, M., Baek, S.W., Sagar, L.K., Fan, J., Hou, Y., Wu, M., Lee, S., Sun, B., Hoogland, S.,

Quintero-Bermudez, R., Ebe, H., Todorovic, P., Dinic, F., Li, P., Kung, H.T., Saidaminov, M.I., Kumacheva, E., Spiecker, E., Liao, L.-S., Voznyy, O., Lu, Z.-H., Sargent, E.H.: Bipolar-shell resurfacing for blue LEDs based on strongly confined perovskite quantum dots. Nat. Nanotechnol. **15**(8), 668–674 (2020)

29. Selopal, G.S., Zhao, H., Wang, Z.M., Rosei, F.: Core/shell quantum dots solar cells. Adv. Func. Mater. **30**(13), 1908762 (2020)
30. Gambaryan, K.M., Boeck, T., Trampert, A., Marquardt, O.: Nucleation chronology and electronic properties of InAs1-x-y Sb x P y graded composition quantum dots grown on an InAs (100) substrate. ACS Appl. Electron. Mater. **2**(3), 646–650 (2020)
31. Hayrapetyan, D.B., Kazaryan, E.M., Sarkisyan, H.A.: Magneto-absorption in conical quantum dot ensemble: possible applications for QD LED. Opt. Commun. **371**, 138–143 (2016)
32. Khachatryan, K.S., Mkrtchyan, M.A., Hayrapetyan, D.B., Kazaryan, E.M., Sarkisyan, H.A.: Adiabatic description of the electroabsorption in strongly prolate and oblate conical quantum dots. Phys. E **134**, 114887 (2021)
33. Kazaryan, E.M., Petrosyan, L.S., Shahnazaryan, V.A., Sarkisyan, H.A.: Quasi-conical quantum dot: electron states and quantum transitions. Commun. Theor. Phys. **63**(2), 255 (2015)
34. Aghekyan, N.G., Kazaryan, E.M., Kostanyan, A.A., Sarkisyan, H.A.: Two electronic states and state exchange time control in spherical nanolayer. Superlattices Microstruct. **50**(3), 199–206 (2011)
35. Baghdasaryan, D.A., Kazaryan, E.M., Sarkisyan, H.A.: Two-electron states and state exchange time control in parabolic quantum dot. Phys. E **58**, 67–72 (2014)
36. Landau, L., Lifshits, E.: Quantum Mechanics. Nauka, Moscow (1989)

Chapter 10
Radiation of Surface Polaritons in Cylindrical Waveguides

A. A. Saharian, L. Sh. Grigoryan, H. F. Khachatryan, and A. S. Kotanjyan

Abstract We investigate the radiation of surface polaritons in a cylindrical waveguide by a charged particle. Two types of motion of the latter are considered: a particle moving parallel to the axis of the cylinder and a particle rotating around the cylinder. The general case is discussed when the cylinder is immersed in a homogeneous medium. Exact expressions are provided for the radiation intensity on a given mode of the cylindrical waveguide. The radiation of surface polaritons is present in the spectral range where the dielectric permittivities of the cylinder and surrounding medium have opposite signs and there is no velocity threshold for their generation. The spectral range of the emitted surface polaritons becomes narrower with decreasing energy of the particle.

10.1 Introduction

The interaction of charged particles with matter results in different types of radiation. The examples, widely investigated both theoretically and experimentally, include the Cherenkov, transition and diffraction radiations. They arise as a consequence of dynamical polarization and further relaxation of the medium by the electromagnetic field of the charge. In problems with two contacting media, the presence of interfaces between them gives rise to new types of channels for the radiation losses by a moving particle. They correspond to the radiation of various types of surface waves. Among those waves, the surface plasmon polaritons (for reviews see Refs. [1–4]) have attracted a great deal of attention in the past decade. The growing interest is motivated by wide range of applications in surface imaging, nanophotonics, spectroscopy, data storage, biosensors, solar cells, communication devices, etc. Surface

Supported by the Science Committee of RA.

A. A. Saharian (✉) · A. S. Kotanjyan
Department of Physics,Yerevan State University, 1 Alex Manoogian Str., 0025 Yerevan, Armenia
e-mail: saharian@ysu.am

A. A. Saharian · L. Sh. Grigoryan · H. F. Khachatryan · A. S. Kotanjyan
Institute of Applied Problems of Physics NAS, 25 Hr. Nersessian Str., 0014 Yerevan, RA, Armenia

D. Blaschke et al. (eds.), *Optics and Its Applications*, Springer Proceedings in Physics 281, https://doi.org/10.1007/978-3-031-11287-4_10

plasmon polaritons are collective excitations of the electromagnetic field and electron subsystem propagating along a metal-dielectric interface. Among the most important properties of these waves we mention here the confining of electromagnetic fields beyond the diffraction limit of light waves and enhancing the local field strengths by orders of magnitude [1, 2]. The surface polariton type waves can also be supported by other materials instead of metals, such as semiconductors, organic and inorganic dielectrics, ionic crystals. An important issue in the investigations of surface polaritons, partially motivated by applications in ultrahigh-speed communications, is the extension of the corresponding frequency range to the terahertz frequencies. This can be done by a suitable choice of the active medium such as doped semiconductors, dielectric-film-coated metal surfaces, textured metal surfaces, graphene surfaces and artificially constructed materials [5–8].

The extensive applications require effective techniques for exciting surface polaritons. The methods used include electron and laser beams, prism and grating coupling, guided photonic modes from waveguides (see references [1–3, 9]). The main part of the previous studies consider the excitation of surface polaritons on planar surfaces. The investigations of the effects induced by the curvature of the interface on the generation and propagation of surface polaritons are worthy of attention in many points of view, from fundamental aspects to practical applications. In the present paper, partly based on our previous research presented in [10, 11], we consider the excitation of surface polaritons on a cylindrical interface between two media by a charged particle. Two cases of the motion will be discussed: a particle uniformly moving along a rectilinear trajectory parallel to the cylinder axis and a particle uniformly rotating around a cylinder along a concentric circular trajectory.

The organization of the paper is as follows. In the next section we describe the scheme for the evaluation of the radiation intensity. The electric field generated by a charge moving along a trajectory parallel to the cylinder axis is presented. The corresponding radiation intensity for surface polaritons is investigated in Sect. 10.3. The electric field and the radiation intensity for a circular motion of a charge around a dielectric waveguide are discussed in Sect. 10.4. The corresponding results are compared with the characteristics in the problem with rectilinear motion. The main results are summarized in Sect. 10.5.

10.2 Electric Field for the Rectilinear Motion

We consider a cylindrical waveguide with dielectric permittivity ε_0 and radius r_c immersed in a homogeneous medium with permittivity ε_1 (magnetic permeabilities will be taken to be unit). In the discussion below the electromagnetic fields will be presented in cylindrical coordinates (r, ϕ, z), where the z-axis is directed along the cylinder axis. If $\mathbf{E}$ is the electric field strength generated by a point charge q in a given state of motion, then the energy losses per unit time can be evaluated in terms of the work done by the electric field on the particle:

$$\frac{dW}{dt} = \int_0^\infty dr \int_0^{2\pi} d\phi \int_{-\infty}^\infty dz\, r\, \mathbf{j} \cdot \mathbf{E}, \tag{10.1}$$

where $\mathbf{j}$ is the current density. First we study the case when the charge follows the rectilinear trajectory $r = r_q$, $\phi = \phi_q$ parallel to the cylinder axis, assuming that $r_q > r_c$. For the motion with a constant velocity v the components of the current density are expressed as

$$j_l = \delta_{l3} \frac{qv}{r} \delta(r - r_q)\delta(\phi - \phi_q)\delta(z - vt), \tag{10.2}$$

with $\delta(x)$ being the Dirac delta function. Here and below the vector component indices $l = 1, 2, 3$ correspond to the cylindrical components (r, ϕ, z).

The electric field generated by the charge is Fourier expanded as

$$\mathbf{E}(t, \mathbf{r}) = \sum_{n=-\infty}^{\infty} e^{in(\phi-\phi_q)} \int_{-\infty}^{\infty} dk_z\, e^{ik_z(z-vt)} \mathbf{E}_n(k_z, r), \tag{10.3}$$

where the Fourier components obey the relation $\mathbf{E}_{-n}(-k_z, r) = \mathbf{E}_n^*(k_z, r)$ and the star stands for the complex conjugate. Note that the angular frequency corresponding to the Fourier mode with given n and k_z is given by $\omega = k_z v$. The electric and magnetic fields can be found by using the electromagnetic field Green tensor from [12]. For the motion described by (10.2), in the evaluation of the energy losses in accordance with (10.1), the component of the electric field along the z-axis is required only. In the region $r > r_q$ the corresponding Fourier component is given by the expression

$$E_{n,3}(k_z, r) = \frac{qk_z}{4\varepsilon_1} \sum_{p=\pm 1} g_n^{(p)} H_n(\lambda_1 r), \tag{10.4}$$

where $H_\nu(x) = H_\nu^{(1)}(x)$ is the Hankel function of the first kind and the coefficients $g_n^{(p)}$ with $p = \pm 1$ are defined by the formula

$$g_n^{(p)} = \left(1 - \beta_1^2\right) J_n(\lambda_1 r_0) + \frac{H_n(\lambda_1 r_0)}{V_n^H}\left[\left(\beta_1^2 - 1\right) V_n^J \right.$$
$$\left. + \sqrt{\beta_1^2 - 1}\frac{2ipk_z}{\pi} \frac{J_n(\lambda_0 r_c)}{r_c \alpha_n} \frac{J_{n+p}(\lambda_0 r_c)}{V_{n+p}^H}\right]. \tag{10.5}$$

Here, $J_\nu(x)$ is the Bessel function,

$$\lambda_j^2 = k_z^2(\beta_j^2 - 1),\ \beta_j^2 = (v/c)^2 \varepsilon_j,\ j = 0, 1, \tag{10.6}$$

the functions V_n^J and V_n^H are defined as

$$V_n^J = J_n(\lambda_0 r_c)\partial_{r_c} J_n(\lambda_1 r_c) - [\partial_{r_c} J_n(\lambda_0 r_c)] J_n(\lambda_1 r_c),$$
$$V_n^H = J_n(\lambda_0 r_c)\partial_{r_c} H_n(\lambda_1 r_c) - [\partial_{r_c} J_n(\lambda_0 r_c)] H_n(\lambda_1 r_c), \tag{10.7}$$

and

$$\alpha_n = \frac{\varepsilon_0}{\varepsilon_1 - \varepsilon_0} + \frac{1}{2}\sum_{l=\pm 1}\left[1 - \frac{\lambda_1}{\lambda_0}\frac{J_{n+l}(\lambda_0 r_c) H_n(\lambda_1 r_c)}{J_n(\lambda_0 r_c) H_{n+l}(\lambda_1 r_c)}\right]^{-1}. \tag{10.8}$$

It can be seen that the coefficients $g_n^{(p)}$ obey the relation $g_{-n}^{(p)} = (-1)^n g_n^{(-p)}$. From here it follows that $E_{-n,3}(k_z, r) = E_{n,3}(k_z, r)$. It can be shown that the Fourier component $E_{n,3}(k_z, r)$, as a function of k_z, is regular at possible zeros of the functions V_n^H and V_{n+p}^H, and it has poles at zeros of the function α_n. We note that the equation $\alpha_n = 0$ determines the eigenmodes of the dielectric cylinder.

The special case $\varepsilon_0 = \varepsilon_1$ corresponds to the problem where the charge moves in a homogeneous medium. In this case $\lambda_0 = \lambda_1$ and we get $V_n^J = 0$, $V_n^H = 2i/\pi r_c$. The function α_n tends to infinity and the second and third terms in the right-hand side of (10.5) become zero. Hence, the first term in the right-hand side of of (10.5) determines the z-component of the electric field in a homogeneous medium with dielectric permittivity ε_1. The corresponding Fourier component is given by

$$E_{n,3}^{(0)}(k_z, r) = \frac{qk_z}{2\varepsilon_1}\left(1 - \beta_1^2\right) J_n(\lambda_1 r_0) H_n(\lambda_1 r), \tag{10.9}$$

The remaining terms in (10.5) correspond to the part in the field that is induced by the cylinder. The expression for the Fourier component $E_{n,3}(k_z, r)$ in the exterior region in the range $r_c < r < r_q$ is obtained from (10.4) by the replacements $J \to H$, $H \to J$ in the part corresponding to the field in homogeneous medium with permittivity ε_1 (in (10.9)). The Fourier components have poles at the zeros of the function α_n. As it has been mentioned before, those zeros determine the eigenmodes of the cylinder. The fields inside the cylinder can be found by using the corresponding expressions for the Green tensor components from [12].

10.3 Radiation Intensity for Surface Polaritons

Substituting (10.2) and (10.3) into (10.1), for the energy losses per unit of path length one gets

$$\frac{dW}{dz} = q^2 {\sum_{n=0}^{\infty}}' \mathrm{Re}\left[\int_0^{\infty} dk_z \frac{k_z}{\varepsilon_1} \sum_{p=\pm 1} g_n^{(p)} H_n(\lambda_1 r_0)\right], \tag{10.10}$$

where the prime on the sign of the summation means that the term $n = 0$ should be multiplied by an additional coefficient 1/2. The expression (10.10) for the energy losses is valid for general case of complex dielectric permittivities for the cylinder and surrounding medium. In what follows we will assume that the permittivities ε_0 and ε_1 are real functions (a small imaginary part for ε_0 will be introduced when considering the contour of the integration over k_z near the poles on the real axis). In this case the only energy losses are in the form of radiation.

In the problem at hand we can have three types of radiation. The first one corresponds to the Cherenkov radiation propagating in the exterior medium. This radiation is present under the Cherenkov condition $\beta_1 > 1$ and propagates along the Cherenkov angle $\theta = \theta_{\text{Ch}} = \arccos(1/\beta_1)$ with respect to the cylinder axis. With that condition the equation $\alpha_n = 0$ has no solutions and the integrand in (10.10) is a regular function on the positive semiaxis. It is easily seen that the part in the energy losses coming from the first term in the righ-hand side of (10.5) coincides with the standard result for the Cherenkov radiation in a homogeneous medium. The last two terms in (10.5) describe the effects of the cylinder on the Cherenkov radiation intensity. Those effects have been investigated in [11] by evaluating the energy flux at large distances from the radiating source. It has been shown that the influence of the cylinder on the spectral distribution of the Cherenkov radiation is essential in the case $\varepsilon_0 > \varepsilon_1$. Under this condition strong narrow peaks appear in the spectral distribution of the radiation intensity coming from the terms in (10.10) with large values of n. The locations of the spectral peaks are related to the zeros of the function that is obtained from the function α_n, given by (10.8), when we replace the Hankel functions by the Neuman functions $Y_n(\lambda_1 r_c)$ and $Y_{n+l}(\lambda_1 r_c)$. In [11] we have analytically estimated the heights and widths of the spectral peaks of the Cherenkov radiation by using the asymptotic expressions of the cylinder functions for large arguments.

For $\beta_1 < 1$ one has $\lambda_1 = i|k_z|\gamma_1$ with $\gamma_1 = \sqrt{1-\beta_1^2}$. In this case the Fourier component (10.4) exponentially decays for large r and the Cherenkov radiation in the exterior medium is absent. Introducing the Macdonald function $K_\nu(x)$ instead of the Hankel functions, the expression for the function (10.8) is presented as

$$\alpha_n = \frac{\varepsilon_0}{\varepsilon_1 - \varepsilon_0} + \frac{1}{2}\sum_{l=\pm1}\left[1 + l\frac{|\lambda_1|}{\lambda_0}\frac{J_{n+l}(\lambda_0 r_c)K_n(|\lambda_1|r_c)}{J_n(\lambda_0 r_c)K_{n+l}(|\lambda_1|r_c)}\right]^{-1}. \tag{10.11}$$

This expression is real for both real and purely imaginary values of λ_0. The parts of the integrand in (10.10) coming from the first term in the right-hand side of (10.5) and from the first term in the square brackets are purely imaginary and those parts are regular functions of k_z. Hence, they do not contribute to the radiation losses. By taking into account this and introducing the Macdonald function, the energy radiated per unit time, given by $I = -v\,dW/dz$, is presented in the form

$$I = -\frac{2q^2 v}{\pi}\sum_{n=0}^{\infty}{}' \,\text{Im}\left[\int_0^\infty dk_z\,\gamma_1 \frac{k_z^2 J_n(\lambda_0 r_c)}{\varepsilon_1 r_c V_n^K \alpha_n}\sum_{p=\pm1}\frac{J_{n+p}(\lambda_0 r_c)}{V_{n+p}^K}K_n^2(k_z\gamma_1 r_q)\right], \tag{10.12}$$

where

$$V_n^K = J_n(\lambda_0 r_c)\partial_{r_c} K_n(k_z\gamma_1 r_c) - [\partial_{r_c} J_n(\lambda_0 r_c)]K_n(k_z\gamma_1 r_c). \tag{10.13}$$

Of course, the fact that the first term in the right-hand side of (10.5) does not contribute to the energy losses corresponds to the absence of the Cherenkov radiation in a homogeneous medium with permittivity ε_1 under the condition $\beta_1 < 1$.

For points on the real semiaxis of k_z, different from the zeros of the function α_n, the integrand in (10.12) is real for both real and imaginary λ_0. Hence, the only nonzero contribution to the energy losses can come from the poles of the integrand been the roots of the equation $\alpha_n = 0$. We will denote those roots with respect to $k_z r_c$ by $k_z r_c = x_{n,s}$, where $s = 1, 2, \ldots$ enumerates the roots for a given n. For the corresponding angular frequencies one has $\omega_{n,s} = x_{n,s} v/r_c$. The eigenmodes of the cylinder, given by $x_{n,s}$, correspond to two types of the waves radiated by the charge. For the first one we have $\beta_0^2 > 1$ and the Cherenkov condition for the material of the cylinder is satisfied. In this case the radial dependence of the electric and magnetic fields inside the cylinder is expressed in terms of the Bessel functions $J_n(\lambda_0 r)$ and $J_{n\pm1}(\lambda_0 r)$ with $\lambda_0 > 0$. This type of modes are referred to as guiding modes. The second type of the waves are radiated when $\beta_0^2 < 1$ and for them $\lambda_0^2 < 0$. The radial dependence of the fields inside the cylinder is expressed in terms of the modified Bessel functions $I_n(|\lambda_0|r)$, $I_{n\pm1}(|\lambda_0|r)$ and they correspond to surface polaritons. Here we are interested in the radiation intensity for the latter type of the waves.

For $\lambda_0^2 < 0$ the function (10.11) that determines the eigenfrequencies $\omega_{n,s} = v x_{n,s}/r_c$ is expressed in terms of the modified Bessel functions:

$$\alpha_n = \frac{\varepsilon_0}{\varepsilon_1 - \varepsilon_0} + \frac{1}{2}\sum_{l=\pm1}\left[1 + \frac{|\lambda_1|}{|\lambda_0|}\frac{I_{n+l}(|\lambda_0|r_c)K_n(|\lambda_1|r_c)}{I_n(|\lambda_0|r_c)K_{n+l}(|\lambda_1|r_c)}\right]^{-1}, \tag{10.14}$$

where $|\lambda_j| = k_z\sqrt{1-\beta_j^2}$. Note that the last term in (10.14) is always positive. It can be shown that the equation $\alpha_n = 0$ has solutions in the spectral range where the dielectric permittivities ε_0 and ε_1 have opposite signs (for dispersion relations corresponding to surface polaritons on cylindrical interfaces see also [13, 14]). This condition for the excitation of surface polaritons on planar boundaries is well known. Here we will consider the case $\varepsilon_0 < 0 < \varepsilon_1$. The roots of the equation $\alpha_n = 0$ with respect to $k_z r_c$ are functions of two combinations of the parameters: $\varepsilon_0/\varepsilon_1$ and β_1. For given values of these combinations we can have one or two roots $x_{n,s}$ ($s = 1$ or $s = 1, 2$). For the modes with $n = 0$ there is a single root in the range $\varepsilon_0 < \varepsilon_0^{(\infty)} \equiv -\varepsilon_1/(1-\beta_1^2)$ and there are no roots for $\varepsilon_0 > \varepsilon_0^{(\infty)}$. In the limit $k_z r_c \to 0$ we have $\varepsilon_0 \to -\infty$. The distribution of the roots as functions of the permittivity ε_0 for the $n = 0$ mode is presented on the left panel of Fig. 10.1 for different values of the ratio v/c (the numbers near the curves). In the limit $\varepsilon_0 \to \varepsilon_0^{(\infty)}$ the single root $k_{n,1} r_c$ tends to infinity. For $n \geq 1$ the surface modes are allowed in the finite range $\varepsilon_0^{(\mathrm{m})} \leq \varepsilon_0 < -\varepsilon_1$ for the dielectric permittivity of the cylinder, where $\varepsilon_0^{(\mathrm{m})}$ is

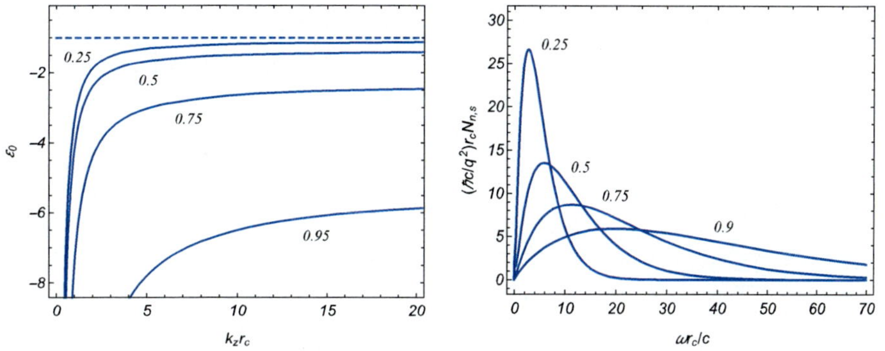

Fig. 10.1 Localization of the eigenvalues for $k_z r_c$ as functions of ε_0 for $\varepsilon_1 = 1$ (left panel) and the spectral distribution of the number of radiated surface polaritons for the mode $n = 0$. The numbers near the curves correspond to the values of the ratio v/c. The graphs on the right panel are plotted for $r_q/r_c = 1.05$

a function of n and v/c. For ε_0 close to $\varepsilon_0^{(\mathrm{m})}$ we have two roots ane there is a single root in the remaining region. In the limit $k_z r_c \to 0$ one has $\varepsilon_0 \to -\varepsilon_1$. For $n \geq 1$ and $v/c \ll 1$ the surface modes are present in the narrow range for the permittivity ε_0 near $\varepsilon_0 = -\varepsilon_1$ with the length of the order β_1^2.

Having specified the properties of the surface modes we return to the radiation intensity given by (10.12). The integrand has poles at the zeros of the function (10.14) and for the evaluation of the integral the contour near the poles needs to be specified. Let us consider α_n as a function of $k_z r_c$ and ε_0: $\alpha_n = \alpha_n(k_z r_c, \varepsilon_0)$. For the derivative of this function with respect to the second argument we have checked that $\partial_{\varepsilon_0}\alpha_n(x_{n,s}, \varepsilon_0) > 0$ for $\lambda_0^2 < 0$. In order to fix the integration contour we note that in physically realistic situations the dielectric permittivity ε_0 has an imaginary part, $\varepsilon_0 = \varepsilon_0' + i\varepsilon_0''$. Expanding the function $\alpha_n(k_z r_c, \varepsilon_0)$ near the zeros for the leading term we get

$$\alpha_n(k_z r_c, \varepsilon_0) \approx \alpha_n'(x_{n,s}) \left(k_z r_c - x_{n,s} + i\varepsilon_0'' b_{n,s}\right), \tag{10.15}$$

where

$$\alpha_n'(x_{n,s}) = \left.\frac{d\alpha_n(k_z r_c, \varepsilon_0')}{d\,(k_z r_c)}\right|_{k_z r_c = x_{n,s}}, \quad b_{n,s} = \frac{\partial_{\varepsilon_0'}\alpha_n(x_{n,s}, \varepsilon_0')}{\alpha_n'(x_{n,s})}. \tag{10.16}$$

Note that in the problem at hand $\omega = k_z v$ and in the presence of dispersion one has $\varepsilon_0 = \varepsilon_0(k_z v)$. If the dispersion is taken into account, the derivative in the definition of $\alpha_n'(x_{n,s})$ is taken with respect to both the arguments. From (10.15) we see that the poles are shifted from the real axis in the complex plane k_z as $k_z r_c = x_{n,s} - i\varepsilon_0''(\omega_{n,s})b_{n,s}$, where $\varepsilon_0''(\omega) > 0$ for $\omega > 0$. By taking into account that for surface polaritons $\partial_{\varepsilon_0'}\alpha_n(x_{n,s}, \varepsilon_0') > 0$, we conclude that the poles $k_z = x_{n,s}/r_c$ in (10.12) should be avoided by semicircles with small radius from above (below) for $\alpha_n'(x_{n,s}) > 0$ ($\alpha_n'(x_{n,s}) < 0$). The integrals along those semicircles are expressed in terms of the corresponding residues. Introducing the radiation intensity $I_{n,s}$ on the angular frequency $\omega_{n,s}$ in accordance with $I = \sum\nolimits_{n=0}^{\infty} \sum_{s} I_{n,s}$, from (10.12) we obtain

$$I_{n,s} = \delta_n \frac{2q^2 v}{\varepsilon_1 r_c^2} \gamma_1 x_{n,s}^2 \left[K_n(\gamma_1 x_{n,s}) \frac{I_n'(\gamma_0 x_{n,s})}{I_n(\gamma_0 x_{n,s})} - K_n'(\gamma_1 x_{n,s}) \right]^{-1} \frac{K_n^2(\gamma_1 x_{n,s} r_q / r_c)}{|\alpha_n'(x_{n,s})|} \times \sum_{p=\pm 1} \left[K_{n+p}(\gamma_1 x_{n,s}) \frac{I_{n+p}'(\gamma_0 x_{n,s})}{I_{n+p}(\gamma_0 x_{n,s})} - K_{n+p}'(\gamma_1 x_{n,s}) \right]^{-1}, \quad (10.17)$$

where $\delta_0 = 1/2$ and $\delta_n = 1$ for $n \geq 1$, and $\gamma_0 = \sqrt{1 - v^2 \varepsilon_0 / c^2}$. Note that $\varepsilon_0 < 0$ and $\gamma_0 > 1$.

We recall that the roots $x_{n,s}$ are completely determined by $\varepsilon_0/\varepsilon_1$ and β_1 and for given values of these parameters the roots do not depend on the cylinder radius. From here it follows that the dependence of the radiation intensity on the cylinder radius enters in the form of the function $K_n^2(\gamma_1 x_{n,s} r_q / r_c)/r_c^2$. In particular, for small values of the cylinder radius the radiation intensity is suppressed by the factor $\exp(-2\gamma_1 x_{n,s} r_q / r_c)/r_c$. Similarly, for large values of r_q the radiation intensity decays as $\exp(-2\gamma_1 x_{n,s} r_q / r_c)/r_q$. For large values of $x_{n,s}$, corresponding to high frequencies, the radiation intensity $I_{n,s}$ is suppressed by the factor $\exp[-2\gamma_1 x_{n,s}(r_q/r_c - 1)]$. The radiation intensity tends to zero in the limit of small values of $x_{n,s}$ as well.

For the numerical example we will consider the radiation on the mode with $n = 0$. In this case the equation $\alpha_n = 0$ for the eigenmodes is simplified to

$$\frac{\varepsilon_0 I_1(\gamma_0 x)}{\gamma_0 I_0(\gamma_0 x)} + \frac{\varepsilon_1 K_1(\gamma_1 x)}{\gamma_0 K_0(\gamma_1 x)} = 0,$$

where $x = k_z r_c$. The roots of this equation are plotted in the left panel of Fig. 10.1 as functions of of the negative permittivity of the cylinder for different values of the ratio v/c (the numbers near the curves). For the permittivity in the region $r > r_c$ we have taken $\varepsilon_1 = 1$. For $n = 0$ the general formula (10.17) is reduced to

$$I_{0,s} = \frac{2q^2 v}{\varepsilon_1 r_c^2} \gamma_1 x_s^2 \left[1 + \frac{I_1(\gamma_0 x_s)}{I_0(\gamma_0 x_s)} \frac{K_0(\gamma_1 x_s)}{K_1(\gamma_1 x_s)} \right]^{-1} \frac{K_1^2(\gamma_1 x_s r_q / r_c)}{K_1^2(\gamma_1 x_s) |\alpha_0'(x_s)|} \times \left[\frac{I_0(\gamma_0 x_s)}{I_1(\gamma_0 x_s)} + \frac{K_0(\gamma_1 x_s)}{K_1(\gamma_1 x_s)} + \frac{1}{\gamma_1 x_s} - \frac{1}{\gamma_0 x_s} \right]^{-1}, \quad (10.18)$$

where $x_s = x_{0,s}$. As it has been mentioned before and it is also seen from Fig. 10.1, in the case $n = 0$ for a given ε_0 we have a single root. On the right panel of Fig. 10.1 we have plotted the quantity $r_c N_{n,s}$ for $n = 0$, where $N_{n,s} = I_{n,s}/(\hbar \omega_{n,s} v)$ is the number of the radiated quanta in the form of the surface polaritons per unit length of the charge trajectory, as a function of the radiation angular frequency $\omega = \omega_{n,s} = x_{n,s} v / r_c$, measured in units of c/r_c. The graphs are plotted for $r_q/r_c = 1.05$ and the numbers near the curves are the values of the ratio v/c. Note that for a given value of the permittivity ε_0 we have a single value for the frequency that is determined from the left panel of Fig. 10.1. The features of the radiation of surface polaritons for the special case of dispersion $\varepsilon_0(\omega) = \varepsilon_\infty - \omega_p^2/\omega^2$, with ε_∞ and ω_p being the background dielectric constant and the plasma frequency, have been discussed in [11].

The radiation on the guiding modes of the cylinder in the setup under consideration have been investigated in [11]. The corresponding waves are radiated under the conditions $\varepsilon_1 < c^2/v^2 < \varepsilon_0$. For the radiation from an electron of the energy 2 MeV and for $\varepsilon_1 = 1$, $\varepsilon_0 = 3.8$, $r_q/r_c = 1.05$, the quantity $(\hbar c/q^2)r_c N_{n,s}$ is of the order 1 for the modes with $n = 1, 2$ and decreases with increasing n. Comparing with the data presented in Fig. 10.1, we see that the number of the radiated quanta for surface polaritons on a given eigenmode of the cylinder can be essentially larger than the corresponding quantity for guiding modes.

10.4 Electric Field and the Radiation Intensity for a Circular Motion

Now we turn to the radiation from a point charge q moving along a circular trajectory of radius r_q around the cylinder with the velocity $v = \omega_0 r_q$. The components of the corresponding current density are expressed as

$$j_l = \frac{q}{r} v \delta_{l2} \delta(r - r_q)\delta(\phi - \omega_0 t)\delta(z). \tag{10.19}$$

For the evaluation of the energy losses in accordance with (10.1) the azimuthal component of the electric field is required. The corresponding Fourier expansion is given by

$$E_2(t, \mathbf{r}) = 2\mathrm{Re}\left[\sum_{n=0}^{\infty}{}' e^{in\phi - i\omega_n t} \int_{-\infty}^{\infty} dk_z\, e^{ik_z z} E_{2,n}(k_z, r)\right], \tag{10.20}$$

where $\omega_n = n\omega_0$. In this case the $n = 0$ mode is time independent and will not contribute to the radiation fields.

By using the Green tensor from [12] for the Fourier component of the required field, with $n \neq 0$, in the region $r > r_q$ we get

$$E_{2,n}(k_z, r) = -\frac{qv}{8\omega_n \varepsilon_1} \sum_p g_{(c)n}^{(p)} \left[\left(\lambda_1^2 + 2k_z^2\right) H_{n+p}(\lambda_1 r) - \lambda_1^2 H_{n-p}(\lambda_1 r)\right], \tag{10.21}$$

where the notations λ_j, $j = 0, 1$, for the circular motion are defined as

$$\lambda_j^2 = \omega_n^2 \varepsilon_j / c^2 - k_z^2, \tag{10.22}$$

and

$$g_{(c)n}^{(p)} = J_{n+p}(\lambda_1 r_q) - \frac{V_{n+p}^J}{V_{n+p}^H} H_{n+p}(\lambda_1 r_q)$$
$$+ip \frac{\lambda_0 J_n(\lambda_0 r_c)}{\pi r_c \alpha_n} \frac{J_{n+p}(\lambda_0 r_c)}{V_{n+p}^H} \sum_{l=\pm 1} \frac{H_{n+l}(\lambda_1 r_q)}{V_{n+l}^H}. \qquad (10.23)$$

Other notations are the same as in (10.7) and (10.8). The part in the field coming from the first term in the right-hand side of (10.23) does not depend on the dielectric permittivity of the cylinder and corresponds to the field in a homogeneous medium with permittivity ε_1 (that is directly seen taking the limit $\varepsilon_0 \to \varepsilon_1$). The Fourier component in the region $r_c < r < r_q$ is obtained from (10.21) by the replacements $J \rightleftarrows H$ in that part.

Substituting (10.19) and (10.20) in (10.1), for the energy losses we get

$$\frac{dW}{dt} = -\frac{q^2 v^2}{2} \mathrm{Re} \left\{ \sum_{n=1}^{\infty} \int_0^{\infty} dk_z \sum_p \frac{g_n^{(c,p)}}{\omega_n \varepsilon_1} \right.$$
$$\left. \times \left[\left(\lambda_1^2 + 2k_z^2 \right) H_{n+p}(\lambda_1 r_q) - \lambda_1^2 H_{n-p}(\lambda_1 r_q) \right] \right\}. \qquad (10.24)$$

The factor λ_1 in the arguments of the Hankel functions, considered as a function of k_z, is defined as $\lambda_1(k_z) = \sqrt{\omega_n^2 \varepsilon_1 / c^2 - k_z^2}$ in the region $0 \leq k_z \leq \omega_n \sqrt{\varepsilon_1}/c$ and as $\lambda_1(k_z) = i\sqrt{k_z^2 - \omega_n^2 \varepsilon_1 / c^2}$ in the integration range $k_z \geq \omega_n \sqrt{\varepsilon_1}/c$. The part of (10.24) in the integration region $0 \leq k_z \leq \omega_n \sqrt{\varepsilon_1}/c$ corresponds to the energy losses in the form of the synchrotron radiation propagating in the exterior medium. In particular, the contribution in that region coming from the first term in the right-hand side of (10.23) corresponds to the radiation in a homogeneous medium with permittivity ε_1.

The spectral-angular distribution of the radiation intensity at large distances from the cylinder has been investigated in [12, 15] by evaluating the energy flux through a cylindrical surface with large radius, coaxial with the dielectric cylinder. It has been shown that under the Cherenkov condition for the velocity $v_c = \omega_0 r_c$ of the charge image on the cylinder surface and for the cylinder dielectric permittivity, $v_c \sqrt{\varepsilon_0}/c > 1$, strong narrow peaks may appear in the angular distribution of the radiation intensity. The locations, heights and widths of those peaks were estimated based on the anlysis of the eigenmode equation for the cylinder. The generalization of the corresponding results for a helical motion of a charge around a cylinder is discussed in [16, 17]. New interesting features of the radiation at large distances arise for a particle rotation around a cylinder with a negative real part of dielectric permittivity [18].

Here we are interested in the radiation losses corresponding to the integration range $k_{\min} < k_z < \infty$, where $k_{\min} = \omega_n \sqrt{\varepsilon_1}/c$. In this range one has $\lambda_1^2 < 0$ and the corresponding contributions to the fields exponentially decay in the exterior medium. Introducing the Macdonald functions instead of the Hankel functions containing in the arguments the factor λ_1, we see that the contribution to the integral coming from the first two terms in the right-hand side of (10.23) is purely imaginary and the corresponding integrand is regular in the integration range. Hence, those terms do not contribute to the radiation losses. For the energy losses in the range $\omega_n\sqrt{\varepsilon_1}/c < k_z < \infty$, corresponding to the radiation on the eigenmodes (em) of the cylinder we get

$$\frac{dW_{(\mathrm{em})}}{dt} = \frac{q^2v^2}{2\pi r_c}\mathrm{Im}\left\{\sum_{n=1}^{\infty}\int_{k_{\min}}^{\infty}dk_z\frac{\lambda_0 J_n(\lambda_0 r_c)}{\varepsilon_1\omega_n\alpha_n}\sum_{l=\pm1}\frac{K_{n+l}(|\lambda_1|r_q)}{V_{n+l}^K}\right.$$
$$\left.\times\sum_{p,j}p\left(k_z^2+j\frac{\omega_n^2}{c^2}\varepsilon_1\right)\frac{J_{n+jp}(\lambda_0 r_c)}{V_{n+jp}^K}K_{n+p}(|\lambda_1|r_q)\right\}. \quad (10.25)$$

Similar to the case of a rectilinear motion, the integrand in (10.25) is real for points different from the poles corresponding to the zeros of the function α_n. These zeros present two types of the eigenmodes for the cylinder. For the first one we have $k_{\min}^2 < k_z^2 < \omega_n^2\varepsilon_0/c^2$ and $\lambda_0^2 > 0$. As a necessary condition for the presence of those modes one gets $\varepsilon_0 > \varepsilon_1$ and they correspond to guiding modes. The radiation intensity for that types of modes has been investigated in [17, 19]. For the second type of modes, presenting the surface polaritons, one has $\lambda_0^2 < 0$ and as a necessary condition one gets $\varepsilon_0\varepsilon_1 < 0$. We are interested in the emission of surface polaritons, radiated in the spectral range, where $\varepsilon_0 < 0 < \varepsilon_1$.

We will denote by $k_z r_c = y_{n,s}$ the roots of the eigenmode equation $\alpha_n = 0$, with α_n defined by (10.14). For the circular motion the factors in the arguments of the modified Bessel functions in (10.14) are defined as

$$|\lambda_0|r_c = \sqrt{x^2 - n^2\beta_{c1}^2\varepsilon_0/\varepsilon_1},\ |\lambda_1|r_c = \sqrt{x^2 - n^2\beta_{c1}^2},\ x = k_z r_c, \quad (10.26)$$

where $\beta_{c1} = v_c\sqrt{\varepsilon_1}/c$. These definitions of λ_j differ from those for the rectilinear motion and, hence, the roots $y_{n,s}$ are different from the roots $x_{n,s}$ considered in Sects. 10.2 and 10.3. The roots $x = y_{n,s}$ are completely determined by two combinations of the parameters, β_{c1} and $\varepsilon_0/\varepsilon_1$. Introducing a small imaginary part for the permittivity ε_0, we can see that in the evaluation of the integral (10.25) the poles $k_z = x_{n,s}/r_c$ should be avoided by small semicircles in the upper half-plane of the complex variable k_z for $\alpha_n'(x_{n,s}) > 0$ and by small semicircles in the lower half-plane for $\alpha_n'(x_{n,s}) < 0$. The integrals are expressed in terms of the residues. The radiation intensity for surface polaritons is obtained as $I = -dW_{(\mathrm{em})}/dt$. It is presented in the form $I = \sum_n I_n$, where for the radiation intensity on a given harmonic n with the frequency $\omega_n = n\omega_0$, after some transformations, we get

$$I_n = \frac{q^2 v^2}{r_c c^2} \sum_s \frac{\omega_n u_{n,s}^{(0)} K_n^{-2}(u_{n,s}^{(1)})}{y_{n,s}^{(1)} u_{n,s} |\alpha_n'(x_{n,s})|} \left[u_{n,s}^{(0)} \frac{I_n'(u_{n,s}^{(0)})}{I_n(u_{n,s}^{(0)})} - u_{n,s}^{(1)} \frac{K_n'(u_{n,s}^{(1)})}{K_n(u_{n,s}^{(1)})} \right]$$
$$\times \left[\sum_p \frac{p K_{n+p}(u_{n,s}^{(1)} r_q / r_c)}{u_{n,s}^{(1)} \frac{I_n'(u_{n,s}^{(0)})}{I_n(u_{n,s}^{(0)})} - u_{n,s}^{(0)} \frac{K_n'(u_{n,s}^{(1)})}{K_n(u_{n,s}^{(1)})} + p n u_{n,s}} \right]^2, \tag{10.27}$$

with the notations

$$u_{n,s}^{(j)} = \sqrt{y_{n,s}^2 - n^2 \varepsilon_j \frac{v_c^2}{c^2}}, \quad u_{n,s} = \frac{u_{n,s}^{(0)}}{u_{n,s}^{(1)}} - \frac{u_{n,s}^{(1)}}{u_{n,s}^{(0)}}. \tag{10.28}$$

For a given value of the rotation frequency, the roots $y_{n,s}$ do not depend on the radius of the rotation orbit. The dependence on the radius enters through the function $K_{n+p}(u_{n,s}^{(1)} r_q/r_c)$ and for large values of the ratio r_q/r_c the radiation intensity decays like $r_q^2 \exp(-2u_{n,s}^{(1)} r_q/r_c)$.

We can find the behavior of the roots $y_{n,s}$ in the asymptotic regions by using the corresponding approximations of the modified Bessel functions in (10.14). For $y_{n,s} \gg n v_c \sqrt{|\varepsilon_j|}/c$ one gets

$$y_{n,s} \approx \frac{n v_c}{c} \sqrt{\frac{\varepsilon_0 \varepsilon_1}{\varepsilon_0 + \varepsilon_1}}. \tag{10.29}$$

This result coincides with the corresponding relation between the projection of the wave vector and frequency for surface polaritons on planar boundaries. In the limit under consideration the wavelength of the surface polariton is small and the effect of the curvature on the dispersion relation is weak. By taking into account that $y_{n,s}$ is large, from (10.29) we conclude that the large values of k_z are realized in the spectral range where the ratio $\varepsilon_0/\varepsilon_1$ is close to -1. The allowed values for $y_{n,s}$ are restricted from below by the condition $y_{n,s} > n\beta_{c1}$. By using the expression (10.14) for α_n it can be seen that for $n = 1$ we have $y_{n,s} \to n\beta_{c1}$ for $\varepsilon_0 \to -\infty$. Combining this with the behavior of the roots for large values of $y_{n,s}$, we conclude that for $n = 1$ one has modes for all values of the cylinder dielectric permittivity in the range $-\infty < \varepsilon_0 < -\varepsilon_1$. The situation is different for the modes with $n > 1$. In this case the eigenmodes corresponding to surface polaritons are present in the finite region $\varepsilon_{0n} < \varepsilon_0 < -\varepsilon_1$, with critical value ε_{0n}, and there are no modes in the region $\varepsilon_0 < \varepsilon_{0n}$. For the mode $n = 1$, the locations of the roots of the eigenvalue equation $\alpha_n = 0$ with respect to the ratio ck_z/ω, $\omega = n\omega_0$, is depicted on the left panel of Fig. 10.2 as functions of the permittivity ε_0 for $\varepsilon_1 = 1$. The numbers near the curves are the values of the ratio v_c/c. As seen, depending on the region for ε_0, for a given ε_0 one can have one, two or three roots of the equation $\alpha_n = 0$.

Let us consider the behavior of the radiation intensity (10.27) in asymptotic regions of the parameters. For the values of dielectric permittivity ε_0 close to the limiting value $-\varepsilon_1$ the eigenvalues $y_{n,s}$ are large and we can replace the modified Bessel functions by

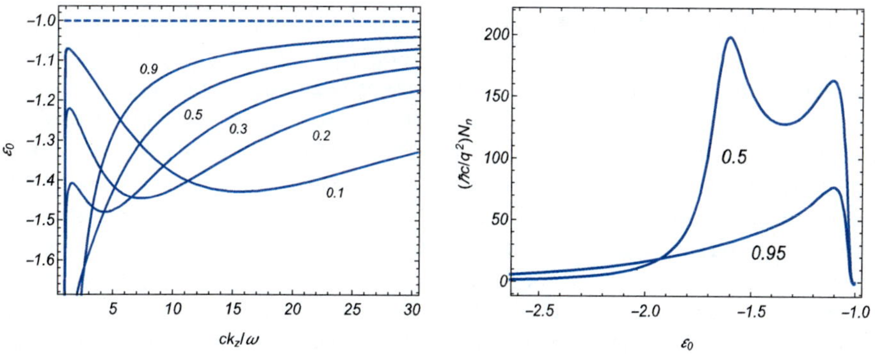

Fig. 10.2 The solutions of the eigenmodes equation $\alpha_n = 0$ with respect to the ratio ck_z/ω_n (left panel) and the number of radiated surface polaritons (right panel) as functions of ε_0 for $\varepsilon_1 = 1$ and $n = 1$. The numbers near the curves correspond to the values of v_c/c on the left panel and to the values for v/c on the right panel. The right panel is plotted for $r_c/r_q = 0.95$

their asymptotic expressions for large arguments. In this way we can see that the radiation intensity is suppressed by the factor $\exp\left[-2n\beta_{c1}\left(r_q/r_c - 1\right)/\sqrt{|\varepsilon_0/\varepsilon_1 + 1|}\right]$. In the spectral range where $\varepsilon_0 \ll -\varepsilon_1$ the surface polaritons are radiated on the harmonic $n = 1$ only. In that range we have a single root $y_{n,s}$ which is close to the lower bound β_{c1} and, hence, the quantity $u_{n,s}^{(1)}$ in (10.28) is small. By using the corresponding asymptotic for the Macdonald function, we see that the radiation intensity decays as $(|\varepsilon_0|/\varepsilon_1)\exp(-\sqrt{|\varepsilon_0|/\varepsilon_1}/\beta_{c1})$ and vanishes in the limit $\varepsilon_0 \to -\infty$. For the modes $n > 1$ the surface polaritons are present in the finite range $\varepsilon_{0n} < \varepsilon_0 < -\varepsilon_1$ of the dielectric permittivity of the cylinder. In the limit $\varepsilon_0 \to \varepsilon_{0n}$ the radiation intensity tends to a nonzero limiting value.

On the right panel of Fig. 10.2 we have displayed the number of the radiated quanta, $N_n = TI_n/(\hbar\omega_n)$, per period of the particle rotation, $T = 2\pi/\omega_0$, for the harmonic $n = 1$ as a function of ε_0. The graphs are plotted for $\varepsilon_1 = 1$, $r_c/r_q = 0.95$, and the numbers near the curves present the values of v/c. As it has been explained by the asymptotic analysis, the radiation intensity tends to zero in the limits $\varepsilon_0 \to -\varepsilon_1$ and $\varepsilon_0 \to -\infty$.

Similar to the case of rectilinear motion, it is of interest to compare the radiation intensities for surface polaritons and guiding modes. Numerical exmaples for the radiation of guiding modes are presented in [19]. In particular, for an electron with energy 2 MeV and for $\varepsilon_1 = 1$, $\varepsilon_0 = 3.74$, $r_c/r_q = 0.99$, the corresponding quantity $(\hbar c/q^2)N_n$ for the mode $n = 1$ is approximately equal to 0.53. Again, we see that the number of quanta for the radiated surface polaritons can be essentially larger.

10.5 Conclusion

Wide applications of surface polaritons as a research tool motivate the investigations of various mechanisms for the control of their excitation and propagation. In the present paper we have considered two geometries for the emission of surface polaritons by charged particles in cylindrical waveguides. In the first setup the charge is uniformly moving along the trajectory parallel to the axis of the cylinder. If the Cherenkov condition is not satisfied in the exterior medium, depending on the spectral range, two types of the waves are generated propagating inside the waveguide. The first one corresponds to guiding modes with oscillatory radial dependence of the electric and magnetic fields. These modes are radiated under the Cherenkov condition for dielectric permittivity of the cylinder. Here, we were interested in the radiation of surface polaritons, emitted in the spectral range where the real parts of the dielectric permittivities for the cylinder and exterior medium have opposite signs. There is no velocity threshold for the generation of this type of modes. The corresponding fields are peaked near the cylinder surface. The radiation intensity is evaluated in terms of the energy losses and for the radiation on a given eigenmode of the cylinder it is expressed as (10.17), where $x_{n,s}$ in the arguments of the modified Bessel functions are determined solving the eigenmode equation for dielectric cylinder. The frequency and the projection of wavevector on the cylinder axis for radiated surface polaritons are expressed as $\omega = \omega_{n,s} = x_{n,s}v/r_c$ and $k_z = x_{n,s}/r_c$. For $n = 0$ one has a single mode in the range $\varepsilon_0 < \varepsilon_0^{(\infty)}$ and there are no surface modes in the region $\varepsilon_0 > \varepsilon_0^{(\infty)}$. In the case $n \geq 1$, a single or two surface modes exist in the finite range $\varepsilon_0^{(\mathrm{m})} \leq \varepsilon_0 < -\varepsilon_1$.

In the second part of the paper we have considered the emission of surface polaritons by a charged particle circulating around a cylinder. Unlike to the case of a rectilinear motion, there exist a lower threshold for the projection of the corresponding wave vector, $|k_z| > \omega_n\sqrt{\varepsilon_1}/c$. The radiation intensity for surface waves on a given harmonic $n = 1, 2, \ldots$ is given by (10.27). The surface polaritons on the harmonic $n = 1$ are emitted for all values of the permittivity ε_0 in the range $\varepsilon_0 < -\varepsilon_1$. For harmonics $n > 1$, one has a critical value ε_{0n} with the absence of surface polaritons in the region $\varepsilon_0 < -\varepsilon_{0n}$. The radiation wavelength decreases with approaching ε_0 to $-\varepsilon_1$ from below. In that range, the wavelength of surface polaritons is much smaller compared to the wavelength of corresponding electromagnetic radiation in free space. In both cases of rectilinear and circular motions the radiation intensity exponentially decreases with increasing distance of the charge trajectory from the cylinder surface. The results presented show that the number of the radiated quanta for surface polaritons can significantly exceed the corresponding number for guiding modes of the cylinder. It should also be noted that the numerical analysis was presented in terms of the ratio r_q/r_c and the absolute values for r_q and for the radius of the cylinder were not fixed. For a given energy of the radiating particle, the spectral range of the radiation is determined by r_q ($\approx r_c$). The features of the radiation described above are not sensitive to the spectral range. The radiation spectrum can be controlled by tuning the waveguide radius.

Acknowledgements A.A.S. and A.S.K. were supported by the Research Project No. 21AG-1C047 of the Science Committee of RA. L.Sh.G. and H.F.K. were supported by the Science Committee of RA, in the frames of the Research Project No. 21AG-1C069.

References

1. Agranovich, V.M., Mills, D.L. (eds.): Surface Polaritons: Electromagnetic Waves at Surfaces and Interfaces. North-Holland Pub. Co., Amsterdam (1982)
2. Maier, S.A.: Plasmonics: Fundamentals and Applications. Springer (2007)
3. Enoch, S., Bonod, N. (eds.): Plasmonics: From Basics to Advanced Topics. Springer (2012)
4. Stockman, M.I., et al.: Roadmap on plasmonics. J. Optics **20**, 043001 (2018). https://doi.org/10.1088/2040-8986/aaa1142
5. Marqués, R., Martín, F., Sorolla, M.: Metamaterials with Negative Parameters: Theory, Design, and Microwave Applications. John Wiley & Sons, Hoboken, NJ (2008)
6. West, P.R., Ishii, S., Naik, G.V., Emani, N.K., Shalaev, V.M., Boltasseva, A.: Searching for better plasmonic materials. Laser Photonics Rev. **4**, 795–808 (2010). https://doi.org/10.1002/lpor.200900055
7. Lindquist, N.C., Nagpal, P., McPeak, K.M., Norris, D.J.: Sang-Hyun Oh: Engineering metallic nanostructures for plasmonics and nanophotonics. Rep. Prog. Phys. **75**, 036501 (2012). https://doi.org/10.1088/0034-4885/75/3/036501
8. Zhang, X., et al.: Terahertz surface plasmonic waves: a review. Adv. Photonics **2**(1), 014001 (2020). https://doi.org/10.1117/1.AP.2.1.014001
9. Han, Zh., Bozhevolnyi, S.I.: Radiation guiding with surface plasmon polaritons. Rep. Prog. Phys. **76**, 016402 (2013). https://doi.org/10.1088/0034-4885/76/1/016402
10. Kotanjyan, A.S., Mkrtchyan, A.R., Saharian, A.A., Kotanjyan, VKh.: Generation of surface polaritons in dielectric cylindrical waveguides. Phys. Rev. Spec. Top. Accel. Beams **22**, 040701 (2019). https://doi.org/10.1103/PhysRevAccelBeams.22.040701
11. Saharian, A.A., Grigoryan, LSh., Grigorian, AKh., Khachatryan, H.F., Kotanjyan, A.S.: Cherenkov radiation and emission of surface polaritons from charges moving paraxially outside a dielectric cylindrical waveguide. Phys. Rev. A **102**, 063517 (2020). https://doi.org/10.1103/PhysRevA.102.063517
12. Grigoryan, L.Sh., Kotanjyan, A.S., Saharian, A.A.: Green function of an electromagnetic field in cylindrically symmetric inhomogeneous medium. Izv. Nats. Akad. Nauk Arm., Fiz. **30**, 239–244 (1995) (Engl. Transl.: J. Contemp. Phys.)
13. Ashley, J.C., Emerson, L.C.: Dispersion relations for non-radiative surface plasmons on cylinders. Surf. Sci. **41**, 615–618 (1974). https://doi.org/10.1016/0039-6028(74)90080-6
14. Khosravi, H., Tilley, D.R., Loudon, R.: Surface polaritons in cylindrical optical fibers. J. Opt. Soc. Am. A **8**, 112–122 (1991). https://doi.org/10.1364/JOSAA.8.000112
15. Kotanjyan, A.S., Khachatryan, H.F., Petrosyan, A.V., Saharian, A.A.: On the features of radiation from charged particle rotating around a dielectric cylinder. Izv. Nats. Akad. Nauk Arm., Fiz. **35**, 115–123 (2000) (Engl. Transl.: J. Contemp. Phys.)
16. Saharian, A.A., Kotanjyan, A.S.: Synchrotron radiation from a charge moving along a helix around a dielectric cylinder. J. Phys. A **42**, 135402 (2009). https://doi.org/10.1088/1751-8113/42/13/135402
17. Kotanjyan, A.S., Saharian, A.A.: Undulator radiation inside a dielectric waveguide. Nucl. Instrum. Methods B **309**, 177–182 (2013). https://doi.org/10.1016/j.nimb.2013.01.076
18. Saharian, A.A., Kotanjyan, A.S., Grigoryan, LSh., Khachatryan, H.F., Kotanjyan, VKh.: Synchrotron radiation from a charge circulating around a cylinder with negative permittivity. Int. J. Mod. Phys. B **34**, 2050065 (2020). https://doi.org/10.1142/S0217979220500654
19. Saharian, A.A., Kotanjyan, A.S., Kotanjyan, VKh.: Synchrotron radiation from a charge on eigenmodes of a dielectric cylinder. J. Contemp. Phys. **54**, 111–116 (2019). https://doi.org/10.3103/S1068337219020014

Chapter 11
Crossing Points in Spectra and Light Absorption in Spheroidal and Cone-Shaped Quantum Dots

V. L. Derbov, A. A. Gusev, O. Chuluunbaatar, L. L. Hai, S. I. Vinitsky, E. M. Kazaryan, and H. A. Sarkisyan

Abstract We briefly review the analysis of the energy spectrum, the envelope eigenfunctions of electron, hole and exciton states, and the direct interband light absorption in cone-shaped and spheroidal impenetrable quantum dots. We apply high-order finite element method and calculation schemes of Kantorovich method in comparison with the adiabatic approximation (in the strong size quantization limit) for solving boundary-value problems that describe axially symmetric quantum dots. We demonstrate the efficiency of the algorithms and software by benchmark calculations of spectral and optical characteristics of the cone-shaped and spheroidal quantum dots and crossing points in their spectra.

Partially supported by the RFBR (grant No. 17-51-44003 Mong_a) and the RUDN University Strategic Academic Leadership Program.

V. L. Derbov
N.G. Chernyshevsky Saratov National Research State University, Saratov, Russia

A. A. Gusev (✉) · O. Chuluunbaatar · S. I. Vinitsky
Joint Institute for Nuclear Research, Dubna, MR 141980, Russia
e-mail: gooseff@jinr.ru

O. Chuluunbaatar
Institute of Mathematics and Digital Technology, Mongolian Academy of Sciences, Ulaanbaatar 13330, Mongolia

L. L. Hai
Ho Chi Minh city University of Education, Ho Chi Minh city, Vietnam

S. I. Vinitsky
Peoples' Friendship University of Russia (RUDN University), 117198 Moscow, Russia

E. M. Kazaryan · H. A. Sarkisyan
Russian-Armenian (Slavonic) University, H. Emin 123, 0051 Yerevan, Armenia

H. A. Sarkisyan
Peter the Great Saint-Petersburg Polytechnic University, Polytechnicheskaya 29, St. Petersburg 195251, Russia

D. Blaschke et al. (eds.), *Optics and Its Applications*, Springer Proceedings in Physics 281, https://doi.org/10.1007/978-3-031-11287-4_11

11.1 Introduction

The study of spectral and optical characteristics of quantum wells, wires and dots with complicated geometry is an urgent problem of both computational and theoretical physics. To solve boundary-value problems (BPVs) that describe corresponding mathematical models, they commonly use finite-difference and variational methods [1], finite element method (FEM) [2–7], Kantorovich method (KM)-reduction to ordinary differential equations(ODEs) [8–14], known in physics as adiabatic method [15], and adapted in the field [16–21].

In this paper, we present a brief review of application of high-order FEM calculation schemes implemented on unstructured grids with triangle elements [4, 5] and KM for solving the BVPs that describe spectral and optical characteristics of cone-shaped and spheroidal quantum dots and crossing points in their spectra *to obtain independent estimates of the applicability range and accuracy of conventional adiabatic approximation (AA) (in the strong size quantization limit), i.e., a diagonal set of the KM-ODEs developed early by our team.* We discuss the application of these methods and appropriate software to the calculation of the energy spectra of electron, hole and exciton states the direct interband light absorption and the light absorption coefficient in ensembles of non-interacting cone-shaped and spheroidal impenetrable quantum dots (QDs).

The paper is organized as follows. In Sect. 11.2 we set the boundary value problem. Section 11.3 presents the basis equations of the Kantorovich method. In Sect. 11.4 we describe the numerical calculations of the energy spectrum of the QDs. Section 11.5 is devoted to the adiabatic approximation. In Sect. 11.6 we give examples of the interband absorption in QDs. In Conclusions we resume the results and discuss the prospects.

11.2 Setting the Problem

Within the effective mass approximation we consider a class of QD models in which the calculation of energy levels and corresponding envelope eigenfunctions is reduced to self-adjoint BVPs for elliptic differential equations [2]

$$(H - \mathcal{E}_t)\,\Psi_t(x) \equiv \left(-\frac{1}{g_0(x)} \sum_{ij=1}^{d} \frac{\partial}{\partial x_i} g_{ij}(x) \frac{\partial}{\partial x_j} + V(x) - \mathcal{E}_t \right) \Psi_t(x) = 0. \quad (11.1)$$

We assume that $g_0(x) > 0$, $g_{ji}(x) = g_{ij}(x)$ and $V(x)$ are real-valued functions, continuous together with their generalized derivatives to a given order in the domain $x \in \bar{\Omega} = \Omega \cup \partial\Omega$ with the piecewise continuous boundary $S = \partial\Omega$, which provides the existence of a nontrivial solution obeying the mixed boundary conditions (BCs) of the first (I) and/or the second kind (II), i.e., Dirichlet and/or Neumann conditions:

$$(I)\ \Psi_t(x)|_S = 0,\ (II)\ \frac{\partial \Psi(x)}{\partial n_D}\Big|_S = 0,\ \frac{\partial \Psi_t(x)}{\partial n_D} = \sum_{ij=1}^{d} (\hat{n}, \hat{e}_i) g_{ij}(x) \frac{\partial \Psi_t(x)}{\partial z_j}, \qquad (11.2)$$

where $\frac{\partial \Psi_t(x)}{\partial n_D}$ is the derivative along the conormal direction, $\hat{n}$ is the outer normal to the boundary of the domain $S = \partial\Omega$, $\hat{e}_i$ is the unit vector of $x = \sum_{i=1}^{d} \hat{e}_i x_i$, and $(\hat{n}, \hat{e}_i)$ is the scalar product in $\mathcal{R}^d$. The eigenfunctions $\Psi_t(x)$ from the Sobolev space $\mathcal{H}_2^{s\geq 1}(\Omega)$, $\Psi_t(x) \in \mathcal{H}_2^{s\geq 1}(\Omega)$, corresponding to the real eigenvalues of energy spectrum $\mathcal{E}$:$\mathcal{E}_1 \leq \mathcal{E}_2 \leq \ldots \leq \mathcal{E}_t \leq \ldots$ satisfy the orthonormality conditions

$$\langle \Psi_t(x) | \Psi_{t'}(x) \rangle = \int_{\Omega} dx g_0(x) \Psi_t(x) \Psi_{t'}(x) = \delta_{tt'}, \quad dx = dx_1 ... dx_d. \qquad (11.3)$$

We solve this problem using high-accuracy finite element schemes, implemented in the appropriate algorithms and programs [4, 5].

11.3 Kantorovich Method

The axially symmetric solutions of the BVP (11.1)–(11.3) at $d = 3$, periodical with respect to the azimuthal angle φ, are sought in the form of a product $\Psi_t(x_f, x_s, \varphi) = \Psi_t^{m\sigma}(x_f, x_s) e^{im\varphi}/\sqrt{2\pi}$, where $m = 0, \pm 1, \pm 2, \ldots$ is the magnetic quantum number and divided into even ($\sigma = +1$) and odd ($\sigma = -1$) reflection parity ones, or marked $\sigma = 0$, i.e., without parity separation. The function $\Psi_t^{m\sigma}(x_f, x_s)$ is solution of the BVP (11.1)–(11.3) at $d = 2$ and satisfies the equation in the 2D domain Ω: $\Omega = \Omega_{x_f}(x_s) \cup \Omega_{x_s} \subset \mathbf{R}^2 \backslash \{0\}$, $\Omega_{x_f}(x_s) = (x_f^{\min}(x_s), x_f^{\max}(x_s))$, $\Omega_{x_s} = (x_s^{\min}, x_s^{\max})$:

$$\left(\hat{H}(x_f, x_s) - 2E_t\right) \Psi_t^{m\sigma}(x_f, x_s) = 0, \quad \hat{H}(x_f, x_s) = \hat{H}_1(x_f; x_s) + \hat{H}_2(x_f; x_s). \qquad (11.4)$$

The Hamiltonian of the slow subsystem $\hat{H}_2(x_f; x_s)$ is expressed as

$$\hat{H}_2(x_f; x_s) = \check{H}_2(x_f; x_s) = -\frac{1}{g_{1s}(x_f; x_s)} \frac{\partial}{\partial x_s} g_{2s}(x_s) \frac{\partial}{\partial x_s} + \check{V}_s(x_f; x_s), \qquad (11.5)$$

and the Hamiltonian of the fast subsystem $\hat{H}_1(x_f; x_s)$ is expressed through the reduced Hamiltonian $\check{H}_f(x_f; x_s)$ and the weighting factor $g_{3s}(x_f; x_s)$:

$$\hat{H}_1(x_f; x_s) = g_{3s}^{-1}(x_f; x_s) \check{H}_f(x_f; x_s), \qquad (11.6)$$
$$\check{H}_f(x_f; x_s) = -\frac{1}{g_{1f}(x_f)} \frac{\partial}{\partial x_f} g_{2f}(x_f) \frac{\partial}{\partial x_f} + \check{V}_f(x_f) + \check{V}_{fs}(x_f; x_s).$$

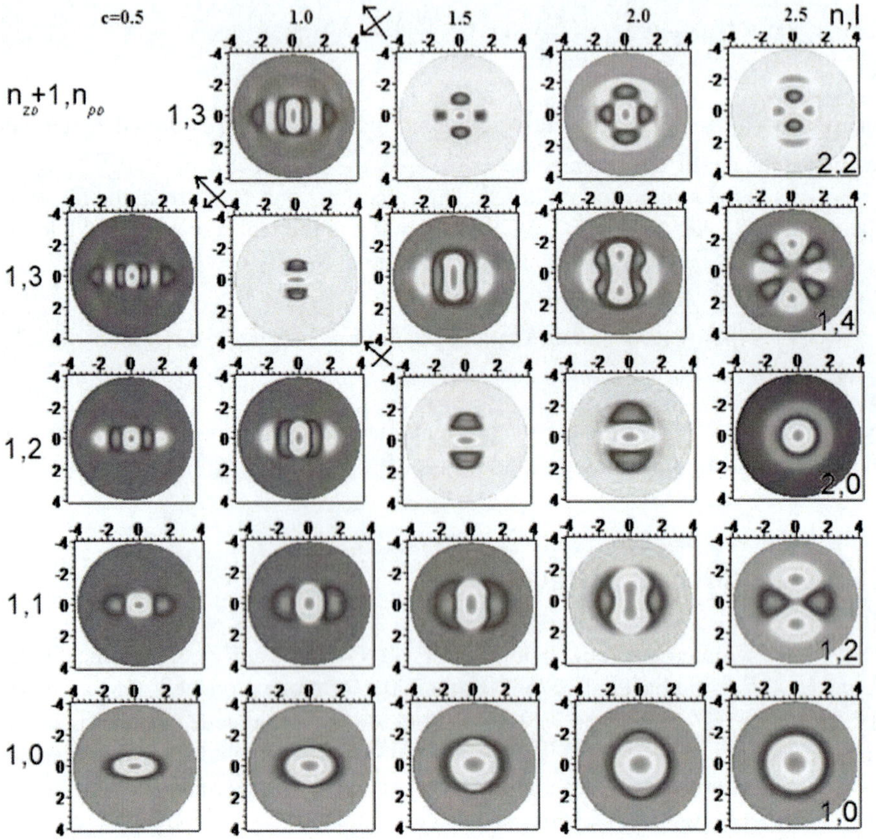

Fig. 11.1 Contour lines of the first five even-parity wave functions $\sigma = +1$ at $m = 0$ in the xz plane of an oblate SQD with the major semiaxis $a = 2.5$ and different values of the minor semiaxis c ($\zeta_{ca} = c/a \in (1/5, 1)$). Arrows indicate the shape transformations of the eigenfunctions when passing through the exact crossing points of pairs of eigenvalues

The choice of fast variable x_f (or slow variable x_s), as well as the interaction potentials, is determined by the QD geometry and construction. The separation of Hamiltonian (11.4) into two Hamiltonians (11.5) and (11.6) with partial derivatives with respect to independent slow variable x_s and fast variable x_f is possible, when one of the QD dimensions is small compared to the other ones. Then the size quantization in the direction x_f (x_s) turns to be much stronger (weaker) compared to that in other directions. This feature allows using adiabatic or diagonal Kantorovich approximation in the strong size quantization limit to estimate the lower-energy part of the QD spectrum. Note, for a convergence of the KM it is sufficient if the Hamiltonian of the fast subsystem has a pure discrete spectrum, while an appropriate choice of the fast and slow variables and parametric basis provides a more high rate of the convergence.

Table 11.1 The values of conditionally fast x_f and slow x_s independent variables, coefficients $g_{is}(x_f; x_s)$, $g_{jf}(x_f)$, and potentials $\check{V}_f(x_f)$, $\check{V}_s(x_f; x_s)$, $\check{V}_{fs}(x_f, x_s)$, in Eqs. (11.4)–(11.6) for OCQD and PCQD, or SQD, OSQD and PSQD in cylindrical (CC) and spherical (SC) coordinates, and oblate & prolate spheroidal (OSC & PSC) coordinates with $(f/2)^2 = \pm(a^2 - c^2)$, + for OSC, − for PSC, f is a focal distance, a and c are semiaxes

	CC		SC	OSC & PSC
	OCQD/OSQD	PCQD/PSQD	SQD	OSQD & PSQD
x_f	z	ρ	η	η
x_s	ρ	z	r	ξ
$g_0(x_f; x_s)$	ρ	ρ	r^2	$(f/2)(\xi^2 \pm \eta^2)$
$g_{1f}(x_f)$	1	ρ	1	1
$g_{2f}(x_f)$	1	ρ	$1 - \eta^2$	$1 - \eta^2$
$g_{1s}(x_f\ x_s)$	ρ	1	r^2	g_0
$g_{2s}(x_s)$	ρ	1	r^2	$\xi^2 \pm 1$
$g_{3s}(x_f\ x_s)$	1	1	r^2	g_0
$\check{V}_f(x_f)$	0	$m^2\rho^2$	$m^2 g_{2f}$	$m^2 g_{2f}$
$\check{V}_s(x_f\ x_s)$	$m^2\rho^2$	0	0	$\mp m^2(g_0 g_{2s})$
$\check{V}_{fs}(x_f, x_s)$	$\check{V}(z, \rho)$	$\check{V}(\rho, z)$	$\check{V}(r, \eta)$	$\check{V}(\xi, \eta)$

Table 11.1 contains the values of conditionally fast x_f and slow x_s independent variables, the coefficients $g_0(x_f; x_s)$, $g_{1s}(x_f; x_s)$, $g_{2s}(x_s)$, $g_{3s}(x_f; x_s)$, $g_{1f}(x_f)$, $g_{2f}(x_f)$, and the reduced potentials $\check{V}_f(x_f)$, $\check{V}_s(x_f; x_s)$, $\check{V}_{fs}(x_f, x_s)$, entering Eqs. (11.4)–(11.6) for QDs: spherical SQD, oblate and prolate spheroidal OSQD and PSQD or cone-shaped OCQD and PCQD in cylindrical coordinates (CC) ($\mathbf{x} = (z, \rho, \varphi)$), spherical coordinates (SC) ($\mathbf{x} = (r, \eta = \cos\theta, \varphi)$), and oblate/prolate spheroidal ($\mathbf{x} = (\xi, \eta, \varphi)$) coordinates [22].

In Table 11.1 for the considered QDs with impenetrable walls the potentials $\check{V}_{fs} = 0$ are zero, since the potential is reformulated below in the form of BCs with respect to the variables x_f and x_s. The solution $\Psi_t^{m\sigma}(x_f, x_s)$ of the problem (11.4)–(11.6) is sought in the form of KM expansion [17]

$$\Psi_t^{m\sigma}(x_f, x_s) = \sum_{j=1}^{j_{\max}} \Phi_j^{m\sigma}(x_f; x_s)\chi_j^{(m\sigma t)}(x_s), \tag{11.7}$$

using as a set of trial functions the eigenfunctions $\Phi_j^{m\sigma}(x_f; x_s)$ of the Hamiltonian $\check{H}_f(x_f; x_s)$ from (11.6), i.e., the solutions of the parametric BVP

$$\left\{\check{H}_f(x_f; x_s) - \check{\lambda}_i(x_s)\right\} \Phi_i^{m\sigma}(x_f; x_s) = 0, \tag{11.8}$$

in the interval $x_f \in \Omega_{x_f}(x_s)$ depending on the conditionally slow variable $x_s \in \Omega_{x_s}$ as a parameter. These solutions obey the boundary conditions

$$\lim_{x_f \to x'_f(x_s)} \left(N_f^{(m\sigma)}(x_s) g_{2f}(x_f) \frac{d\Phi_j^{m\sigma}(x_f; x_s)}{dx_f} + D_f^{(m\sigma)}(x_s) \Phi_j^{m\sigma}(x_f; x_s) \right) = 0 \quad (11.9)$$

at the boundary points $\{x_f^{\min}(x_s), x_f^{\max}(x_s)\} = \partial\Omega_{x_f}(x_s)$, of the interval $\Omega_{x_f}(x_s)$. In Eq. (11.9), $N_f^{(m\sigma)}(x_s) \equiv N_f^{(m\sigma)}$, $D_f^{(m\sigma)}(x_s) \equiv D_f^{(m\sigma)}$, unless specially declared, are determined by the relations $N_f^{(m\sigma)} = 1$, $D_f^{(m\sigma)} = 0$ at $m = 0$, $\sigma = +1$, (or at $\sigma = 0$, i.e., without parity separation), $N_f^{(m\sigma)} = 0$, $D_f^{(m\sigma)} = 1$ at $m = 0$, $\sigma = -1$ or at $m \neq 0$. The eigenfunctions satisfy the orthonormality condition in the same interval:

$$\langle i|j\rangle = \left\langle \Phi_i^{m\sigma} | \Phi_j^{m\sigma} \right\rangle = \int_{x_f^{\min}(x_s)}^{x_f^{\max}(x_s)} \Phi_i^{m\sigma}(x_f; x_s) \Phi_j^{m\sigma}(x_f; x_s) g_{1f}(x_f) dx_f = \delta_{ij}. \quad (11.10)$$

Here $\check{\lambda}_1(x_s) < \ldots < \check{\lambda}_{j_{\max}}(x_s) < \ldots$ is the desired set of real-valued eigenvalues. The corresponding set of potential curves of Eqs. (11.6) is determined by the condition $2E_j(x_s) = g_{3s}^{-1}(x_f; x_s)\check{\lambda}_j(x_s)$. The solutions of the problem (11.8)–(11.10) are calculated in the analytical form [18, 20], or by the program ODPEVP [12].

Substituting the expansion (11.7) into Eq. (11.4) in consideration of (11.8) and (11.10), we get a set of ODEs with respect to unknown $\chi^{(t)}(x_s) = (\varnothing_1^{(t)}(x_s), \ldots, \chi_{j_{\max}}^{(t)}(x_s))^T$

$$\sum_{j=1}^{j_{max}} \left\langle i | g_{1s}(x_f; x_s)[\hat{H}(x_f, x_s) - 2E_t] | j \right\rangle \chi_j^{(m\sigma t)}(x_s) = 0, \quad (11.11)$$

where the matrix elements read as

$$\left\langle i | g_{1s}(x_f; x_s) \hat{H}(x_f, x_s) | j \right\rangle = -\delta_{ij} \frac{d}{dx_s} g_{2s}(x_s) \frac{d}{dx_s} + < i | (\frac{g_{1s}(x_s)}{g_{3s}(x_s)} | j > \check{\lambda}_j(x_s)$$
$$+ < i | g_{1s}(x_s) \check{V}_s(x_s) | j > + g_{2s}(x_s) \left[\mathrm{W}_{ij}(x_s) + \mathrm{Q}_{ij}(x_s) \frac{d}{dx_s} \right] + \frac{d g_{2s}(x_s) \mathrm{Q}_{ij}(x_s)}{dx_s}, \quad (11.12)$$

Here $\mathrm{W}_{ij}(x_s)$, and $\mathrm{Q}_{ij}(x_s)$, $\langle i | g_{1s}(x_s) | j \rangle$ and etc. are given by integrals

$$\langle i | g_{1s}(x_s) | j \rangle = \int_{x_f^{\min}(x_s)}^{x_f^{\max}(x_s)} g_{1f}(x_f) \Phi_i^{m\sigma}(x_f; x_s) g_{1s}(x_f; x_s) \Phi_j^{m\sigma}(x_f; x_s) dx_f, \quad (11.13)$$

$$W_{ij}(x_s) = W_{ji}(x_s) = \int_{x_f^{\min}(x_s)}^{x_f^{\max}(x_s)} g_{1f}(x_f) \frac{\partial \Phi_i^{m\sigma}(x_f; x_s)}{\partial x_s} \frac{\partial \Phi_j^{m\sigma}(x_f; x_s)}{\partial x_s} dx_f,$$

$$Q_{ij}(x_s) = -Q_{ji}(x_s) = -\int_{x_f^{\min}(x_s)}^{x_f^{\max}(x_s)} g_{1f}(x_f) \Phi_i^{m\sigma}(x_f; x_s) \frac{\partial \Phi_j^{m\sigma}(x_f; x_s)}{\partial x_s} dx_f,$$

calculated analytically [18, 20] or by using the program ODPEVP [12]. The solutions $\chi^{(m\sigma t)}(x_s) = \chi^{(t)}(x_s)$ of discrete spectrum $\mathcal{E} = 2E : 2E_1 < 2E_2 \leq \ldots \leq 2E_t \leq \ldots$ obey the BCs at points $x_s^t = \{x_s^{\min}, x_s^{\max}\} = \partial\Omega_{x_s}$ bounding the interval Ω_{x_s}:

$$\lim_{x_s \to x_s^t} \left(N_s^{(m\sigma)} g_{2s}(x_s) \frac{d\chi^{(m\sigma t)}(x_s)}{dx_s} + D_s^{(m\sigma)} \chi^{(m\sigma t)}(x_s) \right) = 0, \tag{11.14}$$

where $N_s^{(m\sigma)} = 1$, $D_s^{(m\sigma)} = 0$ at $m = 0, \sigma = +1$, (or at $\sigma = 0$, i.e without parity separation), $N_s^{(m\sigma)} = 0$, $D_s^{(m\sigma)} = 1$ at $m = 0, \sigma = -1$ or at $m \neq 0$. They obey the orthonormality conditions (11.3) with $g_{1s}(x_f; x_s)$ from Table 11.1 and the KM expansion (11.7), that after integration (11.13) over x_f are reduced to

$$\left\langle \chi^{(m\sigma t)} | \chi^{(m\sigma t')} \right\rangle = \sum_{i,j=1}^{j_{max}} \int_{x_s^{\min}}^{x_s^{\max}} \chi_i^{(m\sigma t)}(x_s) \left\langle i | g_{1s}(x_s) | j \right\rangle \chi_j^{(m\sigma t')}(x_s) dx_s = \delta_{tt'}. \tag{11.15}$$

Solutions of BVP (11.11)–(11.15) are calculated with a given accuracy not vorse then six significant digits by the program KANTBP [9, 11].

Remark 1. Note t is the eigenvalue number in the ascending energy sequence $E_1 \leq E_2 \leq \ldots \leq E_t \leq \ldots$ corresponding to the number v of the eigenvalue $E_{t;j1} \leq E_{t;j2} \leq \ldots \leq E_{t;jv} \leq \ldots$ counted at each $\langle j \rangle = j$, in diagonal approximation of the KM Eqs. (11.11)–(11.12) or a crude AA Eq. (11.16) without diagonal nonadiabatic terms $W_{jj}(x_s)$, where number v determines the number $v-1$ of nodes of the solution $\chi_j^{(v)}(x_s)$ at fixed value j and the quantity $\langle j \rangle = \left\langle \chi^{(t)} | j | \chi^{(t)} \right\rangle$ is the averaged quantum number [10].

11.4 Numerical Calculations of the Energy Spectrum

After the separation of the angular variable φ, the axially symmetric BVP for the electron, hole, and exciton eigenstates in an impenetrable cone-shaped quantum dot (CQD) or spheroidal quantum dot (SQD) is reduced to the BVP (11.1)–(11.3) at $d = 2$ with respect to the radial $x_1 = \rho$ and the axial $x_2 = z$ variables, where $g_0(x) = \rho$, $g_{11}(x) = g_{22}(x) = \rho$, $g_{12}(x) = g_{21}(x) = 0$, $V(x) = V(\rho, z) = m^2/\rho^2 + 2V_C(\rho, z)$, with the BCs at the boundary $\partial\Omega = \partial\Omega_1 \cup \partial\Omega_2$, $\partial\Omega_1 = \{(\rho, z)|z = 0, \rho = \rho_{\max}(z)\}$, $\partial\Omega_2 = \{(\rho, z)|\rho = 0\}$ of the 2D domain $\Omega = \{(\rho, z)|\rho \geq 0, z \geq 0, \rho \leq \rho_{\max}(z)\}$

$$\Psi_t^{(m\sigma)}(\rho, z)|\partial\Omega_2 = 0, \lim_{\rho \to 0} \left(\rho \frac{\Psi_t^{(m\sigma)}(\rho, z)}{\partial\rho} \delta_{m0} + \Psi_t^{(m\sigma)}(\rho, z)(1 - \delta_{m0}) \right) \Bigg| \partial\Omega_1 = 0.$$

Below we restrict ourselves to the case $m = 0$. For CDQ $\rho_{\max}(z) = R(1 - z/H)$, where R is the base radius and H is the height, $\sigma = 0$, for SDQ, $\rho^2/a^2 + z^2/c^2 = 1$, where a and c are the spheroid semiaxes, and $\rho_{\max}(z) = a\sqrt{1 - z^2/c^2}$, $\sigma = +1$.

For oblate and prolate SDQs the BVP (11.1)–(11.3) at $d = 2$ was also solved in the spheroidal coordinates ($x_1 = \xi, x_2 = \eta$), $g_0(x) = (f/2)^2(\xi^2 \pm \eta^2)$, $g_{11}(x) =$

Table 11.2 The first three eigenvalues, $\mathcal{E}_t(n_b, l_{\min})$, $t = 1, 2, 3$, at $m = 0$, $\sigma = +1$ in the units of E_R, for the oblate SQD $a = 2.5$, $c = 0.5$ obtained on the different FEM grid with the maximal element size $l_{\min}$. The number n_b determines the length of the polygonal boundary approximating the boundary of SQD. KM—Kantorovich method with 60 basis functions from [20]. SPH—FEM in spheroidal coordinates on non-uniform grid in the rectangular domain $0 < \xi_0 < c/\sqrt{a^2 - c^2} = 1/2\sqrt{6}$, $0 \leq \eta \leq 1$

$l_{\min}$	0.0625	0.125	0.03125	0.0625	KM	SPH
n_b	0.03125	0.03125	0.015625	0.015625		
t = 1	12.76518	12.76516	12.76490	12.76490	12.764809	12.77105
t = 2	20.04147	20.04143	20.04086	20.04085	20.040651	20.04933
t = 3	29.74910	29.74902	29.74780	29.74779	29.747387	29.75713

$(\xi^2 \pm 1)$, $g_{22}(x) = (1-\eta^2)$, $g_{12}(x) = g_{21}(x) = 0$, $(f/2)^2 = \pm(a^2-c^2)$, f is a focal distance, $V(x) = V(\xi, \eta) = m^2/(\xi^2\pm1)(1-\eta^2)$, and using the Kantorovich method Eqs. (11.11)–(11.15) with $j_{max} = 60$ of basis functions [18].

The comparison of results obtained for eigenvalues of oblate SDQ in the cylindrical coordinates on different FEM grids, in the spheroidal coordinates, and using the (KM) is resented in Table 11.2 [1]. As seen from Table 11.2, the results coincide to five significant digits, and the maximal contribution to the error in cylindrical coordinates is due to the error of approximating the curved boundary by triangle finite elements with rectilinear boundaries.

Figures 11.1 and 11.2 show the lower part of non-equidistant spectrum $\tilde{E}(\zeta_{ca})/E_R = 2E_t$ (or $\tilde{E}(\zeta_{ac})/E_R = 2E_t$) and the eigenfunctions $\Psi_t^{m\sigma}$ from (11.7) for even states oblate (or prolate) SQD at $m = 0$.

For an oblate (or prolate) SQD a correspondence rule holds: $j = n_o = n_{zo}+1 = 2n-(1+\sigma)/2, n = 1, 2, \ldots, n_{\rho o} = (l-|m|-(1-\sigma)/2)/2$ (or $j = n_{\rho p}+1 = n_p = n = n_r+ 1$, $j = 1, 2, \ldots$, $n_{zp} = l-|m|$) between the spherical quantum numbers $(n, l, m, \hat{\sigma})$ of an SQD with radius $r_0 = a = c$ and spheroidal quantum numbers $\{n_\xi = n_r, n_\eta = l-|m|, m, \sigma\}$ of an oblate (or prolate) SQD with the major semiaxis a (or c) and the minor semiaxis c (or a), and the adiabatic set of cylindrical quantum numbers $[n_{zo}, n_{\rho o}, m, \sigma]$ (or $[n_{\rho p}, n_{zp}, m, \sigma]$) upon continuous variation of parameter $\zeta_{ca} = c/a$ (or $\zeta_{ac} = a/c$).

The crossing of similar-parity energy levels in Figs. 11.1 and 11.2 upon the change of symmetry from spherical $\zeta_{ca} = 1$ ($\zeta_{ac} = 1$) to axial, i.e., upon the variation of the parameter $0 < \zeta_{ca} < 1$ ($0 < \zeta_{ac} < 1$), in the BVP with two variables at fixed m for an impenetrable oblate (prolate) SQDs is caused by the possibility to separate the variables $(\xi, \eta\varphi)$ and the additional integral of motion given explicitly [18] with

[1] For electron(e) and hole(h) states $2V_C(\rho, z) = 0$, and for exciton states $2V_C(\rho, z) = -2/\sqrt{\rho^2 + z^2}$, $2V_C(\rho, z) = \tilde{V}_C(\tilde{\rho}, \tilde{z})/E_R$, $\tilde{V}_C(\tilde{\rho}, \tilde{z}) = -2e/(\kappa\sqrt{\tilde{\rho}^2 + \tilde{z}^2})$, where e and m_e are the electron charge and mass, κ is the static permittivity. For GaAs model we use the reduced atomic units, $m_e^* = 0.067m_e$, $m_h^* = m_e^*/0.12$, $\kappa = 13.18$, $a_B = 104$Å, $E_R = 5.275$ meV, i.e., $\mathcal{E} = 2E = \tilde{E}/E_R$, $\Psi(\rho, z) = a_B^{3/2}\tilde{\Psi}^e(\tilde{\rho}, \tilde{z})$, $\rho = \tilde{\rho}/a_B$, $z = \tilde{z}/a_B$, where $\tilde{E}$, $\tilde{V}_C(\tilde{\rho}, \tilde{z})$, $\tilde{\rho}$ and $\tilde{z}$ are dimensioned quantities.

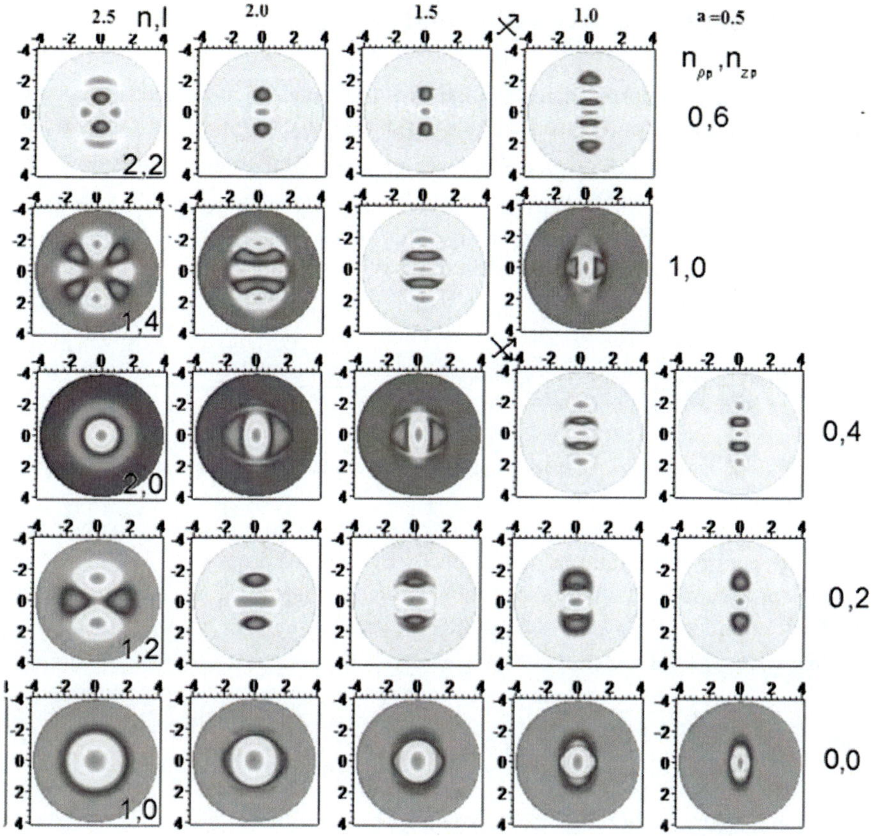

Fig. 11.2 Contour lines of the first five wave functions $\sigma = +1$, $m = 0$ in the xz plane of prolate SQD for the major semiaxis $c = 2.5$ and different values of the minor semiaxis a ($\zeta_{ac} = a/c \in (1/5, 1)$). Arrows show transformations of a shape of eigenfunctions passing thought exact crossing points of pairs of eigenvalues

eigenvalues of a separation parameter having branch points for complex values f of the focal distance or parameter $(f/2)\sqrt{2E}$ of the propagation constant [23, 24]. Thus, the values of parameter $0 < \zeta_{ca} < 1$, or $0 < \zeta_{ac} < 1$, corresponding to crossing points of the eigenvalues $\tilde{E}(\zeta_{ca})/E_R = 2E_t$, or $\tilde{E}(\zeta_{ac})/E_R = 2E_t$ of the lower parts of spectra of oblate or prolate SQDs, strongly restrict the range of applicability of the AA (in the strong size quantization limit), or diagonal approximation of the KM for their estimations.

11.5 Adiabatic Approximation and Size Quantization

For classification and approximate calculation of the spectrum under the size quantization (SQ) conditions for electron (e) hole (h) states the AA is used, $\Psi^{m\sigma}_{jv}(x_f, x_s) = \Phi^{m\sigma}_j(x_f; x_s)\chi^{(m\sigma v)}_j(x_s)$. For prolate SQD $a/c \ll 1$ and CQD with small apex angle $R/H \ll 1$, the parametric spectrum $2E^{m\sigma}_j(z)$ and eigenfunctions $\Phi^{(m\sigma)}_j(\rho; z)$ at $x_f = \rho$, $x_s = z$ of the 'fast' subsystem are solutions of BVP (11.8)–(11.10) at each value of parameter z that are expressed in terms of the cylindrical Bessel function of the first kind [20, 21]

$$2E^{m\sigma}_j(z) = \frac{\alpha^2_{n_{\rho p}+1,|m|}}{\rho_{\max}(z)^2}, \quad \Phi^{m\sigma=0}_j(\rho; z) = \frac{\sqrt{2}J_{|m|}(\rho\sqrt{2E_i(z)})}{\rho_{\max}(z)J_{|m+1|}(\alpha_{n_{\rho p}+1,|m|})},$$

where $\alpha_{n_{\rho p}+1,|m|}$ is the $j = n_p = n_{\rho p}+1$-th positive node of the Bessel function [22] $J_{|m|}(\rho_{\max}(z))$, $j = n_p = n_{\rho p}+1 = 1, 2, \ldots$.

A low part of the spectrum $\mathcal{E}_{t;jv} = 2E_{t;jv}$, and eigenfunctions $\chi^{m\sigma v}_j(z)$ of the 'slow' subsystem are solutions of the BVP of the KM Eqs. (11.11)–(11.15) in diagonal approximation without diagonal nonadiabatic terms $W_{jj}(x_s)$, i.e. in a crude AA, where number $v = n_{zp}+1$ determines the number $v-1 = n_{zp}$ of nodes of the solutions $\chi^{(m\sigma v)}_j(z)$ at fixed j:

$$(-\frac{\partial^2}{\partial z^2} + 2E^{m\sigma}_j(z) - \mathcal{E}_{t;jv})\chi^{(m\sigma v)}_j(z) = 0, \qquad (11.16)$$

that are subjected to orthogonal and normalization conditions

$$\int_{z^{\min}}^{z^{\max}} dz\chi^{m\sigma v}_j(z)\chi^{m\sigma v'}_j(z) = \delta_{vv'} \qquad (11.17)$$

with the BCs $\chi^{m\sigma v}_j(0) = \chi^{m\sigma v}_j(H) = 0$ at $\sigma = 0$ for CQD, or $\chi^{m\sigma v}_j(-c) = \chi^{m\sigma v}_j(c) = 0$ at $\sigma = \pm 1$ for SQD. Considering (11.16), (11.17) in the linear or quadratic approximation for the prolate CQD or SQD leads to the spectrum in the analytical form [20, 21]

$$\mathcal{E}^{CQD}_{(t;n_p n_{zp} m)} = 2E^{(0)}_{n_p m} + 2\varepsilon_{n_{zp}} = \frac{2\alpha^2_{n_p,|m|}}{R^2} - \left(\frac{2\alpha^2_{n_p,|m|}}{R^2 H}\right)^{2/3}\beta_{n_{zp}+1},$$

$$\mathcal{E}^{PSQD}_{(t;n_p n_{zp} m)} = \frac{\alpha^2_{n_p,|m|}}{a^2} + \frac{\alpha^2_{n_p,|m|}}{ac}(2n_{zp}+1),$$

where $\beta_{n_{zp}+1}$ is the v-th negative node of the Airy function of the first kind [22]. For oblate SQD $c/a \ll 1$ and CQD $H/R \ll 1$, using the AA at $x_f = z$, $x_s = \rho$, one has the spectrum and eigenfunctions classified by set $[n_{zo}, n_{\rho o}, m, \sigma]$.

Table 11.3 Comparison of the Coulomb interaction energy $\mathcal{E}_t^{eh}$ between electron and hole, and the size quantization energy $\mathcal{E}_t^{SQ}$ in the units of E_R. Here $H = 10a_B$, $m = 0$. The corresponding values of $\mathcal{E}^{SQ}_{(t;n_p n_{zp} m)}$ (adiabatic calculation) and $\mathcal{E}^{eh}_{(t;n_p n_{zp} m)}$ (first-order perturbation theory) adopted from Ref. [21] are labeled by an asterisk *

R (a_B)	$(t; n_{\rho p}, n_{zp}) = (1; 0, 0)$		$(t; n_{\rho p}, n_{zp}) = (18; 1, 0)$		$(t; n_{\rho p}, n_{zp}) = (2; 0, 1)$	
	$\mathcal{E}_t^{SQ}$	$\mathcal{E}_t^{eh}$	$\mathcal{E}_t^{SQ}$	$\mathcal{E}_t^{eh}$	$\mathcal{E}_t^{SQ}$	$\mathcal{E}_t^{eh}$
0.5	30.42114	–2.64009	142.96526	–4.19801	36.95117	–1.88114
0.5	26.624*	–1.121*	141.542*	–1.113*	34.483*	–0.95*
1.0	8.88906	–1.71413	39.22966	–2.27799	11.93808	–1.25028
1.0	8.359*	–0.902*	36.272*	–1.021*	10.287*	–0.608*
1.5	4.49383	–1.33844	18.79476	–1.66173	6.51718	–0.99660
1.5	4.071*	–0.781*	18.085*	–0.762*	5.193*	–0.503*

Table 11.3 presents the comparison of the energy of Coulomb interaction of exciton $\mathcal{E}_t^{eh} = \mathcal{E}_t^C m_h^*/(m_e^* + m_h^*) - \mathcal{E}_t^{SQ}$ with the size quantization electron energy $\mathcal{E}_t^{SQ}$ at different geometric parameters of a CQD, where exciton energy $\mathcal{E}_t^C$ with the electron-hole reduced mass $m_{eh} = m_e^* m_h^*/(m_e^* + m_h^*)$ in the exciton center-of-mass frame and size quantization electron energy $\mathcal{E}_t^{SQ}$ with the electron mass m_e^* were obtained by solving the BVP (11.1)–(11.3) with Coulomb potential $2V_C(\rho, z)$ and without it, correspondingly. From Table 11.3 the correction energy $\mathcal{E}_t^{eh}$ is always seen to be negative, and with the increasing radius R the relative contribution of Coulomb energy of exciton becomes significant. The comparison with $\mathcal{E}^{SQ}_{(t;n_p n_{zp} m)}$ and $\mathcal{E}^{eh}_{(t;n_p n_{zp} m)}$ calculated using the AA and the perturbation theory [21] show the contribution of nonadiabatic corrections and the applicability of the AA.

Figure 11.3 shows the dependence of the charge carrier energies upon the base radius of fixed-height CQDs and the minor semiaxis of fixed-major semiaxes SQDs, and upon the apex angle of fixed-volume CQDs and SQDs, respectively. Note that each eigenlevel of the 'fast' subsystem has a family of 'slow' subsystem eigenlevels positioned thereupon. For example, for CQD with $\tilde{R} = 0.5a_B$, $\tilde{H} = 10a_B$, the first level ($(t, n_\rho, n_z) = (1, 0, 0)$, $\tilde{E}_1^{SQ}/E_R = 30.42114$) and the eighteenth ($(t, n_\rho, n_z) = (18, 1, 0)$, $\tilde{E}_{18}^{SQ}/E_R = 142.96526$) one belong to the 'fast' subsystem levels. For $\tilde{R} = 1.5a_B$ the first level and the seventh one belong to the 'fast' subsystem, while five levels between them belong to the 'slow' subsystem, etc. The carrier energy is seen to decrease with the increasing base radius $\tilde{R}$ or small semiaxis $\tilde{a}$, because the SQ contribution to the energy decreases. The crossing of the seventh level with the eighth one at $\tilde{R} \approx 1.5a_B$ corresponds to the crossing of the same levels at $\theta_0 \approx \arctan(3/20)$.

Figure 11.3a and b show that for prolate CQD and SDQ the faster growth of energy at small apex angles θ_0 is caused by the size quantization in radial variable ρ and angular variable φ. The slower increase in energy at the apex angles approaching the right angle is caused by the size quantization in longitudinal variable z, i.e., the height. The difference in the rate of energy level variation in the adiabatic domains of variation of geometric parameters is due to the nonuniform scale, since the radius

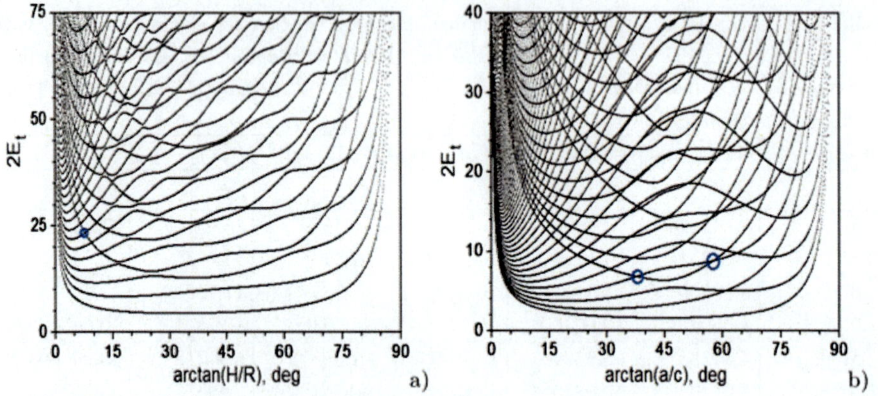

Fig. 11.3 Energy levels $\mathcal{E}_t = 2E_t = \tilde{E}_t^{SQ}/E_R$ at $m = 0$: **a** for prolate and oblate CQDs of equal volume $V = \pi R^2 H/3 = (15/2)\pi a_B^3$ at the base of $H = 10$ and $R = 1.5$ versus the stretch angle $\theta_0 = \arctan(H/R)$ at the base of $\theta_0 \approx 8.53°$, and **b** for prolate and oblate SQDs of equal volume $V = 4\pi a^2 c/3 = (125/6)\pi a_B^3$ at the base of $a = c = 2.5$ versus $\theta_0 = \arctan(a/c)$ at the base of $\theta_0 = 45°$

varies as $(\tan\theta_0)^{2/3}$ and the height as $(\cot\theta_0)^{1/3}$. It is also seen that for the angle values beyond the adiabatic domains there are series of quasicrossings and exact crossings of energy levels for CQD (for comparison, see a discussion of crossing points in the spectra of triangles [25–27]) and SQD, respectively, and its transformation to the oblate QDs.

Remark 2. The above analysis shows the following: the small values of parameters $0 < \zeta_{ca} < 1$ and $H/R \ll 1$, or $0 < \zeta_{ac} < 1$ and $R/H \ll 1$, corresponding to exact crossing and quasicrossings points of eigenvalues in the lower part of the energy spectrum of oblate or prolate QDs strongly restrict the applicability range of the adiabatic approximation (in the strong size quantization limit), or diagonal approximation of Kantorovich method for their estimations.

11.6 Interband Absorption

Consider the direct interband absorption in cone-shaped quantum dots in the regime of strong size quantization, when the Coulomb interaction between an electron and a hole can be neglected. Furthermore, consider the case of a heavy hole with $m_e^* \ll m_h^*$, where m_e^* and m_h^* are the electron and hole effective mass, respectively. Then the absorption coefficient is given by [28]

$$\tilde{K}(\tilde{\omega}^{ph}) = \sum_{\nu\nu'} \tilde{K}_{\nu\nu'}(\tilde{\omega}^{ph}) = \tilde{K}_0 \sum_{\nu\nu'} \left| \int \tilde{\Psi}_\nu^e \tilde{\Psi}_{\nu'}^h d\mathbf{r} \right|^2 \delta(\hbar\tilde{\omega}^{ph} - \tilde{E}_g - \tilde{E}_\nu^e - \tilde{E}_{\nu'}^h), \tag{11.18}$$

where $\tilde{\Psi}^{e(h)}_{\nu(\nu')}$ are envelopes of the electron and hole wave functions, $\nu = [n_{\rho p} n_{zp} m]$ and $\nu' = [n'_{\rho p} n'_{zp} m']$ ($\nu = [n_{zo} n_{\rho o} m]$ and $\nu' = [n'_{zo} n'_{\rho o} m']$) are sets of quantum numbers corresponding to the electron and the heavy hole prolate(oblate) SQD or CQD, respectively, $\tilde{E}_g$ is the band gap of the bulk semiconductor (for GaAs $\tilde{E}_g/E_R = 1.43/(5.27 \cdot 10^{-3})$), $\tilde{\omega}^{ph}$ is the frequency of the incident light, and $\tilde{K}_0$ is proportional to the square of the transition matrix element calculated with Bloch functions [28]. Here the following selection rules for the transitions between the levels with different quantum numbers are valid in the adiabatic classification. In the case of the prolate(oblate) SQD and CQD for the magnetic quantum number the transitions between the levels with $m = -m'$ are allowed. For the prolate(oblate) SQD the transitions between the levels with $n_{\rho p} = n'_{\rho p}$ and $n_{zp} = n'_{zp}$ ($n_{zo} = n'_{zo}$ and $n_{\rho o} = n'_{\rho o}$), respectively, are allowed. For the prolate(oblate)CQD transitions between the levels with $n_{\rho p} = n'_{\rho p}$ ($n_{zp} = n'_{zp}$) are allowed, however there is no selection rule for the axial(radial) quantum numbers n_{zp} ($n_{\rho o}$) and any transitions between different levels are allowed: $n_{zp} \to \forall n'_{zp}$ ($n_{\rho o} \to \forall n_{\rho o}$) respectively, like for prolate (oblate) SQD in uniform electric field [20].

The difference between energy levels for CQD of the same family increases with the increase in the axial quantum number. For example, $\Delta \tilde{E}_{10} = 1.1 E_R$, when $\tilde{R} = 1.5 a_B$ and $\tilde{H} = 10 a_B$ ($n_{\rho p} = 0, m = 0$), and $\Delta \tilde{E}_{10} = 3.4 E_R$, when $\tilde{R} = 1.5 a_B$ and $\tilde{H} = 10 a_B$ ($n_{\rho p} = 1, m = 0$). Note that the transition frequency between these energy levels is $\Delta \tilde{\omega}^{ph}_{10} (n_{\rho p} = 0, m = 0) = 1.43 \cdot 10^{12} \text{sec}^{-1}$ and $\Delta \tilde{\omega}^{ph}_{10} (n_{\rho p} = 1, m = 0) = 4.3 \cdot 10^{12} \text{sec}^{-1}$, which falls into the IR range of spectrum.

In CQD the decrease in the base radius increases the absorption edge energy. It is due to the fact that with the decrease in $\tilde{R}$ the effective width of the bandgap increases due to smaller influence of the CQD walls. The energy levels corresponding to high values of the cone height are located above. Note that the interband transition frequency between the energy levels is $\tilde{\omega}^{ph}_{000} = 5.07 \cdot 10^{-14} \text{sec}^{-1}$ for $\tilde{R} = 0.2 a_B$ and $\tilde{H} = 15 a_B$, which falls into the visible spectral range [21].

For the Lifshits-Slezov distribution, Fig. 11.4 displays the total absorption coefficient $\tilde{K}(\tilde{\omega}^{ph})/\tilde{K}_0$ and the partial absorption coefficients $\tilde{K}_{\nu,\nu}(\tilde{\omega}^{ph})/\tilde{K}_0$, that form the corresponding partial sum (11.18) over a fixed set of quantum numbers ν, ν' at $m = -m' = 0$. In the regime of strong dimensional quantization, the frequencies of the interband transitions ($h \to e$) in GaAS between the levels $n_o = n_{zo} + 1 = 1, n_{\rho o} = 0, m = 0$ for oblate SQD or $n_p = n_{\rho o} + 1 = 1, n_{zp} = 0, m = 0$ for prolate SQD at the fixed values $\tilde{a} = 2.5 a_e$ and $\tilde{c} = 0.5 a_e$ for oblate SQD or $\tilde{a} = 0.5 a_e$ and $\tilde{c} = 2.5 a_e$ for prolate SQD, are equal to $\Delta \tilde{\omega}^{ph}_{100}/(2\pi) = 16.9$ THz or $\Delta \tilde{\omega}^{ph}_{100}/(2\pi) = 33.3$ THz, where $\Delta \tilde{\omega}^{ph}_{100}/(2\pi) = (\tilde{W}_{100,100} - \tilde{E}_g)/(2\pi\hbar)$, $\tilde{W}_{\nu,\nu'} = \tilde{E}_g + \tilde{E}^e_\nu + \tilde{E}^h_{\nu'}$ correspond to the IR spectral region, taking the band gap value $(2\pi\hbar)^{-1} \tilde{E}_g = 346$ THz [20].

With the decreasing semiaxis the threshold energy increases, because the 'effective' band gap width increases, which is a consequence of the dimensional quantization enhancement. Therefore, the above frequency is greater for prolate QD than for oblate QD, because the QD implemented in two directions of the plane (x,y) is effectively larger than that in the direction of the z axis solely at similar values of semiaxes. Higher-accuracy calculations reveal essential difference in the frequency

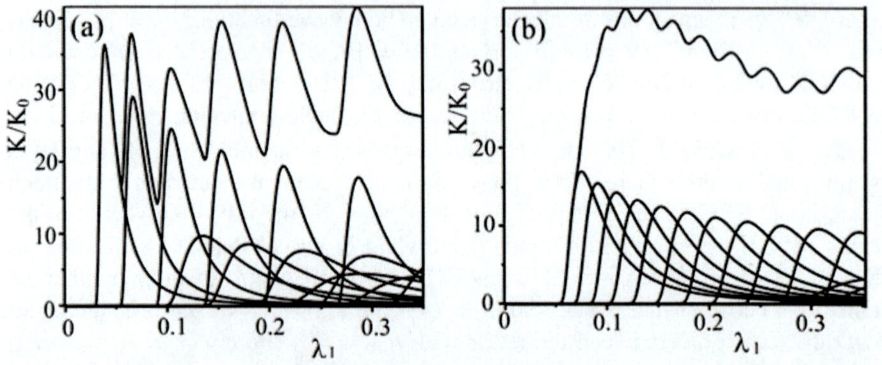

Fig. 11.4 Absorption coefficient K/K_0, Eq. (11.18), consisting of a sum of the first partial contributions versus the energy $\lambda = \lambda_1 = (\tilde{\omega}^{ph} - \tilde{E}_g)/\tilde{E}_g$ of the optical interband transitions for ensembles of GaAs SQDs $(h \to e)$ with the Lifshits-Slezov distribution of the random small semiaxis for an ensemble of **a** oblate SQDs $\bar{c} = 0.5$, $a = 2.5$ and **b** prolate SQDs $\bar{a} = 0.5$, $c = 2.5$

behavior of the AC for interband transitions in the systems of semiconductor oblate or prolate QDs having a distribution of minor semiaxes, which can be used to verify the above models.

11.7 Conclusion

In this paper we briefly review the efficient methods and software for calculating electron, hole and exciton states in axially symmetric QDs by the example of cone-shaped and spheroidal impenetrable QDs. Our analysis shows that the calculation schemes of high-order FEM implemented on unstructured grids together with complementary KM and AA (in the strong size quantization limit) provide useful numerical and analytical tools for describing the energy spectra and *their crossing points that determine the range of AA applicability*, and the optical absorption coefficient in an ensemble of non-interacting axially symmetric QDs.

Further development and application of such approach and software is associated with the investigation of spectral and optical characteristics of quantum wells, wires and dots with complex geometry.

References

1. Harrison, P.: Quantum Wells. Wires and Dots. Theoretical and Computational Physics of Semiconductor Nanostructures. Wiley, New York (2005)
2. Ciarlet, P.: The Finite Element Method for Elliptic Problems. North-Holland Publ. Comp, Amsterdam (1978)

3. Ramdas Ram-Mohan, L.: Finite Element and Boundary Element Applications in Quantum Mechanics. Oxford Univ. Press, New York (2002)
4. Gusev, A. A., Gerdt, V. P., Chuluunbaatar, O., Chuluunbaatar, G. , Vinitsky, S.I., Derbov, V.L., Gozdz, A.: Symbolic-numerical algorithms for solving the parametric self-adjoint 2D elliptic boundary-value problem using high-accuracy finite element method. V.P. Gerdt et al. (Eds.): CASC 2017, LNCS **10490**, 151-166 (2017)
5. Gusev, A. A., Gerdt, V. P., Chuluunbaatar, O., Chuluunbaatar, G. , Vinitsky, S.I., Derbov, V.L., Gozdz, A.: Symbolic-numerical algorithm for generating interpolation multivariate hermite polynomials of high-accuracy finite element method. V.P. Gerdt et al. (Eds.): CASC 2017, LNCS **10490**, 134-150 (2017)
6. Gusev, A.A., Chuluunbaatar, O., Vinitsky, S.I., Derbov, V.L., Hai, L.L., Kazaryan, E.M., Sarkisyan, H.A.: Finite element method for calculating spectral and optical characteristics of axially symmetric quantum dots. SPIE **10717**, 1071712 (2018)
7. Vinasco, J.A., Radu, A., Tiutiunnyk, A., Restrepo, R.L., Laroze, D., Feddi, E., Mora-Ramos, M.E., Morales, A.L., Duque, C.A.: Revisiting the adiabatic approximation for bound states calculation in axisymmetric and asymmetrical quantum structures. Superlattices and Microstructures **138**, 106384 (2020)
8. Kantorovich, L.V., Krylov, V.I.: Approximate Methods of Higher Analysis. Wiley, New York (1964)
9. Chuluunbaatar, O., Gusev, A.A., Gerdt, V.P., Rostovtsev, V.A., Vinitsky, S.I., Abrashkevich, A.G., Kaschiev, M.S., Serov, V.V.: POTHMF: A program for computing potential curves and matrix elements of the coupled adiabatic radial equations for a hydrogen-like atom in a homogeneous magnetic field. Comput. Phys. Commun. **178**, 301–330 (2008)
10. Gusev, A., Vinitsky,S. Chuluunbaatar, O., Gerdt, V., Hai, L. L., Rostovtsev, V.: Symbolic-Numerical Calculations of High-$|m|$ Rydberg States and Decay Rates in Strong Magnetic Fields. V.P. Gerdt et al. (Eds.): CASC 2012, LNCS **7442**, 155-171, (2012)
11. Chuluunbaatar, O., Gusev, A.A., Vinitsky, S.I. and Abrashkevich, A.G.: KANTBP 2.0: New version of a program for computing energy levels, reaction matrix and radial wave functions in the coupled-channel hyperspherical adiabatic approach. Comput. Phys. Commun. **179**, 685–693 (2008)
12. Chuluunbaatar, O., Gusev, A.A., Vinitsky, S.I., Abrashkevich, A.G.: ODPEVP: A program for computing eigenvalues and eigenfunctions and their first derivatives with respect to the parameter of the parametric self-adjoined Sturm-Liouville problem. Comput. Phys. Commun. **181**, 1358–1375 (2009)
13. Derbov V.L., Serov V.V., Vinitsky S.I., Gusev A.A., Chuluunbaatar O., Kazaryan E.M., Sarkisyan A.A.: On Solving the Low-Dimensional Boundary Value Problems of Quantum Mechanics by Kantorovich Method - Reduction to Ordinary Differential Equations. Izvestiya of Saratov University. Physics: Izv. Saratov Univ. (N. S.), Ser. Physics, **10**, iss.1, 4–17 (2010), https://fizika.sgu.ru/en/node/239
14. Vinitsky,S., Gusev, A.A., Chuluunbaatar, O., Derbov,V.L., Zotkina, A.S.: On calculations of two-electron atoms in spheroidal coordinates mapping on hypersphere Proc. of SPIE **9917**, 99172Z (2016). https://doi.org/10.1117/12.2229528
15. Migdal, A.B.: Qualitative Methods in Quantum Theory (A.J. Leggett, Ed.) (1st ed.). CRC Press (1977)
16. Gusev, A.A., Chuluunbaatar, O., Vinitsky, S.I., Kazaryan, E.M., Sarkisyan, H.A.: The application of adiabatic method for the description of impurity states in quantum nanostructures. Journal of Physics: Conference Series **248**, 012047 (2010)
17. Gusev, A.A., Chuluunbaatar, O., Vinitsky, S.I., Derbov, V.L., Kazaryan, E.M., Kostanyan, A.A., Sarkisyan, H.A.: Adiabatic approach to the problem of a quantum well with a hydrogen - like impurity. Phys. Atom. Nucl. **73**, 331–338 (2010)
18. Gusev, A.A., Chuluunbaatar, O., Vinitsky, S.I., Dvoyan, K.G., Kazaryan, E.M., Sarkisyan, H.A., Derbov, V.L., Klombotskaya, A.S., Serov, V.V.: Adiabatic description of nonspherical quantum dot models. Phys. Atom. Nucl. **75**, 1210–1226 (2012)

19. Gusev, A.A., Chuluunbaatar, O., Hai, L,L., Vinitsky, S.I., Kazaryan, E.M., Sarkisyan, H.A., Derbov, V.L.: Spectral and optical characteristics of spheroidal quantum dots. Journal of Physics: Conference Series **393**, 012011 (2012)
20. Gusev, A.A., Hai, L.L., Vinitsky, S.I., Chuluunbaatar, O., Derbov, V.L., Klombotskaya, A.S., Dvoyan, K.G., Sarkisyan, H.A.: Analytical and numerical calculations of spectral and optical characteristics of spheroidal quantum dots. Phys. Atom. Nucl. **76**, 1033–1055 (2013)
21. Hayrapetyan, D.B., Chalyan, A.V., Kazaryan, E.M., Sarkisyan, H.A.: Direct interband light absorption in conical quantum dot. J. Nanomaterials **2015**, 915742 (2015)
22. Abramowitz, M., Stegun, I.A.: Handbook of Mathematical Functions. Dover, New York (1965)
23. Oguchi, T.: Eigenvalues of spheroidal wave functions and their branch points for complex values of propagation constant. Radio Sci. **5**(8–9), 1207–1214 (1970)
24. Skorokhodov, S.L., Khristoforov, D.V.: Calculation of the branch points of the eigenfunctions corresponding to wave spheroidal functions. Comput. Math. Math. Phys. **46**(7), 1132–1146 (2006)
25. Berry, M.V. and Wilkinson, M: Diabolical points in the spectra of triangles. Proc.R. Soc. Lond. A **392**, 15–43 (1984)
26. Pockels, F.: Über die Partielle Differential-Gleichung $\Delta u + k^2 u = 0$ und deren Auftreten in der Mathematischen Physik. B.G. Teubner, Leipzig (1891)
27. McCartin, B.J.: Laplacian Eigenstructure of the Equilateral Triangle. Hikari Ltd., Ruse, Bulgaria (2011)
28. Anselm, A.: Introduction to Semiconductor Theory. Mir, Moscow (1982).in Russian

Chapter 12
Electronic and Magnetic Properties of Laser Dressed Quantum Dot and Ring with Rashba Spin-Orbit Coupling

Vram Mughnetsyan, Aram Manaselyan, Manuk Barseghyan, Albert Kirakosyan, Laura M. Pérez, and David Laroze

Abstract The possibility of the control of energy spectrum and magnetization of circular quantum dots and quantum rings in a transverse magnetic field is considered. The dressing of the confining potential by an intense laser field as well as induction of Rashba spin-orbit interaction by an external electric field is shown to be effective tools for the manipulation of magnetic properties of quantum dots and rings. The temperature effect on magnetization is also considered. It is shown that the intense laser field creates an anisotropy in the system due to which the energy spectrum breaks to groups of states. The Rashba spin-orbit interaction, in its turn, creates several level anticrossings in energy spectrum. Both of these effects have their signatures on magnetization of quantum dots and rings.

12.1 Introduction

Spin-orbit coupling in semiconductor nanostructures is considered as a useful tool for spintronics and quantum information processing since it translates the spatial motion of an electron into rotation of its spin [1–3]. The Rashba spin-orbit interaction (RSOI) in semiconductor quantum dots (QDs) is of late receiving increasing attention because of its relevance to spin transport in low-dimensional electron channels [4]. A major goal of this type of work is to find ways to tune the RSOI field and thereby coherently manipulate electron spins in QDs [5]. Detailed theoretical studies of the

V. Mughnetsyan (✉) · A. Manaselyan · M. Barseghyan · A. Kirakosyan
Department of Solid State Physics, Yerevan State University, Alex Manoogian 1, 0025 Yerevan, Armenia
e-mail: vram@ysu.am

M. Barseghyan
National University of Architecture and Construction of Armenia, Teryan 105, 0009 Yerevan, Armenia

L. M. Pérez
Departamento de Física, FACI, Universidad de Tarapacá, Casilla 7D, Arica, Chile

D. Laroze
Instituto de Alta Investigación, CEDENNA, Universidad de Tarapaca, Casilla 7D, Arica, Chile

D. Blaschke et al. (eds.), *Optics and Its Applications*, Springer Proceedings in Physics 281, https://doi.org/10.1007/978-3-031-11287-4_12

influence of RSOI on the electronic properties of QDs with isotropic and anisotropic confinement potential have already been reported earlier [6–9]. The results have been shown, that the RSOI mainly manifests in multiple level crossings and level repulsions in the energy spectrum. On the other hand by considering the anisotropy of the confinement potential of QD, it has been shown that the Fock-Darwin spectra display a strong signature of RSOI.

Manipulation of electron spin via the RSOI in a quantum ring (QR) would be a promising avenue for quantum information processing [10, 11], as well as for spintronics in reduced dimensions. On the other hand some unique electronic and magnetic properties of QRs subjected to a RSOI effect have been demonstrated [12]. In particular it has been shown that for a nonzero SO coupling, the deviations from the linear behavior of the line-width versus temperature are directly proportional to the strength of the SO coupling.

It is worth to note, that the anisotropy is nearly always present during the fabrication process of QDs and QRs [10, 13]. Whereas, the studies of the influence of THz intense laser field (ILF) on electronic properties of QDs and QRs have shown that the ILF is the useful tool to control the shape anisotropy, and therefore, the physical properties of such systems [14–19].

In this work the possibility of the control of energy spectrum and magnetization of circular QD and QR in a transverse magnetic field are discussed. The dressing of the confining potential by an ILF as well as induction of RSOI by an external electric field is shown to be effective tools for the manipulation of magnetic properties of QDs and QRs. The manuscript is arranged as follows: in Sect. 12.2 the theoretical model is presented, in Sect. 12.3 the results and the corresponding discussion are given, and the conclusions are presented in Sect. 12.4.

12.2 Theoretical Model

In our model we consider the confinement of electron in QD of radius R and QR of inner and outer radii R_1 and R_2, with rectangular potential profile. Namely, for QD we take $V(r) = 0$ when $r \leq R$ and $V(r) = V_0$ when $r > R$ wearies for QR the potential is defined as follows: $V(r) = 0$ when $R_1 \leq r \leq R_2$ and $V(r) = V_0$ when $r < R_1$ or $r > R_2$. We considered here planar structures, in which the sizes in growth direction (z-axis) are much smaller then lateral dimensions ($L_z << R, R_1, R_2$). As a result we can consider the electron motion in the structures as two-dimensional.

The single electron Hamiltonian of the system under the influence of ILF and taking into account RSOI can be written as follows:

$$\left[\frac{1}{2m}\left(\widehat{\mathbf{p}} + \frac{e}{c}\mathbf{A}(t)\right)^2 + V(x, y) + \frac{\alpha}{\hbar}(\boldsymbol{\sigma} \times \mathbf{p})_z + E_Z\right]\Phi(x, y, t) = i\frac{\partial}{\partial t}\Phi(x, y, t), \tag{12.1}$$

where m is the electron effective mass, $\widehat{\mathbf{p}}$ is the two dimensional momentum operator, $\mathbf{A}(t) = \mathbf{A}_{ILF}(t) + \mathbf{A}_m$ is the combined vector potential of the fields, e is the elementary charge, α is the Rashba parameter, $\boldsymbol{\sigma}$ is the operator of spin and $E_Z = g_e\mu_B\sigma_z/2$ is the Zeeman term where g_e is the Lande factor, μ_B is the Bohr magneton.

We consider a linearly polarized laser field in the dipole approximation with the vector potential $\mathbf{A}_{ILF}(t) = A_0\mathbf{e}_x\cos(\omega t)$. The vector potential of the homogeneous magnetic field is $\mathbf{A}_m = (0, Bx, 0)$, where B is the magnetic field induction. Since $\mathbf{A}_{ILF}(t)\cdot\mathbf{A}_m = 0$ and the laser field is only time-dependent [20] the Kramers-Henneberger unitary transformation can be applied [21]. Following the Floquet theory [22–27], one can obtain the following time-independent Schrödinger equation in the high-frequency limit [8, 19]:

$$\left[\frac{1}{2m}\left(\widehat{\mathbf{p}}+\frac{e}{c}\mathbf{A}_m\right)^2 + V_d + \frac{\alpha}{\hbar}(\boldsymbol{\sigma}\times\mathbf{p})_z + E_Z\right]\Phi_d(x,y) = E_d\Phi_d(x,y)\,. \quad (12.2)$$

Here, $V_d = V_d(x, y, \alpha_0)$ shows the time-averaged laser-dressed potential, given by [20]

$$V_d(x, y, \alpha_0) = \frac{\omega_0}{2\pi}\int_0^{2\pi/\omega_0} V(x+\alpha_0 sin(\omega_0 t), y)dt\,, \quad (12.3)$$

where $\alpha_0 = \left(8\pi e^2 I/m^2\epsilon_h^{1/2}c\omega_0^4\right)^{1/2}$ is the so called ILF dressing parameter, I stands for the intensity and ω_0 for the frequency of the ILF and ϵ_h is the high-frequency dielectric constant. The parameter α_0 simultaneously depends on intensity and frequency of the ILF which can be chosen for a broad range in units of KW/cm^2 and terahertz correspondingly [26].

The laser dressed energy spectra and the wave functions of the electron in QD and QR can be calculated using exact diagonalization technique. The eigenfunctions are presented as a linear expansion of the eigenfunctions of the two-dimensional rectangular infinitely high potential well [14, 15, 17]. In our calculations we have used 361 basis states for QD and 625 basis states for QR which is adequate for determining the first few energy eigenvalues with high accuracy.

At zero temperature the magnetization of the QR is defined as $M = -\frac{\partial E_0}{\partial B}$, where E_0 is the ground state energy of the system [28, 29]. In this paper we have calculated the magnetic field dependence of M by evaluating the expectation values of the magnetization operator $\widehat{m} = -\frac{\partial\mathcal{H}}{\partial B}$, where $\mathcal{H}$ is the Hamiltonian of the system with laser dressed potential. We then need to evaluate the expectation values of magnetization operator $\widehat{m}$ using the laser dressed states of the system. We have also studied the temperature dependency of magnetization, following the thermodynamic model discussed in [29]. The temperature dependence of magnetization is evaluated from the following thermodynamic expression

$$M = -\sum_i \frac{\partial E_i}{\partial B} e^{-E_i/kT} \Big/ \sum_i e^{-E_i/kT}\,, \quad (12.4)$$

where the partial derivatives are evaluated as expectation values of operator $\widehat{m}$ for the state i.

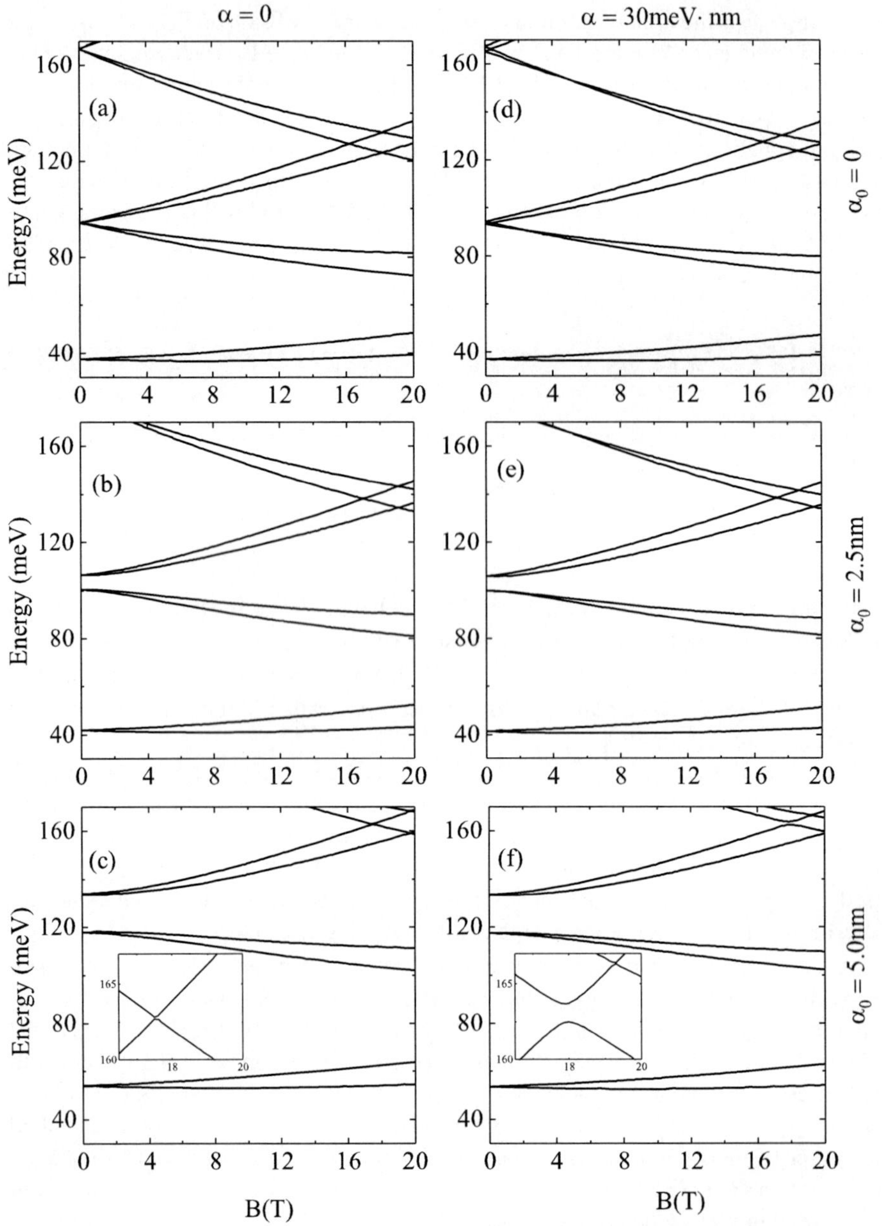

Fig. 12.1 Magnetic field dependence of low-lying energy levels of a QD without and with RSOI. The results are for different values of ILF parameter α_0. The insets demonstrate the shift of level crossing to anti-crossing due to SO coupling

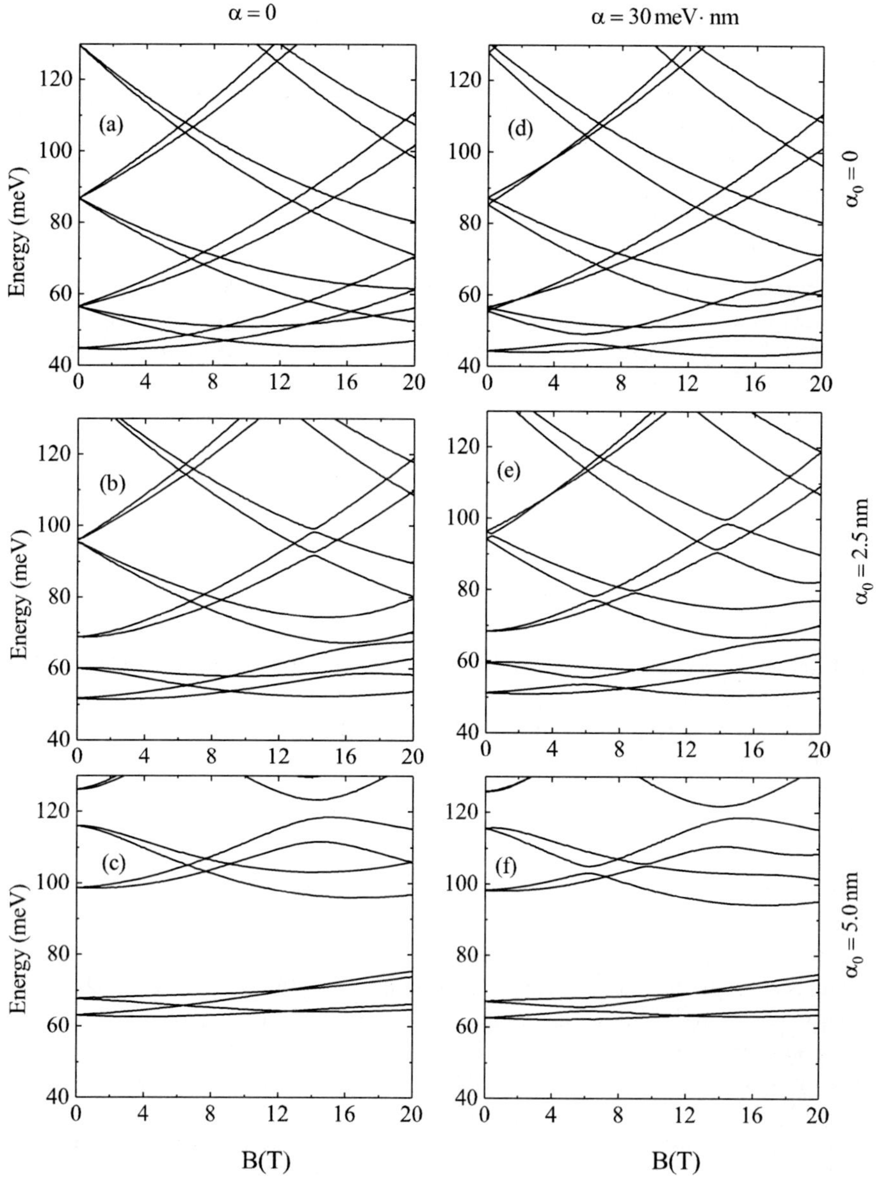

Fig. 12.2 Magnetic field dependence of low-lying energy levels of a QR without and with RSOI. The results are for different values of ILF parameter α_0

12.3 Numerical Results and Discussion

Numerical calculations are carried out for the InAs QD and QR inside undoped GaAs matrix with parameters $m = 0.042m_0$, $V_0 = 280\text{meV}$, $R = 10\text{nm}$, $R_1 = 5\text{nm}$, $R_2 = 15\text{nm}$ respectively.

In Fig. 12.1 the magnetic field dependencies of low-lying energy levels of a QD with and without RSOI are presented for various values of ILF parameter α_0. When the external ILF and the RSOI are absent, we obtain the well known energy spectrum of the cylindrical QD (Fig. 12.1a). In this case, the energy levels of the ground and excited states at zero magnetic field are degenerated. This degeneracy disappears in the case of the shape anisotropy of the system [30]. The ILF applied on a QD creates an anisotropy in the confinement potential (Fig. 12.1b and c) as a result of which the effective length of the confinement along the x direction decreases in the lower part of the QD potential well [18]. It is worth noting that with the increase in α_0, the anisotropy of the QD is strengthened and the degeneracy of the excited states at $B = 0$ partially disappears. As a result the energy spectrum transforms to doublets degenerated at $B = 0$. With the increase of α_0 the distances between neighboring doublets increase. The Rashba SO interaction in its turn couples angular momentum of the electron with spin due to which the degeneracy of energy levels at $B = 0$ disappears even without ILF (Fig. 12.1d). In Fig. 12.1e and f the combined effect of RSOI and ILF on energy spectra is presented. As it can be seen the effect of ILF is much stronger compared with SO interaction. Nevertheless several shifts of level crossing to anti-crossing can be observed due to SO coupling as it is demonstrated in insets of Fig. 12.1c and d.

In Fig. 12.2 the magnetic field dependencies of low-lying energy levels of a QR with and without RSOI are presented for various values of ILF parameter α_0. In this case when the external ILF and the RSOI are absent, we obtain the energy spectrum of the cylindrical QR with well known Aharonov-Bohm oscillations (Fig. 12.2a). Similar with the case of QD, the ILF applied on a QR creates an anisotropy in the confinement potential [17]. Here also with the increase of α_0, the anisotropy of the QR is strengthened due to which the degeneracy of the excited states at $B = 0$ partially disappears. With the increase of α_0 due to the reduced symmetry from C_∞ to C_2, the energy spectrum splits into non-crossing quartets of states which in turn cross repeatedly as B increases. A similar behavior of the energy levels, which can be called 'unusual' AB oscillations, was reported earlier in QRs by other authors that is caused by the effective mass anisotropy [31, 32] and structural distortions in QRs [33]. For $\alpha_0 = 2.5$ nm, only the first group of levels feel the deformation of the potential (Fig. 12.2b). Similarly, for larger values of α_0 more excited states start to feel the deformation of the QR confinement potential, and two periodically crossing groups of levels can be visible in energy spectra (Fig. 12.2c). For the case of QR the effect of RSOI again removes the degeneracy of energy levels at $B = 0$ and creates several level anti-crossings in energy spectra. It should be noted that the effect of SO coupling is stronger for QR compared with QD (Fig. 12.2d–f).

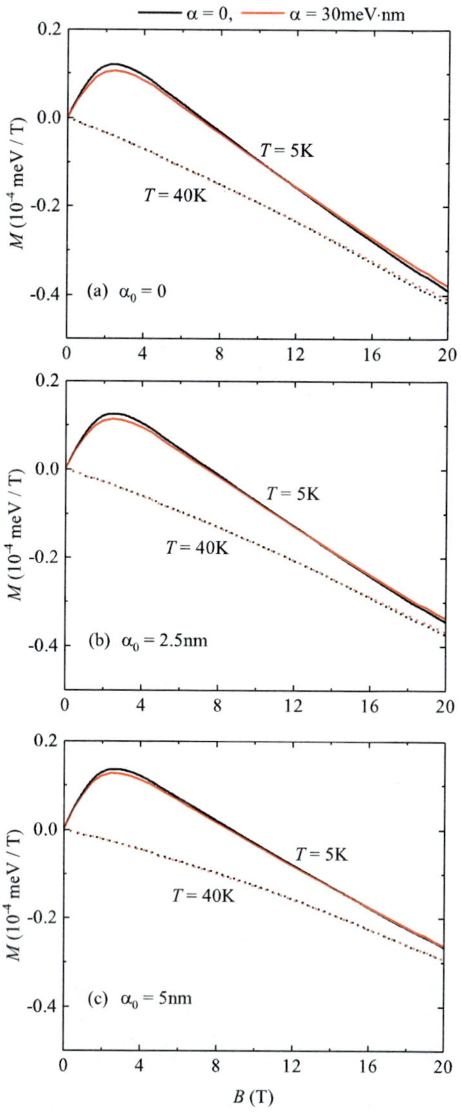

Fig. 12.3 The magnetization of the QD for various values of the RSOI and ILF parameters

In Fig. 12.3 the magnetization of the QD is presented for various values of the RSOI and ILF parameters and for two values of temperature. At low temperatures ($T = 5$K) with the increase of magnetic field the ground state magnetization at first increases and then starts to decrease for larger values of magnetic field. At $B = 0$ the magnetization is equal to zero due to spin degeneration of ground state. For non zero magnetic fields the ground state is with spin +1/2 due to Zeeman splitting, thus we observe a positive value of magnetization. With further increase of magnetic

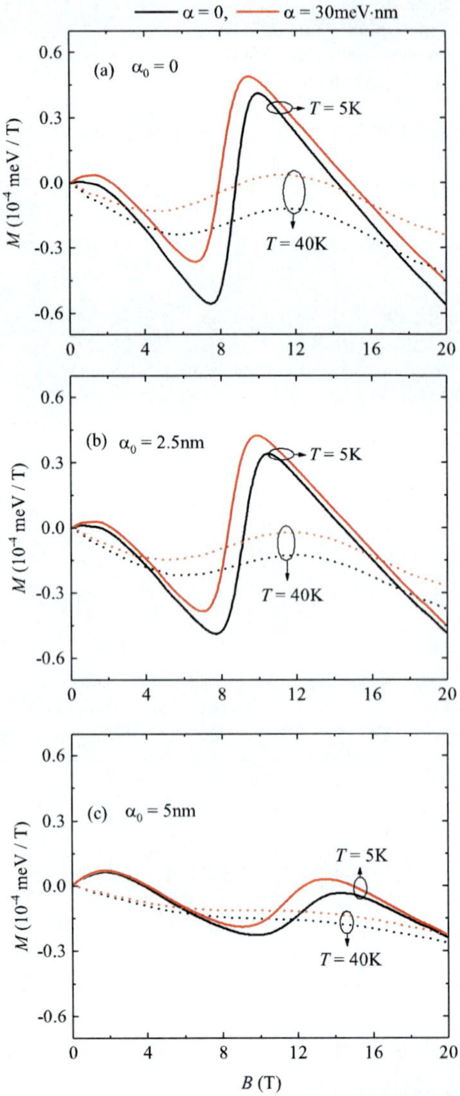

Fig. 12.4 The magnetization of the QR for various values of the RSOI and ILF parameters

field the magnetization decreases linearly due to the diamagnetic effect. With the increase of temperature the ground state mixes with nearest exited states due to which electron populations change. Therefore we have to calculate also the effect of excited states on the magnetization of the system using Eq. (12.4). At higher values of temperature ($T = 40$K) the thermodynamic weights of spin up and spin down states in magnetization are almost equal therefore the magnetization is a decreasing function of magnetic field. As can be seen from the figure, the effect of ILF on magnetization is considerable only for higher values of magnetic field induction and leads to the decrease of the magnetization. The RSOI has a small impact on the magnetization that becomes even smaller with the increase of ILF parameter.

In Fig. 12.4 the magnetization of the QR is presented for various values of the RSOI and ILF parameters and for two values of temperature. At low temperatures the magnetic field dependencies of magnetization for all cases have step like behaviour with several periodic jumps of magnetization (see solid curves on Fig. 12.4), which is a usual picture for semiconductor quantum rings and is the direct signature of Aharonov-Bohm oscillations in QR. In QRs, with increase of the magnetic field, the ground state changes periodically, and after each change of the ground state the angular momentum quantum number increases by one. This change of ground state angular momentum manifests itself in persistent currents and in magnetization of QR [34]. With the increase of temperature, due to the mixing of the ground state with nearest exited states, instead of sharp jumps smooth curves of magnetization can be observed (Fig. 12.4 doted lines). With the increase of laser parameter α_0, due to the broken symmetry of QR from C_∞ to C_2 the first group of states are energetically separated from others (Fig. 12.2c and f). As a result all effects observed in Fig. 12.4b and c weaken. The RSOI in its turn increases the value of magnetization of QR. It should be noted that for given parameters of the considered systems the Rashba effect on magnetization is more pronounced for the case of QR compared with QD. With the increase of ILF the effect of RSOI on magnetization weakens.

12.4 Conclusion

We have studied the combined effect of ILF and RSOI on electronic and magnetic properties of circular QDs and QRs subjected to transverse magnetic field. The temperature effect on magnetization of considered systems is also taken into account. We have demonstrated that the ILF creates an anisotropy due to which the energy spectrum breaks to groups of states for both systems. The RSOI, in its turn, creates several level anticrossings in energy spectrum. Both of these effects have their signatures in magnetization which allows effectively control the magnetic properties of QD and QR.

Acknowledgements The works of V.M., A.M. and M.B. are supported by Armenian State Committee of Science (grants No 21SCG–1C012, 21T-1C247, 21AG–1C008, 20TTWS–1C014). LMP acknowledges financial support from *Convocatoria Nacional Subvención a la Instalación en la Academia Convocatoria Año 2021* with PAI/ANID financing, Grant SA77210040. DL acknowledges financial support from Centers of Excellence with BASAL/ANID financing, Grant AFB180001, CEDENNA.

References

1. Nowack, K.C., Koppens, F.H.L., Nazarov, Yu.V., Vandersypen, L.M.K.: Science **318**, 1430 (2007)
2. Bednarek, S., Szafran, B.: Phys. Rev. Lett. **101**, 216805 (2008)
3. Manchon, A., Koo, H.C., Nitta, J., Frolov, S.M., Duine, R.A.: Nat. Mater. **14**, 871 (2015)
4. Awschalom, D.D., Samarth, N., Loss, D. (eds.): Semiconductor Spintronics and Quantum Computation. Springer, Heidelberg (2002)
5. Kuan, W.H., Tang, C.S., Xu, W.: J. Appl. Phys. **95**, 6368 (2004)
6. Chakraborty, T., Pietiläinen, P.: Phys. Rev. Lett. **95**, 136603 (2005)
7. Pietiläinen, P., Chakraborty, T.: Phys. Rev. B **73**, 155315 (2006)
8. Avetisyan, S., Pietiläinen, P., Chakraborty, T.: Phys. Rev. B **85**, 153301 (2012)
9. Avetisyan, S., Pietiläinen, P., Chakraborty, T.: Phys. Rev. B **88**, 205310 (2013)
10. Fomin, V.M.: Physics of Quantum Rings, 2nd edn. Springer International Publishing, Switzerland (2018)
11. Földi, P., Molnár, B., Benedict, M.G., Peeters, F.M.: Phys. Rev. B **71**, 033309 (2005)
12. Chen, H.Y., Pietiläinen, P., Chakraborty, T.: Phys. Rev. B **78**, 073407 (2008)
13. Sormunen, J., Riikonen, J., Mattila, M., Tiilikainen, J., Sopanen, M., Lipsanen, H.: Nano Lett. **5**, 1541 (2005)
14. Radu, A., Kirakosyan, A.A., Laroze, D., Baghramyan, H.M., Barseghyan, M.G.: J. Appl. Phys. **116**, 09310 (2014)
15. Radu, A., Kirakosyan, A.A., Laroze, D., Barseghyan, M.G.: Semicond. Sci. Technol. **30**, 045006 (2015)
16. Baghramyan, H.M., Barseghyan, M.G., Kirakosyan, A.A., Laroze, D.: Intense terahertz radiation effect on electronic and intraband optical properties of semiconductor quantum rings. In: Fomin, V.M. (Ed.), Physics of Quantum Rings, 2nd edn. Springer International Publishing, Switzerland (2018) (and references therein)
17. Chakraborty, T., Manaselyan, A., Barseghyan, M., Laroze, D.: Phys. Rev. B **97**, 041304(R) (2018)
18. Barseghyan, M.G., Manaselyan, A., Kirakosyan, A.A., Péerez, L.M., Laroze, D.: Physica E **117**, 113807 (2020)
19. Barseghyan, M.G., Baghramyan, H.M., Kirakosyan, A.A., Laroze, D.: Physica E **116**, 113758 (2020)
20. Gavrila, M.: Atoms in Intense Laser Fields. Academic, New York (1992)
21. Henneberger, W.C.: Phys. Rev. Lett. **21**, 838 (1968)
22. Gavrila, M., Kaminski, J.Z.: Phys. Rev. Lett. **52**, 613 (1984)
23. Pont, M., Walet, N.R., Gavrila, M., McCurdy, C.W.: Phys. Rev. Lett. **61**, 939 (1988)
24. Valadares, E.C.: Phys. Rev. B (R) **41**, 1282 (1990)
25. Gavrila, M.: J. Phys. B: At. Mol. Opt. Phys. **35**, R147 (2002)
26. Ganichev, S., Prettl, W.: Intense Terahertz Excitation of Semiconductors. Oxford University Press (2006)
27. Bukov, M., D'Alessio, L., Polkovnikov, A.: Adv. Phys. **64**, 139 (2015)
28. Avetisyan, S., Chakraborty, T., Pietiläinen, P.: Physica E **81**, 334 (2016)
29. Chakraborty, T., Manaselyan, A., Barseghyan, M.: J. Phys.: Condens. Matter **29**, 215301 (2017)
30. Madhav, A.V., Chakraborty, T.: Phys. Rev. B **49**, 8163 (1994)
31. de Sousa, G.O., da Costa, D.R., Chaves, A., Farais, G.A., Peeters, F.M.: Phys. Rev. B **95**, 205414 (2017)
32. Milošević, M.M., Tadić, M., Peeters, F.M.: Nanotechnology **19**, 455401 (2008)
33. Planeles, J., Rajadell, F., Climente, J.I.: Nanotechnology **1**(8), 375402 (2007)
34. Chakraborty, T.: Nanoscopic quantum rings: A new perspective. Adv. Solid State Phys. **43**, 79 (2003)

Chapter 13
Magnetically Induced Atomic Transitions of Alkali Metal Atoms

A. Sargsyan, A. Tonoyan, G. Hakhumyan, and D. Sarkisyan

Abstract Atomic transitions of alkali metals, which in the absence of an external magnetic field have a zero probability, meanwhile in the presence of it have high probabilities are called magnetically induced (MI). Interest in them is due to the high probabilities, which, in a wide range of magnetic fields, exceed the probabilities of ordinary atomic transitions. There are two types MI 1 and MI 2 which exhibit different behavior depending on the magnetic field. Sub-Doppler resolution with a simple single-beam geometry, providing a linear response of atomic media for transmission experiments can be attained using spectroscopic nanocells filled with the Rb and Cs atomic vapors with a thickness of the order of resonant wavelength or half of resonant wavelength. This makes it possible to spectrally resolve a large number of atomic transitions closely spaced in frequency and study individual behavior of MI transitions. Good agreement between the experiment and theoretical calculations is observed, and possible practical applications are noted.

13.1 Introduction

Atomic transitions in alkali metals that are excited by laser radiations which have circular (σ^+, σ^-) or linear (π) polarizations and have zero probability in the absence of a magnetic field but have large probabilities in the presence of a magnetic field are called magnetically induced (MI). MI transitions in atoms of Cs, Rb, K, Na represent a large class of about one hundred atomic transitions exhibiting interesting and important specific features [1–6]. Interest to MI atomic transitions is caused mainly by the fact that probabilities of such transitions in a wide range of magnetic field B can considerably exceed probabilities of ordinary atomic transitions allowed in the absence of magnetic field. In addition, frequency shifts of MI transitions in strong magnetic fields can reach ~30 GHz, which is of practical interest for expanding the frequency range for, e.g., laser-frequency stabilization at frequencies strongly shifted relative to the frequencies of transitions in unperturbed atoms [7].

A. Sargsyan (✉) · A. Tonoyan · G. Hakhumyan · D. Sarkisyan
Institute for Physical Research, 0203, Ashtarak-2, Ashtarak, Armenia
e-mail: sarkdav@gmail.com

D. Blaschke et al. (eds.), *Optics and Its Applications*, Springer Proceedings in Physics 281, https://doi.org/10.1007/978-3-031-11287-4_13

The development of new techniques for sub-Doppler spectroscopy of atomic vapors of alkali metal based on a nanometric thick cell (i.e. nanocell) containing atomic vapors and strong permanent magnets allows successful study of the behavior of hyperfine atomic transitions in the optical domain in a wide range of magnetic fields. A high spectral resolution allows numerous individual transitions between magnetic sublevels to be distinguished and identified, while inside a nanometric thick zone of the atomic vapor column there is high magnetic field uniformity when using strong permanent magnets, whose B-field has a strong gradient.

It is reasonable to divide all MI transitions into two types: type 1 (MI 1) and type 2 (MI 2). Using representation in the form $|F, m_F\rangle$, where F is the total atomic momentum and m_F is its projection, transitions $|F_g, 0\rangle \rightarrow |F_e = F_g, 0'\rangle$ between ground (F_g) and excited (F_e) levels the probabilities of which equal to zero in the absence of magnetic field but experience giant increase with increase in the applied magnetic field, asymptotically approaching a constant value with further increase in the B-field, belong to transitions of the first type (MI 1) (primes denote excited levels). In the present work MI 1 transitions that belong to the D_2 line of the Rb atoms were studied in detail. The total number of the MI 1 transitions in alkali metals in the region of the D_1 and D_2 lines is 28.

Transitions $|F_g, m_F\rangle \rightarrow |F_e, m'_F\rangle$ between ground (F_g) and excited (F_e) levels, where $F_e = F_g \pm 2$ and $m'_F - m_F = 0, \pm 1$, belong to the second type of transitions (MI 2). Probabilities of MI 2 transitions experience giant growth with increase in the B-field. However, this probability tends to zero again upon further increase in the magnetic field $B \gg B_0$, where B_0 is a characteristic magnetic induction $B_0 = A_{hfs}/\mu_B$, where A_{hfs} is the magnetic dipole constant of the ground state of an atom and μ_B is the Bohr magneton [8], was introduced in [9] for quantitative determination of the degree of atom's interaction with magnetic field. For Cs, ^{85}Rb, ^{87}Rb atoms $B_0 = 1.7$ kG, 0.7 kG and 2.4 kG respectively. The total number of the MI 2 transitions in alkali metals in the region of D_2 lines is about 70 transitions.

13.2 Experiment

13.2.1 Experimental Setup

The nanocell (NC) design, which consists of windows made of technical sapphire of 2.4 mm thickness and a vertical side arm -sapphire tube (a metal reservoir) is presented in [10, 11]. The NC is filled with a natural mixture of ^{85}Rb (72.2%) and ^{87}Rb (27.8%). The regions of the NC either $L \simeq \lambda$ or $L \simeq \lambda/2$ are used, where $\lambda = 780$ nm when Rb D_2 line is used, or $\lambda = 852$ nm when Cs D_2 line is used. The NC operated with a specially designed oven (made from non-magnetic materials) with

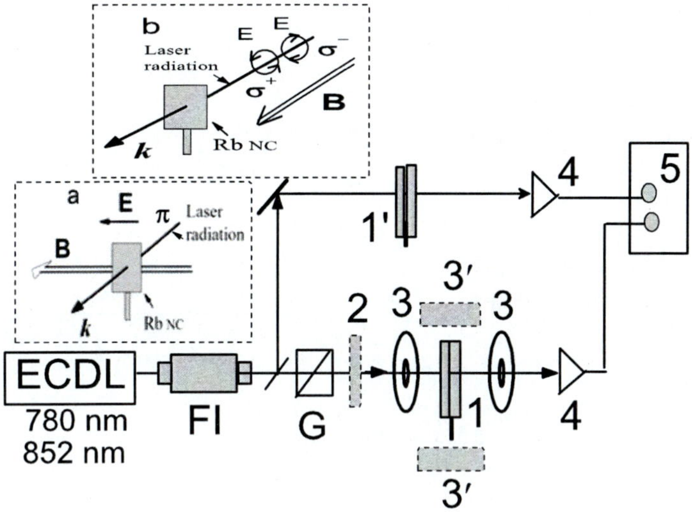

Fig. 13.1 Sketch of the experimental setup. ECDL—diode lasers, lasing wavelengths are either 780 nm or 852 nm, FI—Faraday isolator, 1-main Rb (Cs) nanocell, 1′—an auxiliary Rb (Cs) nanocell (reference), G—Glan polarizer, pair 3 and pair 3′ are—permanent magnets, 4—photodetectors, 5—digital storage oscilloscope, **E**—electrical field of laser radiation, **B**—magnetic field applied to the nanocell. Configurations of the **B**, **E** and **k** are presented in the insets, **a** for this case **B**‖**E**, **b** for this case **B**‖**k**, where **k** is wave vector, and quarter-wave plate 2 is used (it is absent for the case of **a**)

two ports for laser beam transmission. The temperature of the NC reservoir (which contains either Rb or Cs atoms) was $\sim 120\,°C$, but the windows were maintained at a temperature that was by 20 °C higher. At the first stage of the experiment the π-polarized beam of extended cavity diode laser (ECDL, $\lambda = 780$ nm, $P_L = 10$ mW, $\gamma_L < 1$ MHz) resonant with the ^{87}Rb, D_2 transition frequency, is focused (the laser spot diameter is ~0.7 mm) at nearly normal incidence onto the NC 1 with the vapor column thickness $L = \lambda = 780$ nm. To avoid feedback a Faraday insulator is applied. A Glan polarizer (G) is used to purify initial linear radiation polarization of the laser. A part of the laser radiation was directed to the auxiliary NC (1′) and the transmission spectrum of this NC is used as a frequency reference. The transmission signal was detected by a photodiode 4 and was recorded by Tektronix TDS 2014B four-channel storage oscilloscope 5. In order to produce strong magnetic fields two strong permanent magnets pair 3′ is used. Magnetic field B is directed along the laser radiation electric field direction E ($B \| E$). Configuration of the B and E is presented in the inset (a) in Fig. 13.1.

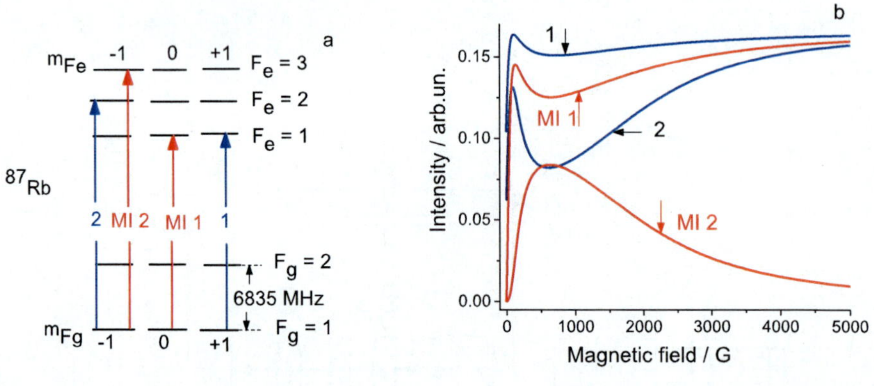

Fig. 13.2 **a** ^{87}Rb, D_2 line, diagram of the atomic transitions $F_g = 1 \rightarrow F_e = 1, 2, 3$ including MI 1 and MI 2 transitions for π-polarization excitation, selection rule is $m'_F - m_F = 0$, **b** the probabilities of the transitions as a function of the B-field

13.2.2 Experimental Results and Discussion

13.2.2.1 Magnetically Induced Transitions MI 1 and MI 2 of the ^{87}Rb Atom D_2 Line

Atomic transitions $F_g = 1 \rightarrow F_e = 0, 1, 2, 3$ between magnetic sub-levels of hyperfine states for the ^{87}Rb, D_2 line in the case of π—polarized laser radiation excitation are depicted in Fig. 13.2a. Note that when $B = 0$ according to the selection rules the $\Delta F = 2$ transitions $F_g = 1 \rightarrow F_e = 3$ (denoted MI 2) as well as atomic transition $F_g = 1, m_F = 0 \rightarrow F_e = 1, m'_F = 0$ (denoted MI 1) are forbidden, while all other presented transitions with $\Delta F = F_e - F_g = 0, \pm 1$ and $\Delta m_F = 0$ are allowed [12]. In Fig. 13.2b the probabilities of these four transitions are presented. As it is seen the probability of the MI 1 transition is zero at zero B-field, then it rapidly increasing with B increase and asymptotically approaching a constant value with further increase of the B-field. As to the probability of MI 2 transition it also zero at zero B-field, then it rapidly increasing with B increase, however the probability tends to zero again upon further increase of the magnetic field when $B \gg B_0$. As to the probabilities of ordinary transitions with numbers 1, 2 at $B = 0$ they are determined by Clebsch-Gordan coefficients, then they are increasing with B increase and asymptotically approaching a constant value when $B \gg B_0$ (so called Pashen-Back regime [9]).

As it was demonstrated earlier in the transmission spectra of the NC when thickness is $L \simeq \lambda$ so called narrow velocity selective optical pumping (VSOP) resonance are formed [2]. This allows one to obtain, identify, and investigate each individual atomic transition between the Zeeman sub-levels in the transmission spectrum of the ^{87}Rb D_2 line in a very wide range of magnetic fields from a few tens up to several kG. Thus, NC was used to implement the method, which ensured the narrowing of atomic transitions in the transmission spectrum $T(\nu)$. To further narrow atomic lines, we

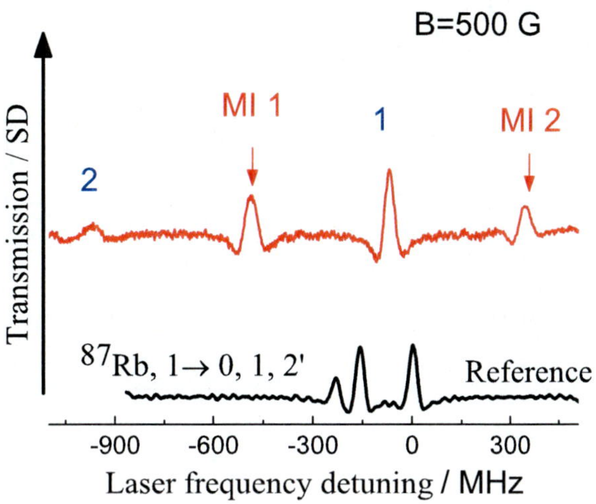

Fig. 13.3 ^{87}Rb, D_2 line, π-polarization, the upper curve is SD transmission spectrum of the nanocell with thickness $L = 780$ nm, atomic transitions $F_g = 1 \rightarrow F_e = 1, 2, 3$, the spectrum contains MI 1 and MI 2 transitions, $B = 500$ G, laser power 50 μW. The lower graph (black line) is the reference spectrum of the ^{87}Rb, $F_g = 1 \rightarrow F_e = 0, 1, 2$ at $B = 0$

performed double differentiation of the transmission spectrum $T''(\nu)$, which ensured the additional significant narrowing of atomic lines in the second derivative (SD) of the spectrum [13]. In Fig. 13.3 SD transmission spectrum when NC thickness is $L \simeq \lambda = 780$ nm, $B = 500$ G is presented by the upper curve. As it is seen the amplitudes of the four VSOP resonances is close to that predicted by the theory presented in Fig. 13.2b for $B = 500$ G.

It is important to note that in the case when laser radiation has circular polarization σ^+ or σ^- only MI 2 transitions can be detected (for this case the MI 1 transitions are absent). In Fig. 13.4a, b atomic transitions frequency shifts and the probabilities are shown for σ^+ circular polarization for $F_g = 1 \rightarrow F_e = 3$ which are indicated by 1–3 in circles (denoted MI 2) and ordinary transitions $F_g = 1 \rightarrow F_e = 2$ which

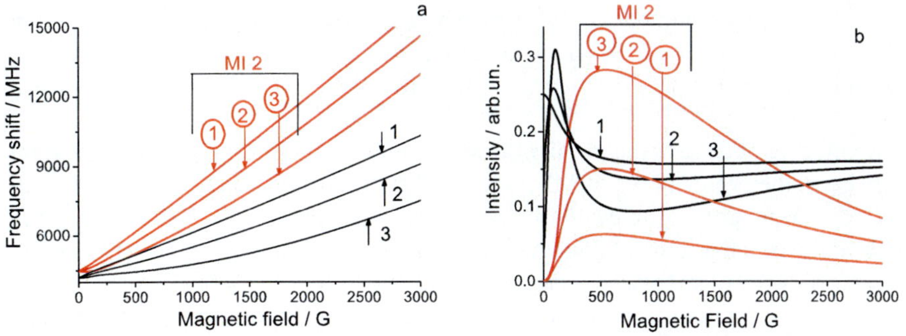

Fig. 13.4 **a** ^{87}Rb, D_2 line, frequency shift of $F_e = 1 \rightarrow F_e = 2, 3$ transitions, when using σ^+-polarized radiation as a function of magnetic induction B, transitions with numbers 1–3 (in circles) are MI 2 transitions, **b** probabilities of MI 2 transitions with numbers 1–3 (in circles) and $F_g = 1 \rightarrow F_e = 2$ transitions for σ^+ polarized radiation as a function of magnetic induction B

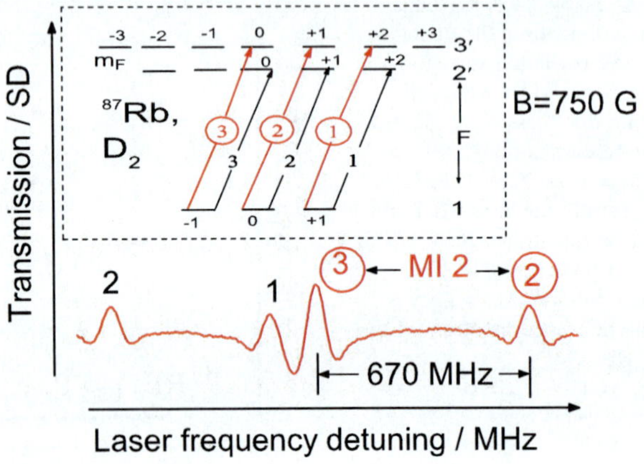

Fig. 13.5 ^{87}Rb, D_2 line, SD transmission spectrum of $F_g = 1 \rightarrow F_e = 2, 3$ transitions for σ^+ circular polarization when NC thickness is $L \simeq \lambda = 780$ nm, transitions labeled 2 and 3 in circles are MI 2 transitions. Diagram of atomic transitions $F_g = 1 \rightarrow F_e = 2, 3$ is presented in the inset

are indicated by 1, 2, 3. Diagram of the atomic transitions $F_g = 1 \rightarrow F_e = 2, 3$ is presented in the inset in Fig. 13.5. In Fig. 13.5 the SD transmission spectrum of $F_g = 1 \rightarrow F_e = 2, 3$ transitions when NC thickness is $L \simeq \lambda = 780$ nm is presented. For this case, as well as for below experiments $\mathbf{B} \| \mathbf{k}$, where $\mathbf{k}$ is wave vector, and quarter-wave plate $\lambda/4$ is used (see the inset (b) in Fig. 13.1.). It is important to note that in the wide range of B-field the probability of MI 2 transition with number 3 in circle is the largest for all transitions of the ^{87}Rb starting from $F_g = 1$.

Although the probability of MI 2 transition with number 3 in circle tends to zero when $B \gg B_0$, it is still detectable at $B = 8$ kG, while demonstrating a giant frequency shift of ~30 GHz from the initial frequency position at $B = 0$, which could be important for practical applications. Note, that MI 2 transitions indicated by 2, 3 in circles recently have been successfully used to obtain the electromagnetically induced transparency phenomenon in strong magnetic fields [14].

It is important to note that we have deduced a general rule applicable for the D_2 lines of all alkali atoms, that is a transition probability is larger for the case of σ^+ (as well as the number of MI 2 transitions is greater) than for σ^- excitation for $F_e - F_g = \Delta F = +2$, whereas the probability (as well as the number of MI 2 transitions is greater) is bigger in the case of σ^- than for σ^+ polarization for $\Delta F = -2$ [5].

Significant change in the probabilities of atomic transitions, in particular, a giant increase in the probabilities of MI 1 and MI 2 transitions, is caused by the effect of "mixing" of magnetic sublevels for ground F_g or excited levels F_e with magnetic sublevels of the nearest transition (the explanation is shown in Fig. 13.6); the "mixing" effect is induced by an external magnetic field [1–5]. A perturbation induced by the external magnetic field only couples the magnetic sublevels with $\Delta m_F = 0$ (indicated in Fig. 13.6 by circular arrows), which must also obey the follows selection rules: $\Delta L = 0$, $\Delta J = 0$, and $\Delta F = 0, \pm 1$ [1–3]. As it is seen in Fig. 13.6 the "mixing" is absent only for the transition $|2, +2\rangle \rightarrow |3', +3'\rangle$, which is called guiding (GT), and whose probability is independent of the B-field [15].

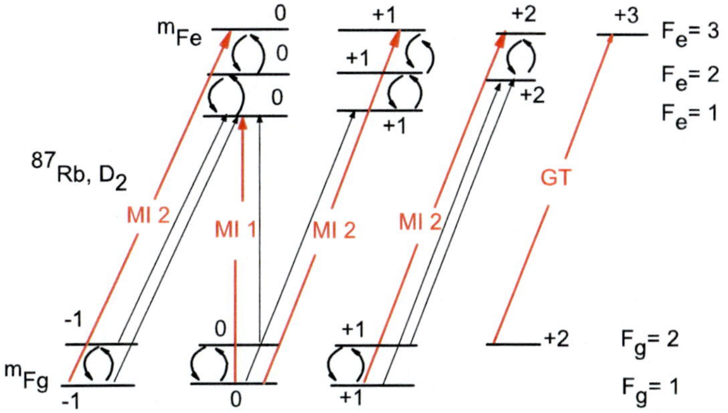

Fig. 13.6 The scheme is illustrating the "mixing" of Zeeman magnetic sublevels. Significant modification in the probabilities of atomic transitions, in particular, a giant increase in the probabilities of MI 1 and MI 2 transitions, is caused by the "mixing" effect of magnetic sublevels of ground F_g or excited levels F_e with magnetic sublevels of the nearest transitions. The "mixing" is absent only for transition called guiding (GT)

13.2.2.2 Magnetically Induced MI 2 Transitions of the ^{85}Rb Atom D_2 Line

Diagram of the transitions $F_g = 2 \rightarrow F_e = 3, 4$ of the ^{85}Rb, including magnetically induced five MI 2 transitions for laser circular polarization σ^+ is shown in the Inset of Fig. 13.7a. MI 2 transitions are indicated by 1–5 in circles. As it is seen at $B > 750$ G the group of MI 2 transitions are separated from the group of ordinary transitions 1–5. The probabilities of MI 2 transitions are presented in Fig. 13.7b. As it is seen in the range of B-field 100–1000 G the probability of the MI 2 indicated by 5 in circle is the biggest among all transitions $F_g = 2 \rightarrow F_e = 3, 4$, which is important for an applications.

In Fig. 13.8 the upper curve (1) shows the SD transmission spectrum when NC thickness is $L \simeq \lambda = 780$ nm, transitions are $F_g = 2 \rightarrow F_e = 3, 4$, including MI 2 transitions indicated by 4, 5 in circles, B-field=1000 G. There is a good agreement of the frequency positions of $F_g = 2 \rightarrow F_e = 3, 4$ transitions with the theory presented in Fig. 13.7a. The middle curve (2) in Fig. 13.8 shows the calculated transmission spectrum and there is a good agreement with the experiment. Note, that MI 2 transitions labeled 4, 5 in circles recently have been successfully used for the electromagnetically induced transparency phenomenon in strong magnetic field [16].

13.2.2.3 Magnetically Induced Transitions MI 2 of the Cs Atom D_2 Line

Diagram of the transitions $F_g = 3 \rightarrow F_e = 4, 5$ of the ^{133}Cs, including magnetically induced five MI 2 transitions for laser polarization σ^+ is shown in the inset of

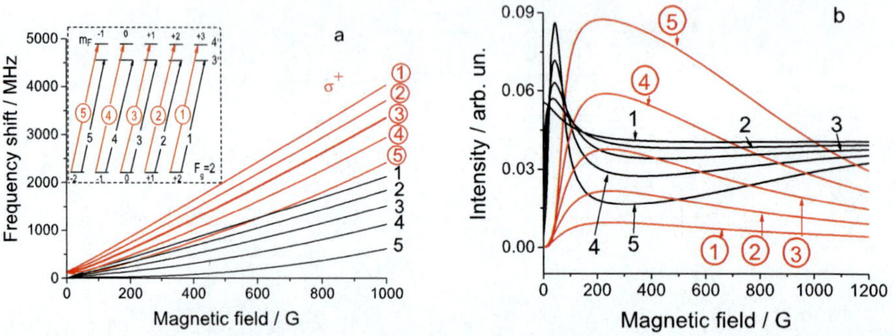

Fig. 13.7 **a** ^{85}Rb, the D_2 line, frequency shift of $F_g = 2 \rightarrow F_e = 3, 4$ transitions labeled 1–5 (in circles) are MI 2 transitions when using σ^+-polarized radiation as a function of magnetic induction B, **b** the probabilities of MI transitions with numbers 1–5 (in circles) and $F_g = 2 \rightarrow F_e = 3$ transitions for σ^+-polarized radiation as a function of B

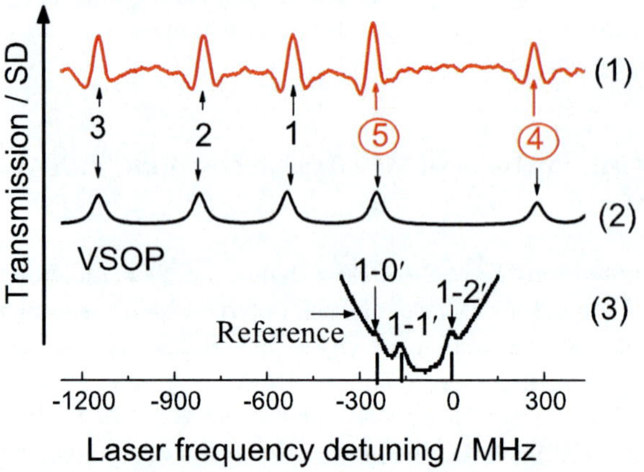

Fig. 13.8 ^{85}Rb, D_2 line the upper curve (1) is SD transmission spectrum when NC thickness is $L \simeq \lambda = 780$ nm, $B = 1$ kG, transitions are $F_g = 2 \rightarrow F_e = 3, 4$, including MI 2 transitions labeled 4, 5 in circles. The middle curve (2)-calculated transmission spectrum, a good agreement with the experiment is seen. The lower graph—reference spectrum of the ^{87}Rb, $F_g = 1 \rightarrow F_e = 0, 1, 2$ transitions, $B = 0$

Fig. 13.9b. The MI 2 transitions are indicated by 1–7 in circles. In the range of B-field 300–3200 G the probability of the MI 2 indicated by 7 in circle is the biggest among all transitions $F_g = 2 \rightarrow F_e = 3, 4$, which is important for applications. The probability of MI 2 transition indicated by 7 in circle tends to zero when $B \gg B_0$, however it is still possible to detect it at $B = 8$ kG, while there is a big frequency shift of ~25 GHz from the initial position at $B = 0$, which could be important for practical applications.

In Fig. 13.10 the calculated spectra for MI 2 transitions which are indicated by 1–7 in circles for polarization σ^+ (are shown in red on the high-frequency wing) and for

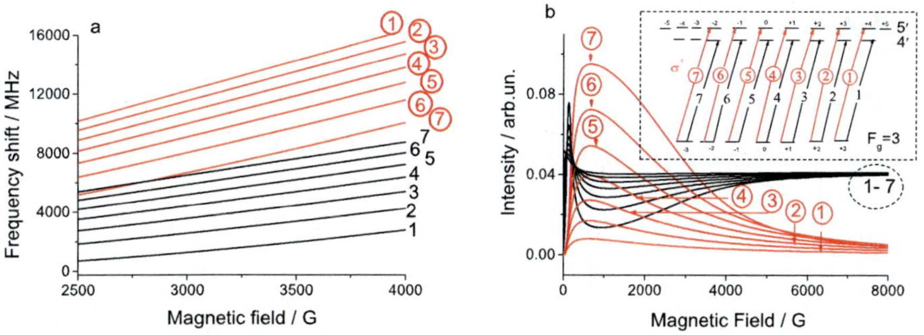

Fig. 13.9 **a** ^{133}Cs, D_2 line, frequency shift of $F_g = 3 \to F_e = 4, 5$ transitions labeled 1–5 (in circles) are MI 2 transitions, when using σ^+-polarized radiation as a function of magnetic induction B, **b** probabilities of MI 2 transitions with numbers 1–7 (in circles) and $F_g = 3 \to Fe = 4$ transitions for σ^+-polarized radiation as a function of B

Fig. 13.10 ^{133}Cs, D_2 line, calculated spectra for MI 2 transitions numbered 1–7 in circles for circular polarization σ^+ (in red on the high-frequency wing) and for circular polarization σ^- (in blue on the low-frequency wing) are presented when B-field is varying from 0.2 kG from the bottom up to 5 kG at the top, MI 2 transition labeled 7 in circle is the strongest transition in the group

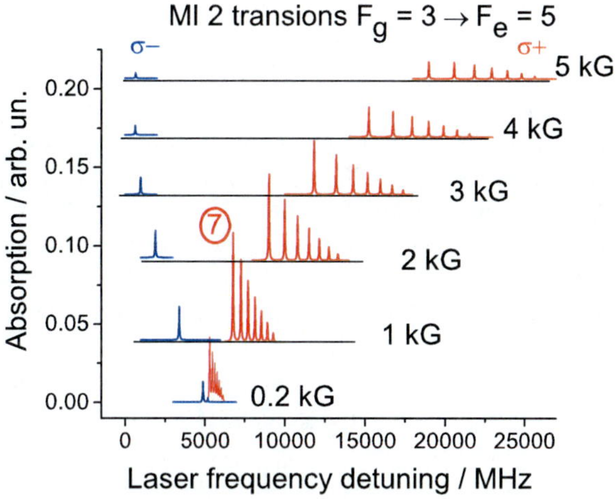

circular polarization σ^- (are shown in blue on the low-frequency wing) are presented when B-field is varying from 0.2 kG from the bottom to up to 5 kG at the top. For clarity, the ordinary transitions 1–7 are not shown. First, it can be seen that using σ^+ polarization there are seven transitions 1–7, while using σ^- polarization there is only one MI 2 transition. Secondly, it is seen the probability of transitions (i.e. the amplitudes) for σ^+ polarization is 2–3 times bigger than that for σ^- polarization. As it is mentioned above that for the case of σ^+ excitation for $F_e - F_g = \Delta F = +2$ the probability of the MI 2 transitions is bigger. The number of MI 2 transitions is also higher than in the case of excitation with σ^- polarization. Third, with an increase in the magnetic field, the MI amplitudes decrease, but their amplitudes become almost the same and MI 2 are located equidistantly in frequency. This behavior is typical for ordinary transitions in Pashen-Back regime, but the difference is that the probabilities

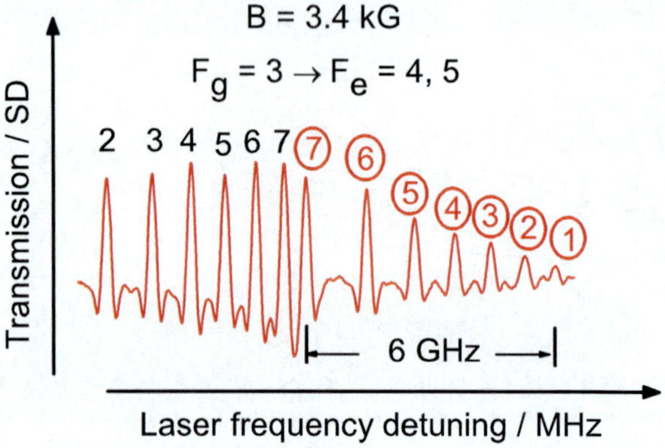

Fig. 13.11 ^{133}Cs, D_2 line, SD transmission spectrum of the NC with $L = \lambda/2 = 426$ nm, transitions $F_g = 3 \rightarrow F_e = 4, 5$, there are seven MI 2 transitions labeled 1–7 in circles and ordinary $F_g = 3 \rightarrow F_e = 4$ transitions labeled 2–7, $B = 3.4$ kG

of some ordinary transitions at $B \gg B_0$ tend to a fixed value, while the probabilities of the MI 2 transitions tends to zero.

As it is predicted theoretically in Fig. 13.9b at $B > 3000$ G the group of MI 2 transitions numbered 1–7 in circles is separated from the group of ordinary transitions numbered 1–7. A narrowband cw external-cavity diode laser with a wavelength of $\lambda = 852$ nm and of 1 MHz-linewidth was used to detect the transmission spectrum. The nanocell with the thickness $L = \lambda/2 = 426$ nm was placed between strong permanent magnets (with a small hole for the transmission of laser radiation) [11]. A quarter-wave plate $\lambda/4$ was used to form radiation with σ^+ circular polarizations Configuration of the B, E and k is presented in the inset (b) in Fig. 13.1, magnetic induction $B = 3.4$ kG, laser power 0.1 mW, and NC temperature is 120 °C. SD transmission spectrum of the NC containing Cs atomic vapor, transitions $F_g = 3 \rightarrow F_e = 4, 5$ is presented in Fig. 13.11. There are seven MI 2 transitions numbered 1–7 in circles and ordinary $F_g = 3 \rightarrow F_e = 4$ transitions numbered 2–7. As it is seen in Figs. 13.3, 13.5, 13.7 and 13.10a, a high spectral resolution with a simple single-beam geometry, providing a linear response of atomic media for transmission experiments can be attained using NC filled with an atomic vapor of alkali metal with a thickness of the order of resonant wavelength, or half of resonant wavelength. This makes it possible to spectrally resolve a large number of atomic transitions closely spaced in frequency. For Cs atoms the MI 1 and MI 2 transitions are considered in paper [17].

13.3 Conclusion

Atomic transitions in alkali metals are considered that have zero probability in the absence of a magnetic field, while have large probabilities in the presence of a magnetic field are called magnetically induced (MI). They are of interest because of their large probabilities, which exceed the probabilities of usual transitions in a wide magnetic field range. Magnetically induced transitions are classified as type-

1 (MI 1) and type-2 (MI 2) and the total number is about 100 transitions. Using nanocells filled with the Rb and Cs atomic vapors with a thickness of the order of resonant wavelength or half of resonant wavelength it makes possible to spectrally resolve a large number of atomic transitions closely spaced in frequency and study individual behavior of MI 1 and MI 2 transitions. Good agreement between the experiment and theoretical calculations is observed. The ^{87}Rb atom MI 2 transitions $F_g = 1 \rightarrow F_e = 3$, ^{85}Rb atom MI 2 transitions $F_g = 2 \rightarrow F_e = 4$ and Cs atom MI 2 transitions $F_g = 4 \rightarrow F_e = 2$ have been successfully implemented and described in the papers [14, 16] to obtain electromagnetically induced transparency phenomena in a strong magnetic fields.

Acknowledgements The work was supported by the Science Committee of RA, in the frames of the research project No 21T-1C005.

References

1. Tremblay, P., Michaud, A., Levesque, M., Thériault, S., Breton, M., Beaubien, J., Cyr, N.: Phys. Rev. A **42**(5), 2766 (1990)
2. Hakhumyan, G., Leroy, C., Mirzoyan, R., Pashayan-Leroy, Y., Sarkisyan, D.: Europhys. J. D **66**, 119 (2012)
3. Sargsyan, A., Tonoyan, A., Hakhumyan, G., Papoyan, A., Mariotti, E., Sarkisyan, D.: Laser Phys. Lett. **11**(5), 055701 (2014)
4. Sargsyan, A., Tonoyan, A., Hakhumyan, G., Sarkisyan, D.: JETP Lett. **106**(11), 700–705 (2017)
5. Tonoyan, A., Sargsyan, A., Klinger, E., Hakhumyan, G., Leroy, C., Auzinsh, M., Papoyan, A., Sarkisyan, D.: Europhys. Lett. **121**(5), 53001 (2018)
6. Sargsyan, A., Klinger, E., Leroy, C., Vartanyan, T.A., Sarkisyan, D.: Opt. Spectrosc. **127**(3), 411–417 (2019)
7. Sargsyan, A., Tonoyan, A., Mirzoyan, R., Sarkisyan, D., Wojciechowski, A.M., Stabrawa, A., Gawlik, W.: Opt. Lett. **39**(8), 2270–2273 (2014)
8. Zentile, M., Keaveney, J., Weller, L., Whiting, D.J., Adams, C.S., Hughes, I.G.: Comput. Phys. Commun. **189**, 162 (2015)
9. Olsen, B.A., Patton, B., Jau, Y.Y., Happer, W.: Phys. Rev. A **84**, 063410 (2011)
10. Sargsyan, A., Klinger, E., Amiryan, A., Tonoyan, A., Sarkisyan, D.: Phys. Lett. A **390**, 127114 (2021)
11. Sargsyan, A., Tonoyan, A., Sarkisyan, D.: JETP Lett. **113**(10), 605–610 (2021)
12. Demtröder, W.: Laser Spectroscopy: Basic Concepts and Instrumentation. Springer, Berlin (2004)
13. Sargsyan, A., Amiryan, A., Pashayan-Leroy, Y., Leroy, C., Papoyan, A., Sarkisyan, D.: Opt. Lett. **44**, 5533 (2019)
14. Sargsyan, A., Tonoyan, A., Papoyan, A., Sarkisyan, D.: Opt. Lett. **44**(6), 1391 (2019)
15. Sargsyan, A., Hakhumyan, G., Papoyan, A., Sarkisyan, D.: JETP Lett. **101**, 303 (2015)
16. Sargsyan, A., Tonoyan, A., Sarkisyan, D.: JETP **133**(1), 16–25 (2021)
17. Sargsyan, A., Tonoyan, A., Vartanyan, T.A., Sarkisyan, D.: Opt. Spectrosc. **128**, 1939–1947 (2020)

Chapter 14
Effect of Molecular and Electronic Geometries on the Electronic Density in FLO-SIC

Simon Liebing, Kai Trepte, and Sebastian Schwalbe

Abstract Recently, Trepte et al. [J. Chem. Phys., vol. 155, 2021] pointed out the importance of analyzing dipole moments in the Fermi-Löwdin orbital (FLO) self-interaction correction (SIC) for cyclic, planar molecules. In this manuscript, the effect of the molecular and electronic geometries on dipole moments and polarizabilities is discussed for non-cyclic molecules. Computed values are presented for water, formaldehyde, and nitromethane. Continuing the work of Schwalbe et al. [J. Chem. Phys. vol. 153, (2020)], we reconfirm that systematic numerical parameter studies are essential to obtain consistent results in density functional theory (DFT) and SIC. In agreement with Trepte et al. [J. Chem. Phys., vol. 155, 2021], DFT agrees well with experiment for dipole moments, while SIC slightly overestimates them. A Linnett double-quartet electronic geometry is found to be energetically preferred for nitromethane.

14.1 Introduction/Motivation

Electronic structure methods have become more important over recent years [1, 2]. These methods can be used to verify experimental observations [3–7]. However, the role of electronic structure methods has changed significantly over the years, as they allow to determine properties which are not easily accessible by experiments [8–11]. Screening for novel materials utilizing purely theoretical and/or computational frameworks saves time, work and money [12–16]. The leading methodology is Kohn-Sham (KS) density functional theory (DFT) [17], based on its suitable accuracy and reasonable numerical effort. Machine learning (ML) strategies are used to speed-

S. Liebing (✉)
JINR Dubna, Bogoliubov Laboratory of Theoretical Physics, 141980 Dubna, Russia
e-mail: science@liebing.cc

S. Liebing · S. Schwalbe
Institute of Theoretical Physics, TU Bergakademie Freiberg, 09599 Freiberg, Germany

K. Trepte
SUNCAT Center for Interface Science and Catalysis, Stanford University, Menlo Park, CA 94025, USA

D. Blaschke et al. (eds.), *Optics and Its Applications*, Springer Proceedings in Physics 281, https://doi.org/10.1007/978-3-031-11287-4_14

up DFT [18] even more or to find novel density functional approximations (DFAs) [19, 20]. The accuracy of novel DFAs [20–22] is getting closer to chemical accuracy.

Some remaining limitations of DFT can be attributed to the so-called self-interaction error (SIE), describing artificial interactions of electrons. The Perdew-Zunger self-interaction correction (PZ-SIC) [23] approximately removes the one-electron SIE. It has a long history of successes and failures [24]. In PZ-SIC, the choice of orbitals is important. Lehtola et al. [25] showed that PZ-SIC suffers from the local minima problem. A recent formulation of PZ-SIC utilizes so-called Fermi-Löwdin orbitals (FLO-SIC) [26–30]. FLO-SIC depends on Fermi-orbital descriptors (FODs) [29] to construct the localized orbitals used for PZ-SIC [31]. These FODs can be imagined as semi-classical electron positions which form an electronic geometry. Recently, Trepte et al. [31] showed that one can guide and classify local minima in PZ-SIC with the help of special sets of FODs that reflect chemical bonding theories of Lewis [32] and Linnett [33, 34]. The latter is known as Linnett's double-quartet (LDQ) theory. While typically one is interested in the *variational* total energy of the system, Trepte et al. [31] proposed to additionally monitor the dipole moment to classify PZ-SIC solutions. The dipole moment is one of the most simple descriptor for the electronic density—the key property in any DFA.

In this work we investigate the influence of molecular and electronic geometries as well as a properly chosen parameter space on the quality of density-related properties in DFT and FLO-SIC. We show that numerical parameters need to be optimized not only for the total energy but also for, e.g., the electric dipole moments and/or polarizabilities. We discuss the results based on small, illustrative and educationally-valuable molecules.

The manuscript is structured as follows. In the first two sections we outline the theoretical background and the computational details, after which we present the major results. In the last section we summarize and conclude our findings.

14.2 Theoretical Background

KS-DFT, see Fig. 14.1, is an approximation to solve the Schrödinger equation.

The total energy of a system is expressed as a functional of the electron density

$$E_{\mathrm{KS}}[n^{\alpha}, n^{\beta}] = T_{\mathrm{s}}[n^{\alpha}, n^{\beta}] + V[n] + J[n] + K_{\mathrm{XC}}[n^{\alpha}, n^{\beta}], \tag{14.1}$$

where $T_{\mathrm{s}}[n^{\alpha}, n^{\beta}]$ is the kinetic energy of the non-interacting system, $V[n]$ is the external potential energy, $J[n]$ is the Coulomb functional, $K_{\mathrm{XC}}[n^{\alpha}, n^{\beta}]$ is the exchange-correlation (XC) functional, n is the electron density, and α and β indicate spin channels.

To compute the XC functional

$$K_{\mathrm{XC}}[n^{\alpha}, n^{\beta}] = \int \varepsilon_{\mathrm{XC}}^{\mathrm{hom}}[n^{\alpha}, n^{\beta}] n(\boldsymbol{r}) F_{\mathrm{XC}} \, \mathrm{d}\boldsymbol{r}, \tag{14.2}$$

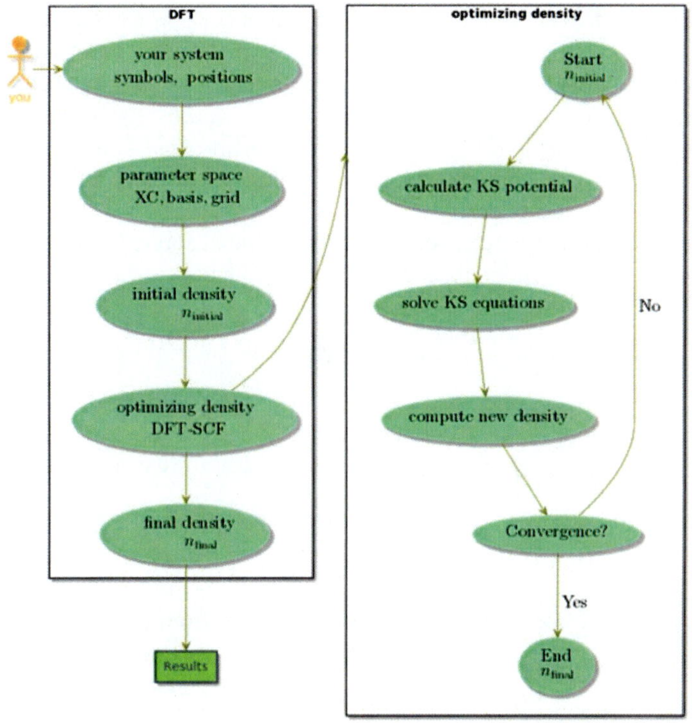

Fig. 14.1 Simplified overview of a DFT calculation

one needs to evaluate an explicit density integral using a numerical quadrature, see Sect. 14.3. Here, $\varepsilon_{\mathrm{XC}}^{\mathrm{hom}}[n^{\alpha}, n^{\beta}]$ is the XC energy-density of the homogeneous electron gas and F_{XC} is an XC enhancement factor.

Several approximations exist for the XC enhancement factor, many of which are available in LIBXC [35]. These approximations lead to artificial interactions of electrons with themselves; this is called self-interaction (SI). The corresponding SI energy comes from an incomplete cancellation of the exchange-correlation energy and the Coloumb energy for one-electron densities n_1^{σ}

$$E_{\mathrm{SI}}[n_1^{\sigma}] = K_{\mathrm{XC}}[n_1^{\sigma}, 0] + J[n_1^{\sigma}]. \tag{14.3}$$

In PZ-SIC, the total E_{KS} is corrected orbital-by-orbital as

$$E_{\mathrm{PZ}} = E_{\mathrm{KS}}[n^{\alpha}, n^{\beta}] + E_{\mathrm{SIC}} = E_{\mathrm{KS}}[n^{\alpha}, n^{\beta}] - \sum_{\sigma}\sum_{i=1}^{N^{\sigma}} E_{\mathrm{SI}}[n_i^{\sigma}]. \tag{14.4}$$

A novel flavor of PZ-SIC is FLO-SIC. This formulation utilizes FODs to construct Fermi orbitals (FO). These FOs are then orthogonalized to become FLOs. The FODs need to be optimized in the employed numerical parameter space using the respective analytical gradients [36]. With energy and gradient expressions at hand one can study, e.g., ionization potentials, atomization energies or barrier heights. However, guided by Trepte et al. [31] our focus is not only on energies, i.e., Eqs. (14.1) and (14.4), but on properties characterizing the density. Thus, having introduced energy expressions for DFT and PZ-SIC, we continue to discuss dipole moments and polarizabilities as fingerprints of the electron density.

Density-related properties can be analyzed using small applied electric fields. The total energy of a system under an external electrical field $\boldsymbol{\varepsilon}$ can be written as

$$E(\boldsymbol{\varepsilon}) = E_0 + \sum_i \mu_i \varepsilon_i + \sum_{ij} \alpha_{ij} \varepsilon_i \varepsilon_j + \mathcal{O}(\boldsymbol{\varepsilon}^3). \tag{14.5}$$

Here, E_0 refers to a ground state energy, e.g., KS-DFT E_{KS} (see Eq. (14.1)) or PZ-SIC E_{PZ} (see Eq. (14.4)). From this energy expression we can derive the electric dipole moment as

$$\mu_i = \left(\frac{\partial E(\boldsymbol{\varepsilon})}{\partial \varepsilon_i} \right)_{\boldsymbol{\varepsilon}=0}. \tag{14.6}$$

Commonly, the dipole moment is directly calculated from the electronic density

$$\boldsymbol{\mu} = \sum_A Z_A \boldsymbol{R}_A - \int d\boldsymbol{r}\, n(\boldsymbol{r})\, \boldsymbol{r}, \tag{14.7}$$

where Z_A, $\boldsymbol{R}_A$, and $n(\boldsymbol{r})$ are nuclear charges and positions and the total electron density, respectively. Note, we only discuss the electric dipole moment in this work and refer to it simply as *dipole moment*. The dipole moment is a measure for the polarity of a system and tells us about the charge separation in this system.

As an example, the molecular electrostatic potential [39] (MEP) and the dipole moment of the H_2O molecule is visualized in Fig. 14.2. The dipole points from the more electronegative O atom to the less electronegative H atoms. This can be directly seen in the coloring scheme of the MEP.

Having the possibility to calculate the total energy of the system under the influence of an external electric field allows to study dipole moments using, e.g., a 2-point finite difference (FD) stencil

$$\mu_{\mathrm{FD},i} = \frac{E(+\varepsilon_i) - E(-\varepsilon_i)}{2\varepsilon_i}. \tag{14.8}$$

One can also derive the electric polarizability α_{ij} from Eq. (14.5) as

$$\alpha_{ij} = \left(\frac{\partial E(\boldsymbol{\varepsilon})}{\partial \varepsilon_i \varepsilon_j} \right)_{\boldsymbol{\varepsilon}=0}. \tag{14.9}$$

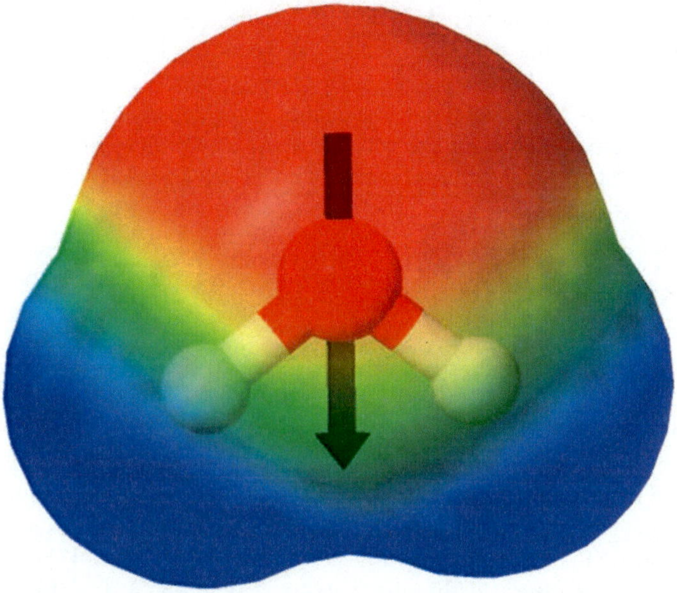

Fig. 14.2 Illustration of the dipole moment $\boldsymbol{\mu}$ in the H_2O molecule using the geometry [37] from CCCBDB [38]. The molecular electrostatic potential (MEP) [39] was calculated using DFT with LDA-PW, aug-pc-3, and a grid=(200,1454) in PySCF. The visualization was done using *Jmol* [40]

The electric polarizability can be calculated using analytical approaches, e.g., solving the coupled perturbed Hartree-Fock (CPHF) equation [41–43]. It describes the tendency of a system to acquire an induced dipole moment in the presence of an external electric field. Similar to the dipole moment we will refer to the electric polarizability as *polarizability*.

With the dipole moment at hand, i.e., Eqs. (14.7) or (14.8), one can calculate the directional components of the polarizability tensor as vector components

$$[\alpha_{\mathrm{FD},ix}, \alpha_{\mathrm{FD},iy}, \alpha_{\mathrm{FD},iz}] = \frac{\boldsymbol{\mu}(+\varepsilon_i) - \boldsymbol{\mu}(-\varepsilon_i)}{2\varepsilon_i}, \tag{14.10}$$

which is a row in Eq. (14.11).

To further simplify the characterization of the density, we introduce scalar values. The vectorial dipole moment $\boldsymbol{\mu} = (\mu_x, \mu_y, \mu_z)$ will be represented as $\mu = |\boldsymbol{\mu}|$, while the tensorial polarizability

$$\alpha_{ij} = \begin{bmatrix} \alpha_{xx} & \alpha_{xy} & \alpha_{xz} \\ \alpha_{yx} & \alpha_{yy} & \alpha_{yz} \\ \alpha_{zx} & \alpha_{zy} & \alpha_{zz} \end{bmatrix} \tag{14.11}$$

will be represented as $\alpha = \mathrm{Tr}(\alpha_{ij})/3$. As shown by Trepte et al. [31], dipole moments are sensitive to the chemical bonding described by the electronic geometry. Moving

FODs changes the dipole moment in FLO-SIC. Dipole moments in FLO-SIC provide insights into whether the electronic density respects the molecular symmetry or not [31].

In practical FLO-SIC calculations, the quality of the density is determined by the used basis set, numerical quadrature, molecular and electronic geometry. Computational details and the used electronic structure codes are discussed next.

14.3 Computational Details

All scripts to produce the data presented in the manuscript are available at https://gitlab.com/opensic/dippo [44]. The calculations were performed with the all-electron Gaussian-type orbital (GTO) codes PySCF [42] and PyFLOSIC2. PyFLOSIC2, see https://gitlab.com/opensic/pyflosic2, is the successor of PyFLOSIC [30]. It offers a cleaner and more modular code structure and can now easily be installed via the Python package manager *pip*. For calculations of real- and complex-valued SIC, i.e., RSIC and CSIC [25, 45, 46], we used the ERKALE code [47]. In previous studies [30, 31] we observed that the pc-n basis sets [48, 49] perform well for DFT as well as SIC calculations. Therefore, for all calculations in this work we use pc-n basis set variants. All codes use the Libxc [35] library, offering access to a vast variety of exchange-correlation functionals. From this library we access LDA-PW [50], PBEsol [51], and r^2SCAN [22]. The used codes are Open-Source codes, meaning they are freely available to anyone [52, 53]. Open-Source codes enable faster code development, re-usable concepts, and versatile tool-boxes.

PySCF and PyFLOSIC2 are written in Python, where only numerically demanding parts in PySCF are written in C. Python is simple and elegant, has a friendly and helpful community, and provides various well-maintained libraries. These are only some reasons why it is easy for students or non-programmers to start coding with Python. This allows to solve even non-trivial tasks, like writing a DFT code from scratch [54] in the limited time of a master thesis when guided and educated with novel strategies [55].

A numerical quadrature [56] is needed to evaluate XC properties in DFT and SIC, see Eq. (14.2) in Sect. 14.2. We will refer to it simply as *grid*. A typical grid consists of a radial and an angular part. Its size is given as a pair of numbers, i.e., the number of radial shells and the number of angular points. SIC requires significantly finer grids than DFT [45, 46, 57]. In analogy to [30, 31], we prune the used grids neither for DFT nor for SIC. This is done as the orbital densities evaluated in SIC are not as smooth as the total density used in DFT [58], thus requiring a finer resolution [46].

FLO-SIC has two major variational degrees of freedom, the density matrix (DM) and the FODs. All FLO-SIC calculations in this work are realized with a two-step FLO-SIC SCF cycle, which follows the idea proposed by [59]. In FLO-SIC, the initial DM and initial molecular coefficients are typically the ones from a DFT calculation. The initial FODs can be generated with various procedures, e.g., Python-based center of mass PyCOM, Fermi-orbital descriptor Monte-Carlo fodMC, or other so-called

FOD generators [29]. All initial FOD configurations used in this work were generated using the FODMC.

Within a full FLO-SIC calculation, the FODs are fully optimized in an inner FOD loop for a given DM. All SIC properties are calculated for this DM and the respective optimized FODs. Based on this, the unified Hamiltonian [30, 60–62] is updated which then provides the next DM in the outer DM loop. The initial FODs in the inner FOD loop are optimized using the *SciPy* L-BFGS-B [63–70] optimization algorithm with a maximum force component threshold of $f_{\mathrm{max,tol}} = 2 \cdot 10^{-4}\ E_{\mathrm{h}}/a_0$. This two-step procedure is repeated until the FOD forces reach $f_{\mathrm{max,tol}}$ and the DM is not changing anymore.

The computational methods introduced in this section have been applied to calculate dipole moments and polarizabilities. The results for water, formaldehyde, and nitromethane are discussed in the next section.

14.4 Results

14.4.1 Sisyphus Rock: The Importance of Grid and Basis Set Size

For the correct description of density-related properties, it is important to converge the used numerical parameters space consisting of grid and basis set. Systematic parameters studies have been performed utilizing CCCBDB molecular geometries [38], and FODs generated with the FODMC in the case of FLO-SIC to exemplify this. The determined trends and optimal values should be transferable to other molecular geometries and other FOD arrangements.

We carried out systematic grid convergence tests varying the number of radial shells with a fixed number of angular points and vice versa. Detailed information can be found at https://gitlab.com/opensic/dippo [44]. The DFT values converge at smaller grids than the respective SIC values, see Table 14.1. For LDA-PW, a value of $N_{\mathrm{rad}} = 200$ for the radial shells gives converged results for both DFT and SIC. The angular dependency for DFT as well as for FLO-SIC is converged at $N_{\mathrm{ang}} = 590$. However, we use a value of $N_{\mathrm{ang}} = 1454$ to resolve all one-electron and total densities accurately.

Note that we investigated the convergence of the density in DFT using the dipole moment as well as the polarizability. Both density fingerprints deliver the same optimal parameters for the tested molecules. Thus, for FLO-SIC we only used the dipole moment to determine optimal parameters.

Having an optimal grid of (200,1454) enabled us to determine a suitable basis set. The convergence of the basis set for water is shown in Fig. 14.3 using DFT and in Fig. 14.4 using FLO-SIC. The aug-pc-3 basis set shows convergence w.r.t. the basis set size. Those parameters are optimal for water, formaldehyde, and nitromethane. However, such convergence checks need to be done for any molecule—a Sisyphus

Table 14.1 Optimal numerical parameters for DFT and FLO-SIC regarding the molecular geometries reported in CCCBDB [38] for H_2O [37], CH_2O [71] and CH_3NO_2 [72]

System	DFT			FLO-SIC		
	n_{rad}	n_{ang}	basis set	n_{rad}	n_{ang}	basis set
H_2O	100	302	aug-pc-3	150	590	aug-pc-3
CH_2O	100	302	aug-pc-3	200	590	aug-pc-3
CH_3NO_2	150	590	aug-pc-3	200	590	aug-pc-3

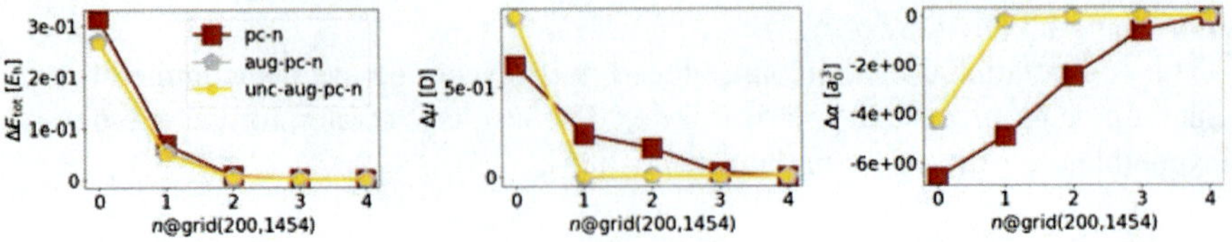

Fig. 14.3 H_2O (DFT, LDA-PW): Convergence of the total energy E_{tot}, the dipole moment μ and the polarizability α w.r.t. increasing basis set size for LDA-PW DFT using PySCF. We used pc-n, aug-pc-n, and unc-aug-pc-n with n=0-4 [48, 49]. Each plot shows the difference to the largest used basis set

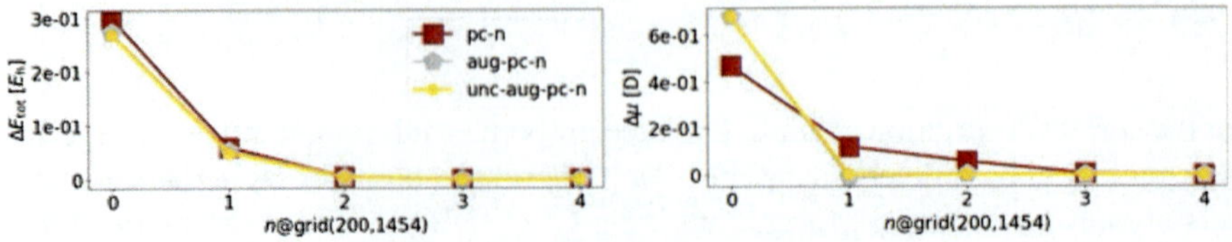

Fig. 14.4 H_2O (FLO-SIC, LDA-PW): Convergence of the total energy E_{tot} and the dipole moment μ w.r.t. increasing basis set size for LDA-PW FLO-SIC using PyFLOSIC2. We used pc-n, aug-pc-n, and unc-aug-pc-n with n=0-4 [48, 49]. Each plot shows the difference to the largest used basis set. Only the density matrix was optimized, while the FODs were fixed

work. Otherwise, the meaning of absolute values for dipole moments or polarizabilities are questionable.

Having established a suitable numerical parameter space, we continue to discuss the influence of molecular geometries on dipole moments and polarizabilities.

14.4.2 Pandora's Box: The Quality of Molecular Geometries Matters

A molecular geometry is needed to perform electronic structure theory calculations; in case of DFT see Fig. 14.1. While such geometries can be optimized within most theories, it is not uncommon to use a fixed molecular geometry to be comparable to other approximations or to simply save computational time.

Table 14.2 Molecular information for H_2O, CH_2O, and CH_3NO_2 with number of atoms, electrons, α and β electrons N_{nuc}, N_{elec}, N_α and N_β. Experimental references for the dipole moments [79] and polarizabilities [38] are provided

System	N_{nuc}	N_{elec}	N_α	N_β	μ_{REF} [D]	α_{REF} [a_0^3]
Water (H_2O)	3	10	5	5	1.85	10.13
Formaldehyde (CH_2O)	4	16	8	8	2.33	18.69
Nitromethane (CH_3NO_2)	7	32	16	16	3.46	32.39

For quantum chemical calculations there exist a vast a variety of seemingly promising databases, such as CCCBDB [38], ChemSpider [73, 74], PubChem [75], and many more. However, the quality of the geometries in these databases can vary [76]. CCCBDB offers access to a vast variety of molecular geometries. In this work, we used the CCCBDB geometries which can be found in the experimental section. Note that those geometries are not necessarily experimental ones. For example, for water [37] the geometry is derived semi-empirically utilizing experimental reference values. PubChem provides molecular geometries calculated using the MMFF94s [77] force field. For other databases such as ChemSpider it is even not that trivial to find the quality of the geometries.

The question of the quality of the molecular geometry might be important for other fields as well. For example, machine learning [78] models might be trained on low quality geometries, which could affect the predictability of the resulting models. For SIC calculations, the quality of the molecular geometry is of great importance, as the orbital densities are sensitive to the underlying molecular geometry. Molecular geometry optimizations are a standard task for commonly used approaches like DFT. However, for more computational demanding methods like FLO-SIC, full geometry optimization require high computational effort. This is caused by the coupled degrees of freedoms of nuclei and FODs [31].

The following results are based on the small, educational systems H_2O, CH_2O, and CH_3NO_2. We summarized some essential molecular information for those molecules in Table 14.2, including experimental reference values for dipole moments and polarizabilities.

The effect of molecular geometries on *density* fingerprints, i.e., dipole moments and polarizabilities, is significant even at the DFT level (see Fig. 14.5). For the used test systems the molecular geometries from CCCBDB provide the best molecular geometries; the dipole moments and polarizabilities are close to the values obtain from a DFT optimized geometry. The basis set size affects the density-related properties significantly, and only the aug-pc-3 basis set provides converged results. The molecular geometries from PubChem and ChemSpider should be used with care, as the dipole moments deviate significantly from values obtained from a DFT optimized geometry.

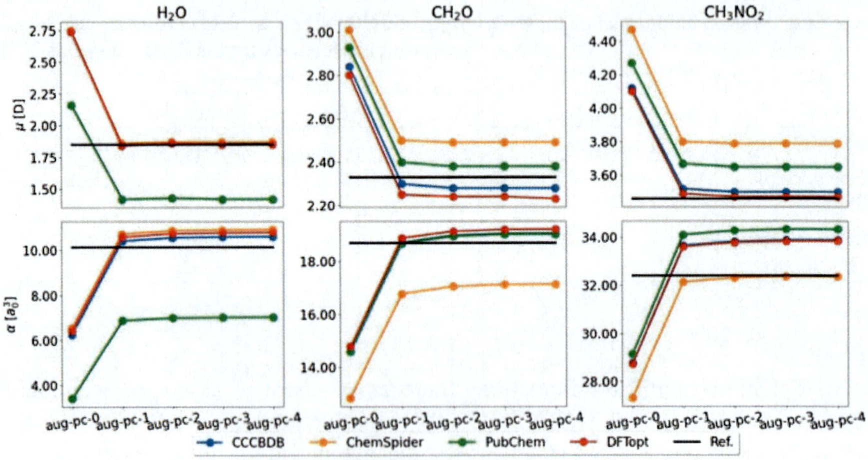

Fig. 14.5 Influence of molecular geometries on density fingerprints, e.g., dipole moments and polarizabilities. For these DFT calculations, LDA-PW was used with a grid=(200,1454) in PySCF. Molecular geometries are taken from common chemical databases. In addition, DFT optimized geometries were used. These DFT geometry optimizations were started from the CCCBDB geometries utilizing ERKALE with aug-pc-3 and grid=(200,1454). The used reference values are provided in Table 14.2

Table 14.3 Finite-difference (FD) errors of polarizabilities in a_0^3 for the chosen step size of $\varepsilon = 10^{-7}$ a.u. utilizing the 2-point FD approximation and LDA-PW DFT. The analytical α as calculated with PySCF were used as reference

Database	aug-pc-0	aug-pc-1	aug-pc-2	aug-pc-3	aug-pc-4
CCCBDB	0.00	0.29	−0.01	−0.01	−0.01
ChemSpider	0.00	−0.08	0.00	−0.02	−0.00
PubChem	−0.01	0.02	0.02	−0.01	0.33
DFTopt	0.00	0.00	−0.03	−0.01	0.01

For all calculations, the finite difference approximations, see Eqs. (14.8) and (14.10), agree well with the analytical results. The mean error for the dipole moments is 0.00 D, while the respective mean errors for the polarizabilities are given in Table 14.3.

Accordingly, for the investigated systems and the employed method the chosen value of 10^{-7} a.u. for the magnitude of $\boldsymbol{\varepsilon}$ regarding the 2-point finite difference approximation delivers reliable numerical results. Note that this finding might not be reproducible for other systems using the same value. Having examined the dependence on the molecular geometry, in Sect. 14.4.3 we proceed to investigate electronic degrees of freedom in FLO-SIC, i.e., DM and FODs.

14.4.3 The Sword of Damocles: Curse and Blessing of Approximations

Approximations are often needed and can be useful to enable the treatment or computation of a specific property at a certain level of theory. However, each approximation needs to be carefully investigated regarding the limits of its predictive power.

Effect of initial FODs Continuing the work of Trepte et al. [31], we show for CH_3NO_2 that it is possible to find several FOD configurations following chemical bonding theories. One possible FOD configuration can be based on Lewis theory of bonding. There, the FODs form one double N=O and one single N-O bond in the -NO_2 group, see Fig. 14.6. Clearly, there exist two identical Lewis configuration where the double and single bonds are exchanged with each other. These Lewis configuration have N-O bond orders of 2 and 1, respectively. Regarding LDQ theory, 2 FODs of one spin channel and 1 FOD of the other spin channel are placed between the N and the O. This leads to bond orders of 1.5 in both N-O bonds. Besides FOD configuration which follow chemical bond theories, other FOD configuration are possible. For example, one can generate a configuration with an over-binding N atom, placing two N=O double bonds in the molecule. We denote this FOD configuration as *other*. Changing the N-O bond order affects the local chemical environment, and with that the resulting SIC solution.

In electronic structure theories there exist several possibilities to treat the spin of the system. In FLO-SIC, we can do restricted calculations where all electrons are paired, $N_\alpha = N_\beta$, and unrestricted calculations where N_α and N_β can vary. In restricted FLO-SIC only one set of FODs is needed, while in unrestricted FLO-SIC two sets of FODs are required.

Given that the calculations for CH_3NO_2 are more computational demanding, we use a grid=(200,1454) and the aug-pc-2 basis set. As seen in Sect. 14.4.1, using this basis set comes with errors in the order of mE_h and mD w.r.t. to the basis set limit. However, the energy difference of the considered FOD configurations, see Fig. 14.6, are in the order of 10^{-2} E_h. The differences in the dipole are in the order of 10^{-1} D. Thus, using the aug-pc-2 basis set should deliver reliable trends.

We performed restricted FLO-SIC calculations for the Lewis configuration of CH_3NO_2. The FODs converge to an electronic geometry which does not follow any bonding theory. The double bond FODs are not lying on the N-O bond axis. Instead, they are distorted towards the respective O atoms. Given this non-symmetric arrangement of FODs, the density of CH_3NO_2 becomes non-symmetric. This leads to an energy of $E_\text{final} = -245.80939$ E_h and a non-symmetric dipole moment with an absolute value of $\mu = 4.06$ D.

The effect of various unrestricted FOD configurations for CH_3NO_2 is shown in Fig. 14.6. The differences in these configurations can already be seen for the initial FODs. Only optimizing the density shows significantly different dipole moments, and only the LDQ value is close to the experimental value. The energy for the initial LDQ arrangement is also the lowest. Upon full optimization of the density

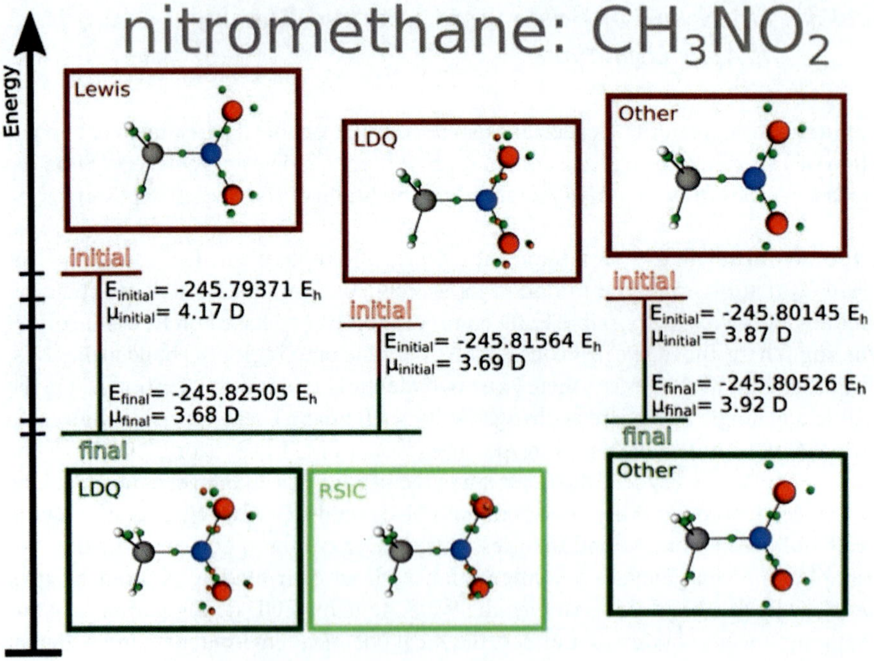

Fig. 14.6 Total energies and dipole moments for various unrestricted FOD configurations for CH_3NO_2, evaluated at the initial and the final FODs. *Other* represents a structure with two N=O double bonds. Any additionally tested structure converged into the LDQ solution. Calculations were performed using PyFLOSIC2 employing LDA-PW, the aug-pc-2 basis set and a grid=(200,1454). In addition, the RSIC center of mass (COM) obtained from ERKALE are shown in the light green box as comparison to the final LDQ FODs. Color code: C-gray, H-white, O-red, N-blue, spin-up FOD/COM - green, spin-down FOD/COM - red. There are situations where spin-up and spin-down FODs/COMs have the same position. Then, only one color is seen

matrix as well as the FODs, the configuration based on LDQ deliver the lowest energy. In the unrestricted calculations, both the initial LDQ and the initial Lewis configuration converge to a final FOD arrangement which can be characterized via LDQ. Interestingly, the *other* FODs stay in the their configuration, but the final energy is higher than for the LDQ FODs.

LDQ not only gives the lowest energy but also the best SIC dipole moment for CH_3NO_2. As a highlight, the center of mass (COM) for the optimal RSIC localized orbitals using ERKALE also reflect the LDQ chemical bonding motif, see the RSIC box in Fig. 14.6. In this section, we reconfirm that the choice of the initial FODs significantly influences dipole moments in FLO-SIC. Next, we describe how the FOD optimization itself can influence dipole moments.

Effect of FOD optimization We continue the discussion for H_2O and CH_2O. These two molecules only have one meaningful FOD configuration, see Fig. 14.7. Having one FOD configuration may allow finding the same local minima in a reproducible

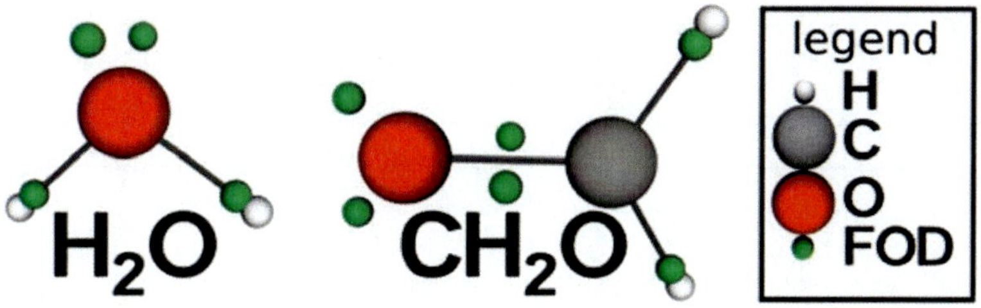

Fig. 14.7 Displaying both the molecular and electronic geometries for H_2O and CH_2O. The picture is generated with the PyFLOSIC2 graphical user interface (GUI). Note, only H_2O and CH_2O are shown as they have only one trivial initial FOD configuration, whereas CH_3NO_2 is more complex; see Fig. 14.6

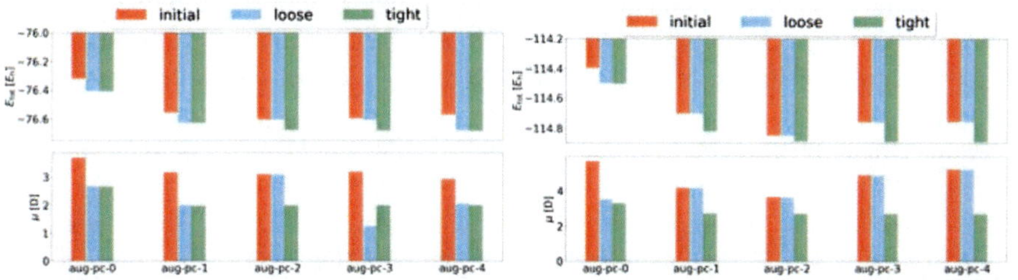

(a) H_2O (b) CH_2O

Fig. 14.8 Effect of FOD optimization vs. basis set size using the LDA-PW XC functional for (**a**) H_2O and (**b**) CH_2O. The tag *initial* represents FODs generated by the FODMC, *loose* represents FODs with a maximal force criterion of $f_{\max,\text{tol}} = 5 \cdot 10^{-3}\ E_\text{h}/a_0$, and *tight* refers to optimized FODs with $f_{\max,\text{tol}} = 2 \cdot 10^{-4}\ E_\text{h}/a_0$. A grid=(200,1454) was used. Note that when *initial* and *loose* give the same result, the initialized FODs are already at a force threshold of $5 \cdot 10^{-3}\ E_\text{h}/a_0$

fashion. Thus, those molecules are promising candidates for systematic FOD convergence studies.

In FLO-SIC it is rather easy to make various approximations. A common approximation is to use fixed FODs which are not optimized for the numerical parameter space of the respective calculation. Optimizing the FODs is, however, critical to obtain reasonable FLO-SIC solutions. Such optimizations are carried out until a specific threshold for the maximum FOD force, $f_{\max,\text{tol}}$, is reached. The influence of the FOD optimization on the total energy and the dipole moment for water and formaldehyde is shown in Fig. 14.8.

FLO-SIC values for the total energy and the dipole moment change drastically when going from an initial set of FODs to optimized FODs. Here, optimized FODs are characterized by $f_{\max,\text{tol}} = 2 \cdot 10^{-4}\ E_\text{h}/a_0$. Stopping the optimization too early, i.e., at $f_{\max,\text{tol}} = 5 \cdot 10^{-3}\ E_\text{h}/a_0$ can lead to insufficiently converged densities and energies. This furthermore leads to an incorrect prediction of trends; increasing the basis set size should smoothly converge the dipole moments, see Fig. 14.5. However,

Table 14.4 Dipole moments for DFT and FLO-SIC using three XC functionals. Using the aug-pc-3 basis set and a grid=(200,1454) in PySCF for DFT and PyFLOSIC2 for FLO-SIC

XC functional	H_2O		CH_2O	
	μ_{DFT}	$\mu_{FLO\text{-}SIC}$	μ_{DFT}	$\mu_{FLO\text{-}SIC}$
LDA-PW	1.86	1.99	2.28	2.72
PBEsol	1.82	1.95	2.23	2.65
r^2SCAN	1.83	1.96	2.31	2.62

with insufficiently optimized FODs these trends can be predicted incorrectly, as can be seen in Fig. 14.8.

Accordingly, global statements and generalization of trends are only valid for optimized FODs in combination with a sufficient basis set and grid. In the tested cases, the aug-pc-3 basis set with a grid=(200,1454) deliver converged results, in analogy to Sects. 14.4.1 and 14.4.2.

As proposed by the authors [31], monitoring the dipole moment is important to classify and analyze SIC solutions. We showed here that converging the total energy is necessary, but might not be sufficient when one aims to study density-related properties, i.e., dipole moments or polarizabilities. All calculations in the previous sections utilized the LDA-PW functional. Next, we discuss how changing the exchange-correlation functional influences dipole moments.

Effect of exchange-correlation functional For H_2O and CH_2O, we find that already LDA-PW describes dipole moments qualitatively correctly for DFT and SIC. However, comparing three exchange-correlation functionals the absolute dipole moments differ, see Table 14.4. Note that for CH_3NO_2 this comparison was not carried, see *Effect of initial FODs* for more information.

All three pure DFAs agree quite well with the experimental dipole reference values, see Table 14.2. FLO-SIC tends to overshot the dipole moment in order of 10^{-1} D w.r.t. the DFT values. This trend has also been observed in the literature, see [31]. For DFT, LDA-PW performs best for H_2O while r^2SCAN agrees the most for CH_2O. In case of FLO-SIC, the best dipoles are given by PBEsol for H_2O and by r^2SCAN for CH_2O.

In the previous sections we have discussed results for FLO-SIC. To investigate the influence of a specific flavor of SIC on dipole moments, we compare FLO-SIC, RSIC, and CSIC in the next section.

Effect of SIC methods To verify the used numerical parameter space, i.e., the aug-pc-3 basis set and a grid=(200,1454), we carried out RSIC and CSIC calculations using ERKALE. There is no significant difference between FLO-SIC and RSIC values, see Table 14.5. This is noteworthy as the results are calculated with two independent electronic structure codes. Thus, the used numerical parameter space is sufficient to deliver reproducible results. SIC suffers from the multiple local-

Table 14.5 Comparison of μ in D using the aug-pc-3 basis set, a grid=(200,1454) and the LDA-PW functional. Molecular structures from the CCCBDB database [38] were used, coming from the following sources: H_2O [37], CH_2O [71] and CH_3NO_2 [72]. DFT values are obtained from PySCF, FLO-SIC values are obtained from PyFLOSIC2, and RSIC and CSIC values come from ERKALE

Molecule	μ_{DFT}	$\mu_{FLO\text{-}SIC}$	μ_{RSIC}	μ_{CSIC}
H_2O	1.86	1.99	1.99	2.06
CH_2O	2.28	2.72	2.72	2.70
CH_3NO_2	3.50	3.67	3.70	3.90

minima problem [25], see Sect. 14.1. Accordingly, we recommend to verify FLO-SIC results with independent SIC methods such as RSIC. For example for nitromethane, see *Effect of initial FODs*, only the LDQ FLO-SIC dipole agrees with the RSIC solution. FLO-SIC, RSIC and CSIC together deliver a consistent SIC description. The advantage of FLO-SIC is the access to bonding information. This information allows to easily classify and further analyze PZ-SIC solutions [31].

In the next section we summarize and conclude our findings.

14.5 Summary and Conclusion

As shown in Sect. 14.4.1, and already stated in earlier works [30, 31, 45], SIC needs finer numerical quadrature meshes in comparison to DFT calculations. Global statements about the predictive power of SIC are only meaningful using very accurate numerical parameter spaces.

In this work we show how density-related properties, i.e., dipole moments and polarizabilities, can help to determine an appropriate numerical parameter space for DFT and SIC. For our investigated molecules, the aug-pc-3 basis set and a numerical grid of (200,1454) is such an appropriate parameter space. Note, even trends can clearly change using other numerical parameters. While it is mandatory to converge the energy it is not necessarily sufficient for the study of density-related properties.

Furthermore, using water, formaldehyde, and nitromethane we show that these density-related properties are not only sensitive to the used numerical parameter space. They are also significantly influenced by the used molecular geometry, see Sect. 14.4.2. Molecular geometries from common chemical database, i.e., CCCBDB, PubChem, or ChemSpider, can deliver very different dipole moments and polarizabilities. Only molecular geometries from the CCCBDB database deliver reasonable trends in comparison to optimized geometries.

Continuing the work of [31], for FLO-SIC we showed that density-fingerprints are also sensitive to the chemical bonding situation introduced by FODs. We demonstrate that the numerical quality of the FODs, represented by their gradients, as well as the choice of initial FODs clearly influences the dipole moments. For molecules with non-trivial bonding situations, e.g., nitromethane, it is highly recommended to use various initial FOD configurations. This allows to reasonably sample the FOD configuration space to determine the most reasonable FOD configuration. Such FOD configurations typically follow chemical bonding theories [31], i.e., Lewis and LDQ. In case where several FOD configurations are possible, FODs based on LDQ are often superior in FLO-SIC. This has been shown in [31] and verified here again for nitromethane in Sect. 14.4.3.

When computational time matters, the simplest density functional approximation, i.e., a local density approximation like LDA-PW, provides reasonable trends for dipole moments in FLO-SIC, as seen in Sect. 14.4.3. If computational time is not limited we recommend to use higher rung functional like PBEsol or r^2SCAN to verify and further analyze determined trends.

We want to emphasis the importance of Open-Science and Open-Source developments, as this work would not be possible without them.

Acknowledgements S. Schwalbe has been funded by the Deutsche Forschungsgemeinschaft (DFG, German Research Foundation)—Project ID 421663657—KO 1924/9-2. S. Liebing wants to express his gratitude for Prof. Jens Kortus enabling the authors to work on this topic. All authors thank M. Sc. Wanja T. Schulze for his contributions to the PyFLOSIC2 code. We thank both referees, as their comments significantly improved the content and shape of our manuscript. This work is part of our OpenSIC project, and we thank all members for fruitful discussions. We thank the ZIH Dresden for computational time and support.

References

1. Becke, A.D.: Perspective: Fifty years of density-functional theory in chemical physics. J. Chem. Phys. **140**(18), 18A301 (2014)
2. Verma, P., Truhlar, D.G.: Status and challenges of density functional theory. Trends Chem. **2**(4), 302–318 (2020)
3. Förster, S., Hahn, T., Loose, C., Röder, C., Liebing, S., Seichter, W., Eißmann, F., Kortus, J., Weber, E.: Synthesis and characterization of new derivatives of azulene, including experimental and theoretical studies of electronic and spectroscopic behavior. J. Phys. Org. Chem. **25**(10), 856 (2012)
4. Pfaff, U., Hildebrandt, A., Schaarschmidt, D., Hahn, T., Liebing, S., Kortus, J., Lang, H.: Di- and triferrocenyl (hetero) aromatics: Synthesis, characterization,(spectro-) electrochemistry, and calculations. Organometallics **31**(19), 6761 (2012)
5. Seidel, N., Hahn, T., Liebing, S., Seichter, W., Kortus, J., Weber, E.: Synthesis and properties of new 9, 10-anthraquinone derived compounds for molecular electronics. New J. Chem. **37**(3), 601 (2013)
6. Trepte, K., Schaber, J., Schwalbe, S., Drache, F., Senkovska, I., Kaskel, S., Kortus, J., Brunner, E., Seifert, G.: The origin of the measured chemical shift of ^{129}Xe in UiO-66 and UiO-67 revealed by DFT investigations. Phys. Chem. Chem. Phys. **19**, 10020–10027 (2017)
7. Trepte, K., Schwalbe, S., Schaber, J., Krause, S., Senkovska, I., Kaskel, S., Brunner, E., Kortus, J., Seifert, G.: Theoretical and experimental investigations of ^{129}Xe NMR chemical shift isotherms in metal-organic frameworks. Phys. Chem. Chem. Phys. **20**, 25039–25043 (2018)
8. Trepte, K., Schwalbe, S., Seifert, G.: Electronic and magnetic properties of DUT-8(Ni). Phys. Chem. Chem. Phys. **17**, 17122–17129 (2015)

9. Rühlig, K., Abylaikhan, A., Aliabadi, A., Kataev, V., Liebing, S., Schwalbe, S., Trepte, K., Ludt, C., Kortus, J., Büchner, B., et al.: Ni^{II} formate complexes with bi-and tridentate nitrogen-donor ligands: synthesis, characterization, and magnetic and thermal properties. Dalton Trans. **46**(12), 3963–3979 (2017)
10. Taubert, F., Schwalbe, S., Seidel, J., Hüttl, R., Gruber, T., Janot, R., Bobnar, M., Gumeniuk, R., Mertens, F., Kortus, J.: Thermodynamic characterization of lithium monosilicide (LiSi) by means of calorimetry and DFT-calculations. Int. J. Mater. Res. **108**(11), 942–958 (2017)
11. Schwalbe, S., Gruber, T., Trepte, K., Taubert, F., Mertens, F., Kortus, J.: Mechanical, elastic and thermodynamic properties of crystalline lithium silicides. Comput. Mater. Sci. **134**, 48–57 (2017)
12. Schwalbe, S., Trepte, K., Seifert, G., Kortus, J.: Screening for high-spin metal organic frameworks (MOFs): density functional theory study on DUT-8(M_1, M_2) (with M_i = V,..., Cu). Phys. Chem. Chem. Phys. **18**, 8075–8080 (2016)
13. Friedrich, R., Usanmaz, D., Oses, C., Supka, A., Fornari, M., Nardelli, M.B., Toher, C., Curtarolo, S.: Coordination corrected ab initio formation enthalpies. NPJ Comput. Mater. **5**(1), 1–12 (2019)
14. Mehl, M.J., Ronquillo, M., Hicks, D., Esters, M., Oses, C., Friedrich, R., Smolyanyuk, A., Gossett, E., Finkenstadt, D., Curtarolo, S.: Tin-pest problem as a test of density functionals using high-throughput calculations. Phys. Rev. Mater. **5**(8), 083608 (2021)
15. Trepte, K., Schwalbe, S.: porE: A code for deterministic and systematic analyses of porosities. J. Comput. Chem. **42**(9), 630–643 (2021)
16. Friedrich, R., Ghorbani-Asl, M., Curtarolo, S., Krasheninnikov, A.V.: Data-driven quest for two-dimensional non-van der Waals materials. Nano Lett. **22**, 989 (2022)
17. Kohn, W., Sham, L.J.: Self-consistent equations including exchange and correlation effects. Phys. Rev. **140**(4A), A1133 (1965)
18. Ellis, J.A., Fiedler, L., Popoola, G.A., Modine, N.A., Stephens, J.A., Thompson, A.P., Cangi, A., Rajamanickam, S.: Accelerating finite-temperature Kohn-Sham density functional theory with deep neural networks. Phys. Rev. B **104**(3), 035120 (2021)
19. Brown, K., Maimaiti, Y., Trepte, K., Bligaard, T., Voss, J.: MCML: Combining physical constraints with experimental data for a multi-purpose meta-generalized gradient approximation. J. Comput. Chem. **42**, 2004–2013 (2021)
20. Kirkpatrick, J., McMorrow, B., Turban, D.H., Gaunt, A.L., Spencer, J.S., Matthews, A.G., Obika, A., Thiry, L., Fortunato, M., Pfau, D., et al.: Pushing the frontiers of density functionals by solving the fractional electron problem. Science **374**(6573), 1385–1389 (2021)
21. Sun, J., Ruzsinszky, A., Perdew, J.P.: Strongly constrained and appropriately normed semilocal density functional. Phys. Rev. Lett. **115**(3), 036402 (2015)
22. Furness, J.W., Kaplan, A.D., Ning, J., Perdew, J.P., Sun, J.: Accurate and numerically efficient r^2SCAN meta-generalized gradient approximation. J. Phys. Chem. Lett. **11**, 8208–8215 (2020)
23. Perdew, J.P., Zunger, A.: Self-interaction correction to density-functional approximations for many-electron systems. Phys. Rev. B **23**(10), 5048 (1981)
24. Perdew, J.P., Ruzsinszky, A., Sun, J., Pederson, M.R.: Paradox of self-interaction correction: How can anything so right be so wrong? Adv. At. Mol. Opt. Phys. **64**, 1–14 (2015)
25. Lehtola, S., Head-Gordon, M., Jónsson, H.: Complex orbitals, multiple local minima, and symmetry breaking in Perdew-Zunger self-interaction corrected density functional theory calculations. J. Chem. Theory Comput. **12**(7), 3195–3207 (2016)
26. Pederson, M.R., Ruzsinszky, A., Perdew, J.P.: Communication: Self-interaction correction with unitary invariance in density functional theory. J. Chem. Phys. **140**(12), 121103 (2014)
27. Pederson, M.R., Baruah, T.: Self-interaction corrections within the Fermi-orbital-based formalism. Adv. At. Mol. Opt. Phys. **64**, 153 (2015)
28. Yang, Z.-H., Pederson, M.R., Perdew, J.P.: Full self-consistency in the Fermi-orbital self-interaction correction. Phys. Rev. A **95**, 052505 (2017)
29. Schwalbe, S., Trepte, K., Fiedler, L., Johnson, A.I., Kraus, J., Hahn, T., Peralta, J.E., Jackson, K.A., Kortus, J.: Interpretation and automatic generation of Fermi-orbital descriptors. J. Comput. Chem. **40**(32), 2843–2857 (2019)

30. Schwalbe, S., Fiedler, L., Kraus, J., Kortus, J., Trepte, K., Lehtola, S.: PyFLOSIC: Python-based Fermi-Löwdin orbital self-interaction correction. J. Chem. Phys. **153**(8), 084104 (2020)
31. Trepte, K., Schwalbe, S., Liebing, S., Schulze, W.T., Kortus, J., Myneni, H., Ivanov, A.V., Lehtola, S.: Chemical bonding theories as guides for self-interaction corrected solutions: multiple local minima and symmetry breaking. J. Chem. Phys. **155**(22), 224109 (2021)
32. Lewis, G.N.: The atom and the molecule. J. Am. Chem. Soc. **38**(4), 762–785 (1916)
33. Linnett, J.W.: Valence-bond structures: A new proposal. Nature **187**(4740), 859 (1960)
34. Linnett, J.W.: A modification of the Lewis-Langmuir octet rule. J. Am. Chem. Soc. **83**(12), 2643–2653 (1961)
35. Lehtola, S., Steigemann, C., Oliveira, M.J.T., Marques, M.A.L.: Recent developments in LIBXC–A comprehensive library of functionals for density functional theory. SoftwareX **7**, 1 (2018)
36. Pederson, M.R.: Fermi orbital derivatives in self-interaction corrected density functional theory: applications to closed shell atoms. J. Chem. Phys. **142**(6), 064112 (2015)
37. Hoy, A.R., Bunker, P.R.: A precise solution of the rotation bending Schrödinger equation for a triatomic molecule with application to the water molecule. J. Mol. Spectrosc. **74**(1), 1–8 (1979)
38. Johnson, R.D.: NIST Computational Chemistry Comparison and Benchmark Database. NIST Standard Reference Database Number 101 (2020)
39. Scrocco, E., Tomasi, J.: Electronic molecular structure, reactivity and intermolecular forces: an euristic interpretation by means of electrostatic molecular potentials. Adv. Quantum Chem. **11**, 115–193 (Elsevier, 1978)
40. Jmol: an open-source java viewer for chemical structures in 3d. http://www.jmol.org/ (2022)
41. Handy, N.C., Schaefer, H.F., III.: On the evaluation of analytic energy derivatives for correlated wave functions. J. Chem. Phys. **81**(11), 5031–5033 (1984)
42. Sun, Q., Zhang, X., Banerjee, S., Bao, P., Barbry, M., Blunt, N.S., Bogdanov, N.A., Booth, G.H., Chen, J., Cui, Z.-H., Eriksen, J.J., Gao, Y., Guo, S., Hermann, J., Hermes, M.R., Koh, K., Koval, P., Lehtola, S., Li, Z., Liu, J., Mardirossian, N., McClain, J.D., Motta, M., Mussard, B., Pham, H.Q., Pulkin, A., Purwanto, W., Robinson, P.J., Ronca, E., Sayfutyarova, E.R., Scheurer, M., Schurkus, H.F., Smith, J.E.T., Sun, C., Sun, S.-N., Upadhyay, S., Wagner, L.K., Wang, X., White, A., Whitfield, J.D., Williamson, M.J., Wouters, S., Yang, J., Yu, J.M., Zhu, T., Berkelbach, T.C., Sharma, S., Sokolov, A.Y., Chan, G.K.-L.: Recent developments in the PySCF program package. J. Chem. Phys. **153**(2), 024109 (2020)
43. Smith, J.E.T.: Modern Multireference Electronic Structure Theory. PhD thesis, https://scholar.colorado.edu/concern/graduate_thesis_or_dissertations/9306t035k. Accessed Feb 2022 (2020)
44. Liebing, S., Trepte, K., Schwalbe, S.: Dippo-(dip)oles and (po)larizabilities. https://doi.org/10.5281/zenodo.6246152 (2022)
45. Lehtola, S., Jónsson, H.: Variational, self-consistent implementation of the Perdew-Zunger self-interaction correction with complex optimal orbitals. J. Chem. Theory Comput. **10**(12), 5324 (2014)
46. Lehtola, S., Jónsson, E.Ö., Jónsson, H.: Effect of complex-valued optimal orbitals on atomization energies with the Perdew-Zunger self-interaction correction to density functional theory. J. Chem. Theory Comput. **12**, 4296–4302 (2016)
47. Lehtola, J., Hakala, M., Sakko, A., Hämäläinen, K.: ERKALE–A flexible program package for X-ray properties of atoms and molecules. J. Comput. Chem. **33**(18), 1572 (2012)
48. Jensen, F.: Polarization consistent basis sets: principles. J. Chem. Phys. **115**, 9113 (2001)
49. Jensen, F.: Polarization consistent basis sets II: estimating the Kohn-Sham basis set limit. J. Chem. Phys. **116**(17), 7372 (2002)
50. Perdew, J.P., Wang, Y.: Accurate and simple analytic representation of the electron-gas correlation energy. Phys. Rev. B **45**, 13244–13249 (1992)
51. Perdew, J.P., Ruzsinszky, A., Csonka, G.I., Vydrov, O.A., Scuseria, G.E., Constantin, L.A., Zhou, X., Burke, K.: Restoring the density-gradient expansion for exchange in solids and surfaces. Phys. Rev. Lett. **100**, 136406 (2008)

52. Oliveira, M.J.T., Papior, N., Pouillon, Y., Blum, V., Artacho, E., Caliste, D., Corsetti, F., de Gironcoli, S., Elena, A.M., García, A., García-Suárez, V.M., Genovese, L., Huhn, W.P., Huhs, G., Kokott, S., Küçükbenli, E., Larsen, A.H., Lazzaro, A., Lebedeva, I.V., Li, Y., López-Durán, D., López-Tarifa, P., Lüders, M., Marques, M.A.L., Minar, J., Mohr, S., Mostofi, A.A., O'Cais, A., Payne, M.C., Ruh, T., Smith, D.G.A., Soler, J.M., Strubbe, D.A., Tancogne-Dejean, N., Tildesley, D., Torrent, M., Yu, V.W.: The CECAM electronic structure library and the modular software development paradigm. J. Chem. Phys. **153**(2), 024117 (2020)
53. Lehtola, S., Karttunen, A.: Free and open source software for computational chemistry education. ChemRxiv, https://doi.org/10.33774/chemrxiv-2021-hr1r0-v2 (2021)
54. Schulze, W.: Domain-averaged fermi holes: a self-interaction correction perspective. Master Thesis, https://doi.org/10.13140/RG.2.2.27958.42568/2 (2021)
55. Ismail-Beigi, S., Arias, T.: New algebraic formulation of density functional calculation. Comput. Phys. Commun. **128**(1–2), 1–45 (2000)
56. Lebedev, V.I., Laikov, D.N.: A quadrature formula for the sphere of the 131st algebraic order of accuracy. Dokl. Math. **59**(3), 477–481 (1999)
57. Vydrov, O.A., Scuseria, G.E.: Effect of the Perdew-Zunger self-interaction correction on the thermochemical performance of approximate density functionals. J. Chem. Phys. **121**(17), 8187–8193 (2004)
58. Shahi, C., Bhattarai, P., Wagle, K., Santra, B., Schwalbe, S., Hahn, T., Kortus, J., Jackson, K.A., Peralta, J.E., Trepte, K., Lehtola, S., Nepal, N.K., Myneni, H., Neupane, B., Adhikari, S., Ruzsinszky, A., Yamamoto, Y., Baruah, T., Zope, R.R., Perdew, J.P.: Stretched or noded orbital densities and self-interaction correction in density functional theory. J. Chem. Phys. **150**(17), 174102 (2019)
59. Karanovich, A., Yamamoto, Y., Jackson, K.A., Park, K.: Electronic structure of mononuclear Cu-based molecule from density-functional theory with self-interaction correction. J. Chem. Phys. **155**(1), 014106 (2021)
60. Harrison, J.G., Heaton, R., Lin, C.C.: Self-interaction correction to the local density Hartree-Fock atomic calculations of excited and ground states. J. Phys. B-At. Mol. **16**(12), 2079 (1983)
61. Pederson, M.R., Heaton, R.A., Lin, C.C.: Local-density Hartree-Fock theory of electronic states of molecules with self-interaction correction. J. Chem. Phys. **80**, 1972 (1984)
62. Lehtola, S., Jónsson, H.: Unitary optimization of localized molecular orbitals. J. Chem. Theory Comput. **9**(12), 5365–5372 (2013)
63. Broyden, C.G.: The convergence of a class of double-rank minimization algorithms 1 general considerations. IMA J. Appl. Math. **6**(1), 76 (1970)
64. Fletcher, R.: A new approach to variable metric algorithms. Comput. J. **13**(3), 317 (1970)
65. Goldfarb, D.: A family of variable-metric methods derived by variational means. Math. Comput. **24**(109), 23 (1970)
66. Shanno, D.F.: Conditioning of quasi-Newton methods for function minimization. Math. Comput. **24**(111), 647 (1970)
67. Nocedal, J.: Updating quasi-Newton matrices with limited storage. Math. Comput. **35**(151), 773 (1980)
68. Liu, D.C., Nocedal, J.: On the limited memory BFGS method for large scale optimization. Math. Program. **45**(1–3), 503 (1989)
69. Byrd, R.H., Lu, P., Nocedal, J., Zhu, C.: A limited memory algorithm for bound constrained optimization. SIAM J. Sci. Comput. **16**(5), 1190 (1995)
70. Zhu, C., Byrd, R.H., Lu, P., Nocedal, J.: Algorithm 778: L-BFGS-B: Fortran subroutines for large-scale bound-constrained optimization. ACM Trans. Math. Softw. **23**(4), 550 (1997)
71. Gurvich, L.V., Veits, I.V., Alcock, C.B.: Thermodynamics Properties of Individual Substances, vol 1 and 116. Hemisphere Pub. Co. (1989)
72. Hellwege, K.H., Hellwege, A.M.: Structure Data of Free Polyatomic Molecules. Springer Verlag (1976)
73. ChemSpider. https://www.chemspider.com/. Accessed Jan 2022
74. Pence, H.E., Williams, A.: Chemspider: an online chemical information resource. J. Chem. Educ. **87**(11), 1123–1124 (2010)

75. Kim, S., Chen, J., Cheng, T., Gindulyte, A., He, J., He, S., Li, Q., Shoemaker, B.A., Thiessen, P.A., Yu, B., Zaslavsky, L., Zhang, J., Bolton, E.E.: PubChem in 2021: new data content and improved web interfaces. Nucleic Acids Res. **49**, D1388–D1395 (2020)
76. Williams, A.J., Ekins, S., Tkachenko, V.: Towards a gold standard: regarding quality in public domain chemistry databases and approaches to improving the situation. Drug Discov. Today **17**(13–14), 685–701 (2012)
77. Halgren, T.A.: MMFF VI. MMFF94s option for energy minimization studies. J. Comput. Chem. **20**(7), 720–729 (1999)
78. Fiedler, L., Shah, K., Bussmann, M., Cangi, A.: A deep dive into machine learning density functional theory for materials science and chemistry. arXiv:2110.00997 (2021)
79. Nelson, R.D., Jr., Lide, D.R., Jr., Maryott, A.A.: Selected values of electric dipole moments for molecules in the gas phase. Tech. Rep, National Standard Reference Data System (1967)

Chapter 15
Approximate Solutions of a Kinetic Theory for Graphene

D. B. Blaschke, V. V. Dmitriev, N. T. Gevorgyan, B. Mahato, A. D. Panferov, S. A. Smolyansky, and V. A. Tseryupa

Abstract The effective mass approximation is analysed in a nonperturbative kinetic theory approach to strong field excitations in graphene. This problem is highly actual for the investigation of quantum radiation from graphene, where the collision integrals in the photon kinetic equation are rather complicated functionals of the distribution functions of the charge carriers. These functions are needed in the explicit analytic definition as solutions of the kinetic equations for the electron-hole excitations in the presence of a strong electromagnetic field. In the present work it is shown that the suggested approach is rather effective in a certain range of nonlinearity parameters. In the standard massive quantum electrodynamics the usability of the analogical approximation is limited to a very narrow region of parameters of the external field.

D. B. Blaschke · B. Mahato
Institute for Theoretical Physics, University of Wroclaw, 50-204 Wroclaw, Poland
e-mail: david.blaschke@uwr.edu.pl

D. B. Blaschke
Bogoliubov Laboratory for Theoretical Physics, Joint Institute for Nuclear Research, Dubna, RU 141980, Russia

National Research Nuclear University (MEPhI), Moscow, RU 115409, Russia

V. V. Dmitriev · A. D. Panferov · S. A. Smolyansky (✉) · V. A. Tseryupa
Saratov State University, Saratov, RU 410026, Russia
e-mail: smol@sgu.ru

N. T. Gevorgyan
A.I. Alikhanyan National Science Laboratory (YerPhI), 0036 Yerevan, Armenia

Byurakan Astronomical Observatory, National Academy of Sciences, 0213 Byurakan, Armenia

S. A. Smolyansky
Tomsk State University, Tomsk, RU 634050, Russia

D. Blaschke et al. (eds.), *Optics and Its Applications*, Springer Proceedings in Physics 281, https://doi.org/10.1007/978-3-031-11287-4_15

15.1 Introduction

It is well known that the interaction of carriers in graphene with an external electromagnetic field is strongly nonlinear with respect to the parameters of the external field [1–3] and is nonanalytic in the fine structure constant [4, 5]. This renders impossible applications of perturbation theory unjustified and stimulates the search for nonperturbative approaches. Paradigms are here exactly solvable quantum field theory models. However, these solutions in graphene are limited to narrow classes of external field models (constant electric field [6], Eckart's potential [7]). Alternative cases are founded either on direct application of the basic equations of motion [8, 9] or on the nonperturbative kinetic theory [1–3, 10] constructed in analogy to the standard strong field QED [11–14].

A higher level of description of the graphene excitations is connected with the investigation of different mechanisms of radiation. In the simplest case the question is about the emission of a quasiclassical electromagnetic field by the plasma currents in graphene [15, 16]. There is also the radiation of a quantized field that is the result of the direct interaction of the charge carriers with the photon field. A consistent realization of this approach is based on a truncation procedure of the Bogoliubov-Born-Green-Kirkwood-Yvon (BBGKY) chain of equations for the correlation functions of the electron-hole-photon—system [3]. This leads to a closed system of kinetic equations (KEs) with collision integrals (CIs) of non-Markovian type in the electron-hole and photon sectors and with the Maxwell equation for the acting inner (plasma) field. As far as the evolution of the electron-hole plasma is accompanied by high-frequency quantum oscillations, in the standard strong field quantum electrodynamics known as Zitterbewegung, any solution of this KE system turns out very susceptible to the selection not only of the model and the parameters of the external field but also to the necessary roughening in the process of calculating the physical quantities.

Below we will consider this problem on the examples of two approximative methods for the solution of the basic KEs describing the production of *eh*-pairs under the action of an external field: the low density approximation [17] and the method of the asymptotic decompositions [18] (Sect. 15.2). The ideas for these approaches are borrowed from the standard QED. In the considered case of the massless theory, an essential role plays the effective mass approximation [19]. The results of analytical calculations of the basic functions of the kinetic theory are compared among themselves and with the exactly solvable model (Sect. 15.3). The considered examples show that the introduction of the effective electromagnetic mass allows to obtain rather simple expansions for the distribution functions of the electron-hole plasma in graphene for time dependent electric fields with different pulse shape.

15.2 The Basic KE and Its Approximate Solutions

The basic KE for description of excitation and evolution of the electron-hole plasma in graphene under the action of an external quasiclassical spatially uniform time-dependent electric field with the vector potential $A^{\mu} = (0, A^{(1)}(t), A^{(2)}(t), 0)$ in the Hamiltonian gauge $A^{(0)} = 0$ can be written in the integro-differential form [1–3, 10]

$$\dot{f}(\mathbf{p}, t) = \frac{1}{2}\lambda(\mathbf{p}, t) \int_{t_0}^{t} \lambda(\mathbf{p}, t')[1 - 2f(\mathbf{p}, t')] \cos\theta(\mathbf{p}; t, t')dt' \qquad (15.1)$$

or as the equivalent system of ordinary differential equations

$$\dot{f}(\mathbf{p}, t) = \frac{1}{2}\lambda(\mathbf{p}, t)u(\mathbf{p}, t), \qquad (15.2)$$

$$\dot{u}(\mathbf{p}, t) = \lambda(\mathbf{p}, t)\left[1 - 2f(\mathbf{p}, t)\right] - \frac{2\varepsilon(\mathbf{p}, t)}{\hbar}v(\mathbf{p}, t), \;\; \dot{v}(\mathbf{p}, t) = \frac{2\varepsilon(\mathbf{p}, t)}{\hbar}u(\mathbf{p}, t).$$

Here $f(\mathbf{p}, t)$ is the distribution function of charge carriers introduced by taking into account the electroneutrality $f(\mathbf{p}, t) = f_e(\mathbf{p}, t) = f_h(-\mathbf{p}, t)$, while the auxiliary functions $u(\mathbf{p}, t)$ and $v(\mathbf{p}, t)$ are defined as [2]

$$u(\mathbf{p}, t) = i\left[f^{(+)}(\mathbf{p}, t) - f^{(-)}(\mathbf{p}, t)\right], \; v(\mathbf{p}, t) = \left[f^{(+)}(\mathbf{p}, t) + f^{(-)}(\mathbf{p}, t)\right] \quad (15.3)$$

via the anomalous averages

$$f^{(+)}(\mathbf{p}, t) = \langle \mathrm{in}|a^{+}(\mathbf{p}, t)b^{+}(-\mathbf{p}, t)|\mathrm{in}\rangle, \;\; f^{(-)}(\mathbf{p}, t) = \langle \mathrm{in}|b(-\mathbf{p}, t)a(\mathbf{p}, t)|\mathrm{in}\rangle, \quad (15.4)$$

where a^{+} (a) and b^{+} (b) are the creation (annihilation) operators for electrons and holes, respectively. The excitation function $\lambda(\mathbf{p}, t)$ in the low energy model is determined as

$$\lambda(\mathbf{p}, t) = \frac{ev_F^2[E^{(1)}(t)P^2 - E^{(2)}(t)P^1]}{\varepsilon^2(\mathbf{p}, t)}, \qquad (15.5)$$

where $v_F = 10^6$ m/s is the Fermi velocity for electrons in graphene, $E^{(k)} = -\frac{1}{c}\dot{A}^{(k)}(t)$ is the field strength $(k = 1, 2)$, $P^k = p^k + \frac{e}{c}A^k(t)$ is the quasi-momentum (generalized momentum) and $\varepsilon(\mathbf{p}, t) = v_F\sqrt{P^2}$ is the quasi-energy. The electron charge is $-e$. Finally, the quantity $\theta(\mathbf{p}; t, t')$ in the KE (15.1) is the phase,

$$\theta(\mathbf{p}; t, t') = \frac{2}{\hbar}\int_{t'}^{t} dt''\varepsilon(\mathbf{p}, t''). \qquad (15.6)$$

The KE (15.1) is an integro-differential equation of the non-Markovian type with a fastly oscillating kernel. There is an integral of motion [1, 2]

$$(1 - 2f)^2 + u^2 + v^2 = \mathrm{const}, \qquad (15.7)$$

where the constant is fixed with the corresponding initial condition.

The KEs (15.1), (15.2) for the case of graphene were obtained in the works [1–3, 10] by the Bogoliubov method of canonical transformations that can by realized in the massless $D = 2 + 1$ theory in an explicit form. On the other hand the system (15.2) can be reduced from the general system of twelve KEs in the standard QED [20].

The massless low energy model of graphene with the lightlike dispersion law $\varepsilon(\mathbf{p}, t)$ leads to the absence of the critical field that is characteristic for massive QED and results in a specific feature of the momentum dependence of the excitation function $\lambda(\mathbf{p}, t)$ (15.5): this function decreases in the ultraviolet area, $\lambda(\mathbf{p}, t) \sim 1/P \to 0$ at $P \to \infty$ and is singular in the infra-red area $\lambda(\mathbf{p}, t) \to \infty$ at $P \to 0$. The last distinction also leads to a nonanalytic structure of the theory in its dependence on the coupling constant and to the absence of the standard perturbation theory.

Let us write also the differential equation of the third order that is equivalent to the system of equations (15.2),

$$\dddot{g} + \ddot{g}\left[\varepsilon\lambda\left(\frac{1}{\varepsilon\lambda}\right)' - \frac{\dot{\lambda}}{\lambda}\right] + \dot{g}\left[4\varepsilon^2 + \lambda^2 - \varepsilon\lambda\left(\frac{\dot{\lambda}}{\varepsilon\lambda^2}\right)'\right] + g\varepsilon\lambda\left(\frac{\lambda}{\varepsilon}\right)' = 0, \tag{15.8}$$

where $g = 1 - 2f$. In Eq. (15.8) the $'$ denotes also the time derivative.

At the present time, an exact solution of the KEs (15.1), (15.2) is not known. However, below we will assume that the well-known exact solutions of the Dirac equation for a constant electric field and the Eckart potential are at the same time solutions of the KEs (15.1), (15.2). This assumption gives a basis for comparing these exact solutions with the known approximate solutions of the KEs (15.1), (15.2) and to construct then some new classes of approximate solutions.

In order to estimate the effectivity of the approximate solutions, we will compare them to the exact solutions (analytical and numerical). Such a comparison will be made on the level of an integral macroscopic quantity. The number density of pairs $n(t)$ will be considered as the simplest quantity of such type,

$$n(t) = \frac{N_f}{(2\pi\hbar)^2} \int f(\mathbf{p}, t) d\mathbf{p}, \tag{15.9}$$

where $N_f = 4$ is the number of flavours. The integration allows here to smoothen out some insignificant details in the momentum dependence of the distribution functions in different approximations.

As the next step, we will consider two approximate methods of solving the KEs (15.1), (15.2). To this end, we will consider the case of a linearly polarized electric field $A^{(1)} = 0,\ A^{(2)}(t) = A(t)$.

15.2.1 Low Density Approximation

This approximation corresponds to the limit $f \ll 1$ in the r.h.s. KE (15.1). It was introduced in the strong field vacuum production of charged particles [17] and was used many times in strong field QED and in the kinetic theory of excitations in graphene. It leads to the quadrature formula [1, 2]

$$f(t) = \frac{1}{4}\left[\int_{-\infty}^{t} dt' \lambda_s(t')\right]^2 + \frac{1}{4}\left[\int_{-\infty}^{t} dt' \lambda_c(t')\right]^2, \tag{15.10}$$

where

$$\lambda_c(t) = \lambda(t)\cos\theta(t, -\infty), \quad \lambda_s(t) = \lambda(t)\sin\theta(t, -\infty). \tag{15.11}$$

In particular, it follows from Eq. (15.10), that $f(\mathbf{p}, t) \geq 0$.

In the low-density approximation, the two last KEs of the system (15.2) separate from it,

$$\dot{u} = \lambda - \frac{2\varepsilon}{\hbar}v, \ \dot{v} = \frac{2\varepsilon}{\hbar}u. \tag{15.12}$$

Then the distribution function f can be found from the first equation of the system (15.2) alone.

The system of Eq. (15.12) corresponds to the ordinary differential equation,

$$\frac{\hbar}{2\varepsilon}\frac{d}{dt}\left(\frac{\hbar}{2\varepsilon}\frac{du}{dt}\right) + u - \frac{\hbar}{2\varepsilon}\frac{d}{dt}\left(\frac{\hbar\lambda}{2\varepsilon}\right) = 0 \tag{15.13}$$

or

$$\mathcal{D}^2 u + u - \mathcal{D}\left(\frac{\hbar\lambda}{2\varepsilon}\right) = 0, \ \mathcal{D} = \frac{\hbar}{2\varepsilon}\frac{d}{dt}. \tag{15.14}$$

15.2.2 Effective Electromagnetic Mass Approximation

This well known approximation [19] was used already in the case of the harmonic model of an external field in the analysis of the radiation effects in the electron-positron plasma [21] and in the electron-hole plasma in graphene [3]. Below we will consider a generalization of this approach to other models of the external field.

The idea of the method is that the time dependent value of the square of the kinetic momentum $P^2(t)$ in the definition of the quasienergy $\varepsilon(\mathbf{p}, t)$ gets substituted by corresponding time average $\prec P^2(t) \succ$, where the symbol $\prec ... \succ$ means the averaging procedure over a characteristic time of the external field. In the case of the linearly polarized electric field we obtain

$$\prec P_2^2(t) \succ = p_2^2 + (e/c)^2 \prec A^2(t) \succ, \tag{15.15}$$

if $\prec A(t) \succ = 0$. Implying the substitution

$$P_2^2(t) \to \prec P_2^2(t) \succ, \tag{15.16}$$

in the square of quasienergy $\varepsilon^2(\mathbf{p}, t) = v_F^2[p_1^2 + P_2^2(t)]$, one can introduce the longitudinal (with respect to external field) effective electromagnetic mass

$$(e/c)^2 \prec A^2(t) \succ = m_*^2 v_F^2, \tag{15.17}$$

or

$$m_*^2 = \frac{e^2}{c^2 v_F^2} \frac{1}{2T} \int_{-T}^{T} dt A^2(t). \tag{15.18}$$

Thus, this approximation corresponds to change

$$\varepsilon(\mathbf{p}, t) \to \varepsilon_*(\mathbf{p}) = v_F \sqrt{p_1^2 + (p_2^2 + m_*^2 v_F^2)}. \tag{15.19}$$

The appearance of the longitudinal mass is a reflection of the anisotropy of the system stipulated by the presence of the external field and leads to a reduction of the mobility of charge carriers along the direction of the action of the external field. In the limiting case $p_2 \ll m_* v_F$, we obtain a strong anisotropic momentum dependence of the quasienergy,

$$\varepsilon_*(\mathbf{p}) = v_F \sqrt{m_*^2 v_F^2 + p_1^2}. \tag{15.20}$$

The approximation of the effective mass (15.16), (15.17) is valid in field models with square integrable functions $A(t)$ only.

The transition amplitude (15.5) in the effective mass approximation in the linearly polarized external field will be

$$\lambda_*(\mathbf{p}, t) = -\frac{e v_F^2 p_1}{\varepsilon_*^2(\mathbf{p})} E(t) \equiv \Lambda(\mathbf{p}) E(t), \tag{15.21}$$

where $p_1 = \cos\varphi$ and φ is the polar angle between the vectors $\mathbf{p}$ and $\mathbf{E}(t)$.

The problem of evaluating the distribution function is now brought to the calculation of the integral

$$J(\mathbf{p}, t) = \int_{-\infty}^{t} dt' E(t') \cos\theta_*(\mathbf{p}; t', -\infty), \tag{15.22}$$

where $\theta_*(\mathbf{p}; t', -\infty)$ is defined by the relation (15.6) with the replacement $\varepsilon(\mathbf{p}, t) \to \varepsilon_*(\mathbf{p})$. The analogous integral with the replacement $\cos\theta_* \to \sin\theta_*$ is equal to zero, if $E(t) = E(-t)$. The distribution function (15.10) will then be

$$f(\mathbf{p}, t) = \frac{1}{4}\Lambda^2(\mathbf{p}) J^2(\mathbf{p}, t). \tag{15.23}$$

According to the relations (15.21) and (15.23), the anisotropy of the distribution function $\sim \cos^2\varphi$ is universal and does not depend on the selection of the external field model. Another distinctive feature of the amplitude (15.21) is the singular slit of the surfaces $\lambda(\mathbf{p}, t)$ on the plane (p_1, p_2) along the axis $p_1 = 0$ (or $\varphi = \pi/2$) for $p_2 \neq 0$, i.e. $\lambda(\mathbf{p}, t)|_{p_1=0} = 0$, and $\lambda(\mathbf{p}, t)|_{p_1\neq 0} \neq 0$. As it follows from the KEs (15.1) and (15.2), this peculiarity is reproduced also in the distribution function, $f(\mathbf{p}, t)|_{p_1=0} = 0$. This means that quasiparticles are not created in the directions of the external field action in the strict sense. Figure 15.1 demonstrates this on the example of the Sauter impulse. This slit is evident in the figures shown below in the case of the massless version of the theory. The introduction of a mass results in a widening of this slit and in the appearance of the energy gap. Let us remark, that the exact solution of the problem [7] in the case of the Sauter pulse field has the singular line $p_1 = 0$ in the case of the linearly polarized external field. The presence of this infinitely thin slit is not reflected in calculations of the integral "observable" macroscopical averages of the type of the pair number density (15.9).

15.2.3 *Method of Asymptotic Decompositions*

This method is adopted from the standard strong field QED [18]. We consider now the dimensionless excitation amplitude in the exact case (15.5)

$$\Lambda(\mathbf{p}, t) = e\hbar v_F^2 E(t) p^1 / 2\varepsilon^3(\mathbf{p}, t) \tag{15.24}$$

and in the effective mass approximation

$$\Lambda_*(\mathbf{p}, t) = e\hbar v_F^2 E(t) p^1 / 2\varepsilon_*^3(\mathbf{p}, t). \tag{15.25}$$

In contrast to $\Lambda(\mathbf{p}, t)$, the amplitude $\Lambda_*(\mathbf{p}, t)$ (15.25) is limited everywhere,

$$\Lambda_*(\mathbf{p}, t) \leq \Lambda_*^{max} = \frac{e\hbar E_0}{3\sqrt{3} m_*^2 v_F^3}, \tag{15.26}$$

where Λ_*^{max} is the maximal value of the amplitude (15.25) in the point of time where $E(t) = E_0$ (see Fig. 15.1).

In order to clarify the physical meaning of the parameter Λ_*^{max} given in (15.26), let us consider the case of a harmonic field, where the momentum of the electromagnetic field is equal to $p_A = (e/c)A(t) \sim eE_0/\omega$. It corresponds to the contribution of the electromagnetic field in the quasienergy $\varepsilon_A = v_F p_A = eE_0 v_F/\omega$. Then the relation (15.26) can be rewritten as

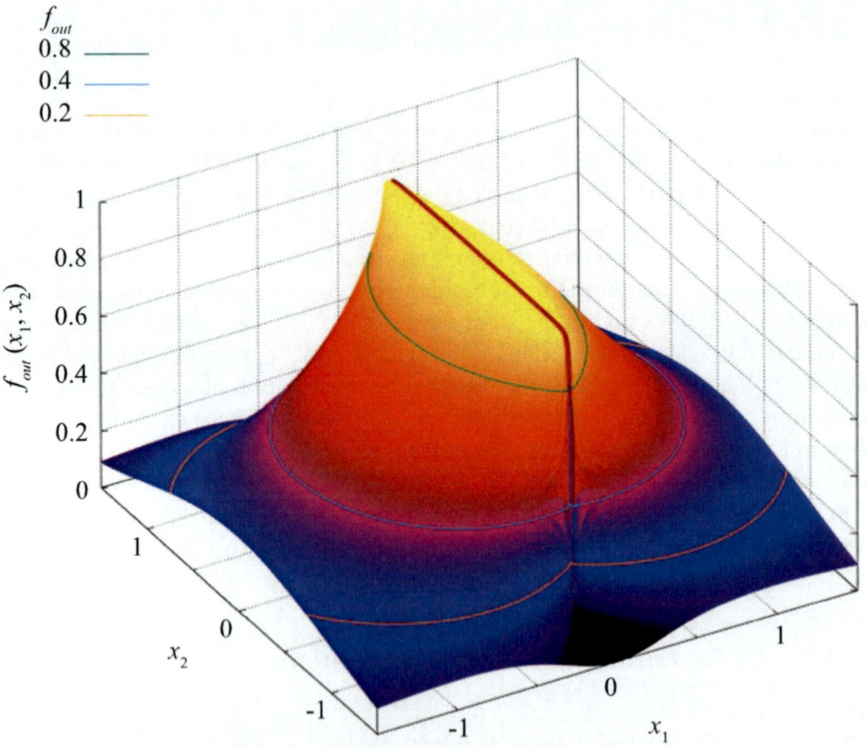

Fig. 15.1 The residual distribution function $f_{out}(x_1, x_2)$ for the Sauter field model with $E = 1000$ V/cm, $\kappa = 10^{12}$ Hz, where $x_{1,2} = p_{1,2}/m_* v_F$ are dimensionless momenta

$$\Lambda_*^{max} = \frac{2\hbar\omega^2}{3\sqrt{3}eE_0 v_F} \sim \frac{\hbar\omega}{\varepsilon_A}, \tag{15.27}$$

which corresponds to the ratio of the energy of the absorbed quant of external field to the part of energy of quasiparticle acquired as a result of acceleration in this field.

In the case of a sufficiently large low-frequency external field, $\hbar\omega \ll eE_0 v_F$, one can search a solution of the KE system (15.2) by means of an asymptotic decomposition of the functions $f,\ u,\ v$ for the small parameter $\Lambda_*(\mathbf{p}, t) \ll \Lambda_*^{max} \ll 1$,

$$f = \sum_{n=0} f_n, \quad u = \sum_{n=0} u_n, \quad v = \sum_{n=0} v_n. \tag{15.28}$$

Substituting these decompositions in the KE system (15.2) and equating the contributions of the same orders, one can obtain the leading terms of the asymptotic decompositions (15.28) as

$$f_4 = \frac{1}{4}\Lambda_*^2(\mathbf{p},t) = \frac{e^2 v_F^4 E^2(t) p_1^2}{16\varepsilon_*^6}, \tag{15.29}$$

$$u_3 = \frac{1}{2\varepsilon_*}\dot{\Lambda}_*(\mathbf{p},t) = \frac{e v_F^2 \dot{E}(t) p_1}{4\varepsilon_*^4}, \quad v_2 = \Lambda_*(\mathbf{p},t).$$

Formally, these expressions have the same form as the analogous results in standard QED which were obtained in the framework of the asymptotic decompositions of the functional series in $E_0/E_c \ll 1$, where $E_c = m^2c^3/e\hbar$ is the critical field. In a similar way one can obtain the post-leading terms.

The obtained asymptotic solutions of the KE (15.2) can be used for estimating the convergences of the integral macroscopical physical values (e.g., the densities of the conduction and polarization currents an so on) and also in analytical calculations in theory of radiation and other transport phenomena.

Let us consider now the realization of the effective mass approximation for different external field models.

15.3 Approximate Solutions of KEs for Different External Field Models in Graphene

15.3.1 The Sauter Pulse

The field of this pulse is given by

$$A(t) = -(cE_0/\kappa)\tanh\kappa t, \; E(t) = E_0/\cosh^2\kappa t\,. \tag{15.30}$$

It is a classical example of the external field model leading to an exact solution of the basic equations of motion of QED [11]. The analogous solution for the massless graphene model was obtained in the work [7].

The effective electromagnetic mass (15.18) in this model is

$$m_* = eE_0/v_F\kappa. \tag{15.31}$$

The corresponding integral (15.22) is

$$J(\mathbf{p},t) = \frac{E_0}{\kappa}\int_{-\infty}^{\kappa t} dx \frac{\cos[\Omega(\mathbf{p})/\kappa]x}{\cosh^2 x}, \tag{15.32}$$

where $\Omega(\mathbf{p}) = 2\varepsilon_*(\mathbf{p})/\hbar$. The corresponding distribution function $f(\mathbf{p},t)$ is defined then according to the Eq. (15.23). The vacuum polarization functions $u(\mathbf{p},t)$ and $v(\mathbf{p},t)$ can be restored then with help of the Eq. (15.2). In the asymptotic limit $t \to \infty$ it follows that

$$J_{out}(\mathbf{p}) = \frac{2E_0}{\kappa} \frac{\pi\Omega(\mathbf{p})/2\kappa}{\sinh[\pi\Omega(\mathbf{p})/2\kappa]}. \tag{15.33}$$

Then the distribution function (15.23) in the out-state will be

$$f_{out}(\mathbf{p}) = \left[\frac{eE_0 v_F^2 p_1}{\kappa\varepsilon_*^2(\mathbf{p})} \frac{\pi\Omega(\mathbf{p})/2\kappa}{\sinh[\pi\Omega(\mathbf{p})/2\kappa]}\right]^2. \tag{15.34}$$

The corresponding expression for the pair number density (15.9) written in terms of the dimensionless momentum $x_k = p_k/m_* v_F$ is

$$n_{out} = \frac{N_f}{\pi}\left(\frac{\eta^2\kappa}{8\pi v_F}\right)^2 \int_0^\infty \frac{dx\, x^3}{1+x^2} \frac{1}{\left[\sinh\left(\frac{\eta}{2}\sqrt{1+x^2}\right)\right]^2}, \tag{15.35}$$

where

$$\eta = \frac{2\pi m_* v_F^2}{\hbar\kappa} = \frac{2\pi e E_0 v_F}{\hbar\kappa^2}. \tag{15.36}$$

The point $\eta = 1$ separates two domains: the domain $\eta < 1$, where the energy of quasiparticles acquired in the external field $eE_0 v_F/\kappa$ is less than the energy of an absorbed quant of the field $\hbar\kappa$, and the domain $\eta > 1$, where the field acceleration mechanism dominates.

Some results of the numerical comparison of the exact and approximate (15.35) dependencies $n_{out}(\eta)$ are given in Fig. 15.2. From here it follows that the effective

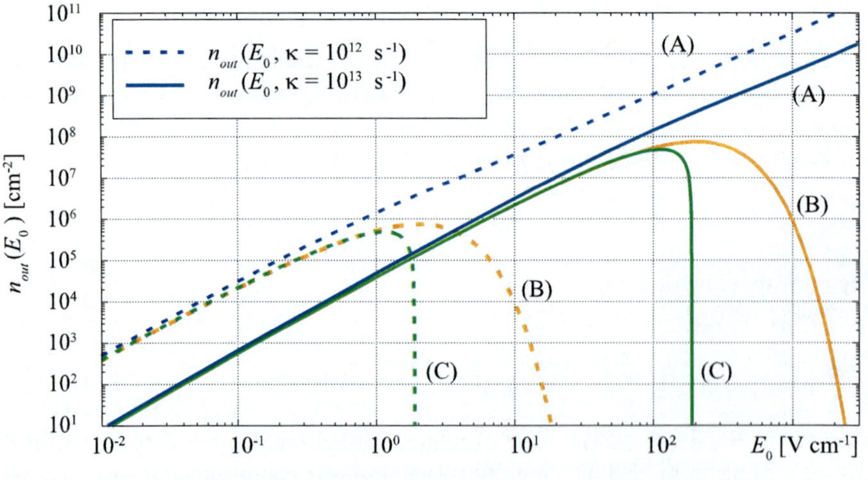

Fig. 15.2 Sauter field model for two values $\kappa = 10^{12}\ \text{s}^{-1}$ and $\kappa = 10^{13}\ \text{s}^{-1}$. Lines labelled (A) are numerical solutions of the KE (15.1), lines (B) show the approximation of Eq. (15.35), and lines (C) belong to the approximation of Eq. (15.37)

electromagnetic mass approximation is valid in the domain $\eta \leq 1$ and can be dubbed "slow switching" with $\kappa \gg (2\pi e E_0 v_F/\hbar)^{1/2}$. Assuming that $\eta \ll 1$ on the r.h.s. of Eq. (15.35), one can obtain

$$n_{out} = -\frac{N_f}{\pi}\left(\frac{eE_0}{2\hbar\kappa}\right)^2\left[1+\frac{1}{2}\ln\frac{1}{24}+\ln\eta\right], \quad \eta \ll 1. \tag{15.37}$$

This corresponds to the result obtained in [7].

15.3.2 The Gaussian Pulse

This field model

$$A(t) = -\sqrt{\frac{\pi}{2}}cE_0\tau\,\mathrm{erf}\left(\frac{t}{\sqrt{2}\tau}\right), \quad E(t) = E_0\exp(-t^2/2\tau^2) \tag{15.38}$$

results in the effective mass

$$m_* = \frac{eE_0\tau\pi^{1/4}}{v_F}\left(\sqrt{2}\,\mathrm{erf}(1/\sqrt{2})e^{-1/2}+\sqrt{\pi}\,\mathrm{erf}^2(1/\sqrt{2})/2-\mathrm{erf}(1)\right)^{1/2}$$
$$\sim 0.5257\,\frac{eE_0\tau}{v_F}. \tag{15.39}$$

The distribution function will be

$$f(\mathbf{p},t) = \frac{1}{2}\left[\frac{eE_0v_F^2\tau p_1}{\varepsilon_*^2(\mathbf{p})}\,I\left(\frac{t}{\sqrt{2}\tau},\sigma\right)\right]^2, \tag{15.40}$$

where $\sigma = \Omega(\mathbf{p})\tau$ and

$$I\left(\frac{t}{\sqrt{2}\tau},\sigma\right) = \int_0^{t/\sqrt{2}\tau} dx\cos(\sqrt{2}\sigma x)e^{-x^2}. \tag{15.41}$$

A simple result follows from Eqs. (15.40) and (15.41) in the asymptotic case $t \to \infty$,

$$f_{out}(\mathbf{p}) = \frac{\pi}{2}\left\{\frac{eE_0\tau v_F^2 p_1}{\varepsilon_*^2(\mathbf{p})}\exp\left[-\frac{2\varepsilon_*^2\tau^2}{\hbar^2}\right]\right\}^2. \tag{15.42}$$

From here one can find after simple calculations the pair number density in the out-state,

$$n_{out} = -N_f\left(\frac{e\tau E_0}{4\hbar}\right)^2\left[e^{-\xi^2}+(1+\xi^2)\mathrm{Ei}(-\xi^2)\right], \tag{15.43}$$

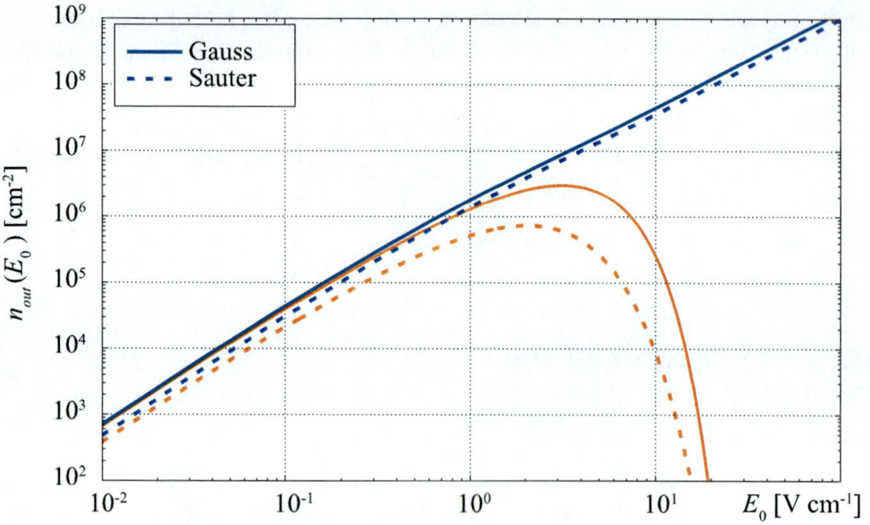

Fig. 15.3 The pair number densities n_{out} for the numerical (upper lines) and approximate (lower lines) solutions for the Sauter ($\kappa = 10^{12}$ s^{-1}) and Gaussian ($\tau = 10^{-12}$ s) pulses

where $\xi = 2m_* v_F^2 \tau/\hbar$ and $\mathrm{Ei}(-\xi^2)$ is the exponential integral function. This result in the region $\xi \ll 1$ corresponds to Eq. (15.37) for the case of the Sauter pulse.

In Fig. 15.3 we compare the behaviour of the pair number densities n_{out} for the exact and approximate solutions for the Sauter ($\kappa = 10^{12}$ s^{-1}) and Gauss ($\tau = 10^{-12}$ s) pulses.

15.3.3 The Harmonic Field Model

The harmonic field model

$$A(t) = -(cE_0/\omega)\sin\omega t, \;\; E(t) = E_0 \cos\omega t; \tag{15.44}$$

corresponds to the effective mass

$$m_* = eE_0/\sqrt{2}v_F\omega. \tag{15.45}$$

The distribution function in this field model was obtained in the work [3],

$$f(\mathbf{p}, t) = f^{(0)}(\mathbf{p}) + f^{(2)}(\mathbf{p}, t), \tag{15.46}$$

where the function

$$f^{(0)}(\mathbf{p}) = \frac{(e\hbar E_0 v_F^2 p_1)^2(4\varepsilon_*^2 + \hbar^2\omega^2)}{8\varepsilon_*^4(4\varepsilon_*^2 - \hbar^2\omega^2)^2} \tag{15.47}$$

corresponds to a stationary background distribution while the function

$$f^{(2)}(\mathbf{p}, t) = \frac{(e\hbar E_0 v_F^2 p_1)^2}{8\varepsilon_*^4(4\varepsilon_*^2 - \hbar^2\omega^2)} \cos 2\omega t \tag{15.48}$$

corresponds to the breathing mode on the doubled frequency of the external field. The residual functions u, v can be reconstructed using Eqs. (15.46)–(15.48) and the KE system (15.2)

$$u(\mathbf{p}, t) = \frac{e\hbar^2 v_F^2 p_1}{\varepsilon_*^2(4\varepsilon_*^2 - \hbar^2\omega^2)} \dot{E}(t), \tag{15.49}$$

$$v(\mathbf{p}, t) = \frac{2e\hbar^2 v_F^4 p_1}{\varepsilon_*(4\varepsilon_*^2 - \hbar^2\omega^2)} E(t). \tag{15.50}$$

The distribution function (15.46)–(15.47) corresponds to the first and third harmonics of the current density and radiation spectrum of the plasma oscillations [3]. These results are valid in the case

$$\hbar\omega^2/(\sqrt{2}eE_0 v_F) < 1. \tag{15.51}$$

This limitation holds also for other field models, if the quantity ω is interpreted as the corresponding characteristic frequency of the field alteration.

A general feature of the two outlined approximate approaches is the E^2 - proportionality of all the resulting distribution functions, $f(\mathbf{p}, t) \sim (eE_0)^2$. This feature was obtained in the work [22] on the basis of an analysis of the numerical solutions of the corresponding KEs in standard QED, see also [23].

The effectiveness of the low density approximation in the standard strong field QED for rather weak fields $E_0 \ll E_c$ has been investigated in the work [24]. The additional introduction of the effective mass approximation results herein a strong restriction of the domain of applicability of the method.

15.4 Conclusion

In the present work we have outlined a simple and rather general approach to obtain approximate solutions for the distribution functions of charge carriers in graphene based on nonperturbative KEs. This was achieved by a combination of the low density approximation and the concept of an effective electromagnetic mass. Such an approach is effective for a rather wide class of external field models with the parameters limited by the relation (15.51).

The considered approximation is of particular interest for the investigation of such complicated nonlinear single-photon effects in graphene as the emission (absorption) and annihilation (photoproduction) and the more complex two-photon processes. Such kind of nonlinear phenomena in graphene became accessible for experimental verification recently, see [15, 16]. The first step in this direction was done in the work [3], where the effect of quantum radiation was predicted which is identified well on the background of the quasiclassical radiation of the plasma currents.

Acknowledgements The work of D.B. was supported by the Russian Federal Program "Priority-2030". N.G. received support from Volkswagen Foundation (Hannover, Germany) under collaborative research grant No. 97029. B.M. acknowledges a stipend from the International Max Planck Research School for "Many-Particle Systems in Structured Environments" at the Max-Planck Institute for Physics of Complex Systems (Dresden, Germany).

References

1. Smolyansky, S.A., Panferov, A., Blaschke, D., Gevorgayn, N.: Nonperturbative kinetic description of electron-hole excitations in graphene in a time dependent electric field of arbitrary polarization. Particles **2**, 208–230 (2019). https://doi.org/10.3390/particles2020015
2. Smolyansky, S.A., Panferov, A.D., Blaschke, D.B., Gevorgyan, N.T.: Kinetic equation approach to graphene in strong external fields. Particles **3**(2), 456–476 (2020). https://doi.org/10.3390/particles3020032
3. Gavrilov, S.P., Gitman, D.M., Dmitriev, V.V., Panferov, A.D., Smolyansky, S.A.: Radiation problems accompanying carrier production by an electric field in the graphene. Universe **6**, 205 (2020). https://doi.org/10.3390/universe6110205
4. Vozmediano, M.A.H., Katsnelson, M.I., Guinea, F.: Gauge fields in graphene. Phys. Rep. **496**, 109–148 (2010). https://doi.org/10.1016/j.physrep.2010.07.003
5. Lewkowicz, M., Kao, H.C., Rosenstein, B.: Signature of the Schwinger pair creation rate via radiation generated in graphene by a strong electric current. Phys. Rev. B **84**, 035414 (2011). https://doi.org/10.1103/PhysRevB.84.035414
6. Gavrilov, S.P., Gitman, D.M., Yokomizo, N.: Dirac fermions in strong electric field and quantum transport in graphene. Phys. Rev. D **86**, 125022 (2012). https://doi.org/10.1103/PhysRevD.86.125022
7. Klimchitskaya, G.L., Mostepanenko, V.M.: Creation of quasiparticles in graphene by a time-dependent electric field. Phys. Rev. D **87**, 125011 (2013). https://doi.org/10.1103/PhysRevD.87.125011
8. Ishikava, K.L.: Nonlinear optical response of graphene in time domain. Phys. Rev. B **82**, 201402 (2010). https://doi.org/10.1103/PhysRevB.82.201402
9. Ishikava, K.L.: Electronic response of graphene to an ultrashort intense terahertz radiation pulse. New J. Phys. **15**, 055021 (2013). https://doi.org/10.1088/1367-2630/15/5/055021
10. Panferov, A.D., Churochkin, D.V., Fedotov, A.M., Smolyansky, S.A., Blaschke, D.B., Gevorgyan, N.T.: Nonperturbative kinetic description of e-h excitations in graphene due to a strong, time-dependent electric field. In: Proceedings of the Ginzburg Centennial Conference on Physics, May 29–June 3, 2017, Moscow. http://gc2.lpi.ru/proceedings/panferov.pdf. Accessed 16 March 2020
11. Grib, A.A., Mamaev, S.G., Mostepanenko, V.M.: Vacuum Quantum Effects in Strong Fields. Friedmann Laboratory Publishing, St. Petersburg, Russia (1994)
12. Bialynicky-Birula, I., Gornicki, P., Rafelski, J.: Phase space structure of the Dirac vacuum. Phys. Rev. D **44**, 1825 (1991). https://doi.org/10.1103/PhysRevD.44.1825

13. Schmidt, S.M., Blaschke, D., Röpke, G., Smolyansky, S.A., Prozorkevich, A.V., Toneev, V.D.: A Quantum kinetic equation for particle production in the Schwinger mechanism. Int. J. Mod. Phys. E **7**, 709–718 (1998). https://doi.org/10.1142/S0218301398000403
14. Kluger, Y., Mottola, E., Eisenberg, J.: Quantum Vlasov equation and its Markov limit. Phys. Rev. D **58**, 125015 (1998). https://doi.org/10.1103/PhysRevD.58.125015
15. Bowlan, P., Martinez-Moreno, E., Reimann, K., Elsaesser, T., Woerner, M.: Ultrafast terahertz response of multilayer graphene in the nonperturbative regime. Phys. Rev. B. **89**, 041408 (2014). https://doi.org/10.1103/PhysRevB.89.041408
16. Baudisch, M., Marini, A., Cox, J.D., et al.: Ultrafast nonlinear optical response of Dirac fermions in graphene. Nat. Commun. **9**, 1018 (2018). https://doi.org/10.1038/s41467-018-03413-7
17. Schmidt, S.M., Blaschke, D., Röpke, G., Prozorkevich, A.V., Smolyansky, S.A., Toneev, V.D.: Non-Markovian effects in strong-field pair creation. Phys. Rev. D **59**, 094005 (1999). https://doi.org/10.1103/PhysRevD.59.094005
18. Mamaev, S.G., Trunov, N.N.: Vacuum polarization and particle production in a non-stationary homogeneous electromagnetic field. Sov. J. Nucl. Phys. **30**, 677 (1979)
19. Berestetskii, V.B., Lifshitz, E.M., Pitaevskii, L.P.: Quantum Electrodynamics. Pergamon, Oxford (1982)
20. Aleksandrov, I.A., Dmitriev, V.V., Sevostyanov, D.G., Smolyansky, S.A.: Kinetic description of vacuum e+e- production in strong electric fields of arbitrary polarization. Eur. Phys. J. Spec. Top. **229**, 3469–3485 (2020). https://doi.org/10.1140/epjst/e2020-000056-1
21. Smolyansky, S.A., Fedotov, A.M., Dmitriev, V.V.: Kinetics of the vacuum e-e+ plasma in a strong electric field and problem of radiation. Mod. Phys. Lett. A **35**, 2040028 (2020). https://doi.org/10.1142/S0217732320400283
22. Kravtcov, K.Y., Dmitriev, V.V., Levenets, S.A., Panfyorov, A.D., Smolyansky, S.A., Juchnowski, L., Blaschke, D.B.: The choice of the optimal approximation in the kinetic description of the vacuum creation of electron-positron plasma in strong laser fields. In: Derbov, V.L., Postnov, D.E. (eds.) SARATOV FALL MEETING 2017, Laser Physics and Photonics XVIII; and Computational Biophysics and Analysis of Biomedical Data IV, vol. 10717, p. 1071702 (2018). https://doi.org/10.1117/12.2306171
23. Blaschke, D.B., Juchnowski, L., Otto, A.: Kinetic approach to pair production in strong fields—two lessons for applications to heavy-ion collisions. Particles **2**(2), 166–179 (2019). https://doi.org/10.3390/particles2020012
24. Fedotov, A.M., Gelfer, E.G., Korolev, K.Yu., Smolyansky, S.A.: On the kinetic equation approach to pair production by time-dependent electric field. Phys. Rev. D **83**, 025011 (2011). https://doi.org/10.1103/PhysRevD.83.025011

Chapter 16
Possibility of Creating a Low-Cost Robot Assistant for Use in General Medical Institutions During the COVID-19 Pandemic

V. Chekubasheva, O. Glukhov, O. Kravchuk, Y. Levchenko, E. Linnyk, and V. Rohovets

Abstract This article analyzes the experience and prospects of creating a robot assistant for medical institutions using in the modern conditions of Ukraine. The main directions of fundamental and applied research are proposed, taking into account budget equipment. Recently, the creation of a system for monitoring the condition of patients in difficult epidemiological conditions has become of particular relevance in the world. It involves the use of an integrated autonomous control system for robotic patients to work with patients in hospitals. For medical institutions, the task of equipping with modern pieces of equipment is also relevant. Autonomous systems carry out independent processing of medical premises and instruments, especially in infectious diseases departments and also premises with increased requirements for cleanliness. One of the possible solutions to this problem is the use of medical autonomous robotic assistants.

16.1 Introduction

Today, slowing down the spread of the infectious disease caused by the SARS-CoV-2 virus is one of the priorities of biomedical research. The main centers that have a significant impact on this are the infectious diseases departments of hospitals. Junior medical personnel is at the greatest risk due to prolonged direct contact with the infected people, whose task is to take all the main medical parameters of patients and monitor patients' conditions. At the same time, due to the huge influx of infected people, medical facilities require constant disinfection to reduce the risk of infection. Including rooms with increased requirements for cleanliness, such as surgical rooms. The performance of such work by personnel exposes them to a constant high risk of infection and requires a lot of time.

V. Chekubasheva (✉) · O. Glukhov · O. Kravchuk · Y. Levchenko · E. Linnyk · V. Rohovets
Kharkiv National University of Radio Electronics, 14 Nauka Ave, 61166 Kharkov, Ukraine
e-mail: valeriia.chekubasheva@nure.ua

D. Blaschke et al. (eds.), *Optics and Its Applications*, Springer Proceedings in Physics 281, https://doi.org/10.1007/978-3-031-11287-4_16

The solution to these problems can be an autonomous robotic system that replaces part of the functions of junior medical staff and thereby reduces the time of interaction with sick patients. Such systems using frees you from daily everyday activities and allows you to focus on more seriously ill patients.

At the moment, robotic systems are being widely implemented in various fields: medicine [1], agriculture [2], logistics [3], industrial and space sectors [4], military industry [5], etc. The use of cross solutions from different areas makes it possible to improve the existing [6] robotic system and equip it with a set of devices for performing specialized tasks. In addition, the use of already known [7, 8] and available technologies can reduce the cost of a robotic system compared to existing analogs [9–11].

Thus, the purpose of this work is to create a budget autonomous robotic system with a replaceable set of blocks for various functional purposes to perform certain medical tasks. Namely: disinfection of premises, assistance in the delivery of tests and drugs, in the removal of basic medical indications of patients.

16.2 Methods and Materials

Pathophysiological mechanisms of COVID-19 involve inflammation, fever, hypoxia, electrolytes, acid-base balance disorder, shock, and other basic pathological processes. According to existing studies and literature reports, patients with COVID-19 have been found to often suffer from the following dysfunctions: dyspnea, fever, fatigue, poor appetite, tachycardia, and decreased oxygen-carrying capacity. That is why the primary parameters that nursing staff control when bypassing patients are saturation, body temperature, pressure, pulse, diuresis, and respiratory rate [12]. The first four earlier noted parameters measurements can be carried out based on a robotic system using sensors of certain functionality [13].

Often the patient's condition remains stable during the bypass, but in just a few hours it can deteriorate sharply. In overcrowded hospitals, when patients are placed literally in corridors and other departments that are not equipped for emergency needs, it is extremely difficult to conduct a high-quality and timely examination of each patient. Because doctors have to spend a lot of time on especially seriously ill people. In addition, it is medical workers who are most at risk, not only from infection but also from professional overtime work and nervous strain. In this way, a functional diagram of the robotic platform was developed based on the above provisions (see Fig. 16.1).

This article proposes a development that combines the functions of processing rooms for various needs, as well as daily patient surveys, which improves the capabilities and equipment of hospital wards, can record a standard number of polled parameters with their sorting. Also, this device can be used as local delivery of medicines and food to patients. This system will help reduce the workload on them and compensate for the shortage of personnel.

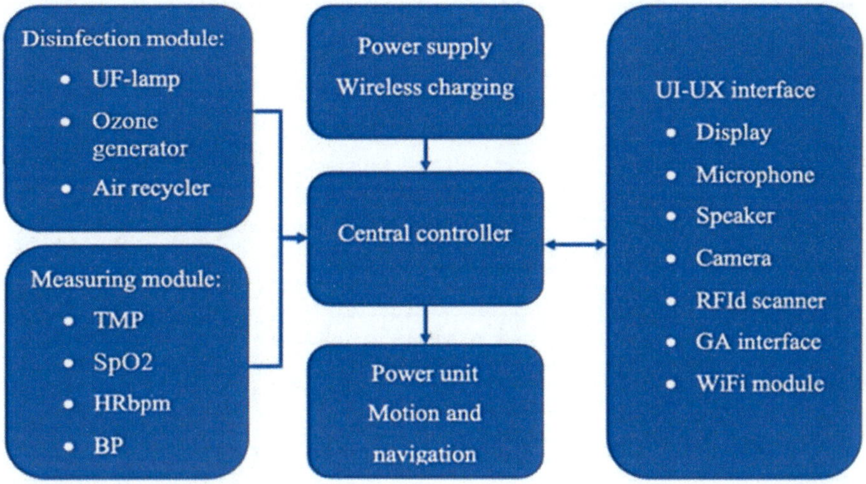

Fig. 16.1 The functional scheme

16.3 Results and Their Analysis

The system is modular and allows you to combine blocks depending on the task. The robot consists of 7 blocks (see Fig. 16.2): a mobile platform for autonomous movement; battery pack; a processing unit for a room without a person present, represented by an ozone generator; sensor unit for monitoring human parameters; a block for placing things necessary for medical personnel, patient test results, and drugs; a tablet for visualizing the attending physician and an interface for managing the system; block for processing premises with the presence of people.

16.3.1 Navigation Algorithm, the Motion Platform

TurtleBot 3 Waffle Pi [14] is used as a mobile platform as the best option to provide the necessary functions: reliability, functionality (support for ROS), optimal dimensions (281mm/306mm/141mm), high transported weight (maximum payload 30 kgs). This has the following parameters: implementation on Raspberry Pi 3, maximum translational velocity is 0.26 m/s and rotational velocity is 104.27 deg/s, presence of lidars, camera module, gyroscope, and accelerometer to implement the vSLAM method [15] to optimize the movement of the robot. The device is equipped with a camera that allows you to effectively analyze the space, remembering the location of objects on the territory and using the ceiling for orientation. This format builds a more accurate floor plan that can be viewed in a mobile application. Difference from the first type: the user can interact with the map. The application provides access to

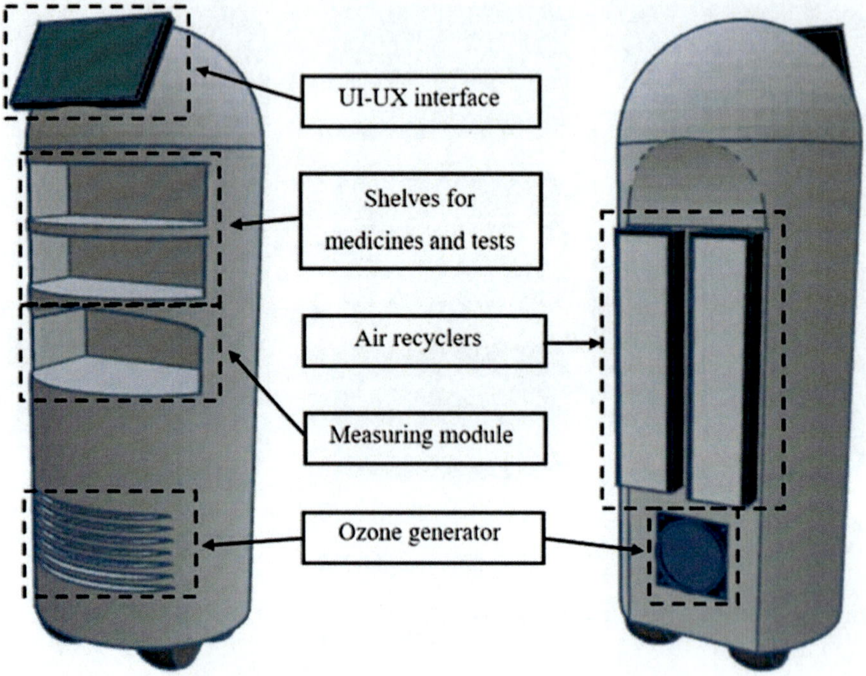

Fig. 16.2 The 3D model image of robotic assistant

building virtual walls and designating specific places for local processing [16]. It is also necessary to use methods [17] for determining the presence of people in the room.

16.3.2 Power Supply Design

The main method of charging the batteries of the assistant robot is an induction charging station, which is provided using two insulated coils. The maximum power transmitted by the charging station is 24 W, while the efficiency is 86%, the time for a full charge is 5–6 hours. The main advantage of this system is the minimal participation of a person in ensuring the process of charging the device. Power is supplied through shielded coils. Before power is applied, the transmitter sends a test signal to determine the receiver, after which a connection is established. The receiver tells the transmitter how much power to send and when (see Fig. 16.3). In addition, UV lamps are placed on the docking station to ensure the treatment of the external surfaces of the system.

To ensure maximum efficiency of charging stations it is necessary to ensure accurate positioning of the mating and receiving parts with an error of no more than $5mm$.

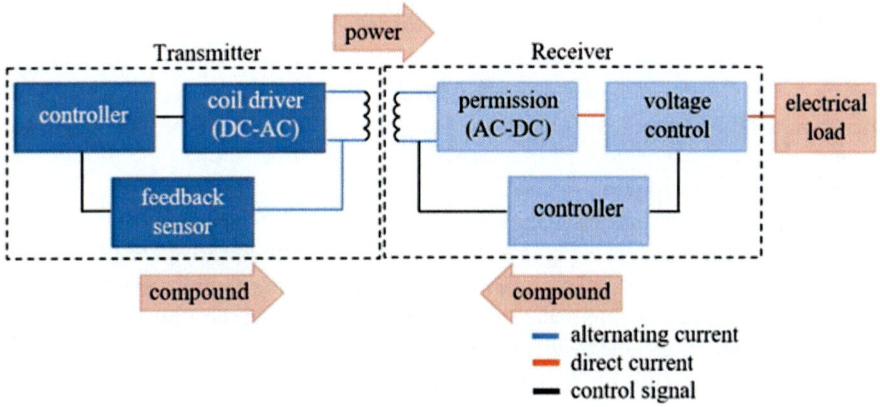

Fig. 16.3 Block diagram of an induction charging station

To accomplish this task, an induction positioning system is used based on two Hall sensors at the charging station and neodymium magnets attached to the robot body.

The system also provides for charging from a 220/110 V network. This allows you to charge the device in less than 2 h. This option is appropriate if there is a need to urgently prepare an assistant for operation. To provide a different input voltage, a dual-circuit power part of the battery charge controller is used.

16.3.3 Disinfection Equipment

The ozone generator uses allows for a higher degree of purification compared to a bactericidal recycler (90% for recirculators [18] and 99% for ozone generator [19]). First, ozone is more effective at inactivating viruses than any other treatment, while leaving no chemical waste. Daily ozone exposures increase mortality and respiratory morbidity rates. In short-term pulmonary function studies, lung inflammation, lung permeability, respiratory symptoms, increased medication usage, morbidity, and mortality [20]. Therefore, it is important to use ozonation strictly indoors without people.

The best solution for the premises treatment is to use ultraviolet bactericidal recirculators. They are safer in comparison with open-type ultraviolet and quartz emitters because they do not irradiate the entire room and people in it, but the air that is passed through the mechanism. Ultraviolet glass in recirculators transmits part of the UV spectrum radiation that is harmful to pathogenic microorganisms, but delays ozone-forming rays [21]. UV radiation destroys the DNA of bacteria and microorganisms.

16.3.4 Health Parameters Measurement Unit

The device is capable of measuring the following parameters: temperature, heart rate, saturation, and pressure. This unit is a retractable platform with a recess under the patient's arm for ease of use of the system.

The human body temperature recording is carried out by a non-contact thermometer MLX90614. It has an error measurement of 0.02 °C. It is necessary to take into account the temperature of the area where the measurement is carried out. The skin temperature on the wrist is 30.4 $^{\circ}C$. In addition, the measurement is carried out taking into account the possible decrease in blood circulation in the limbs of patients. However, for this block, the task is not to accurately determine the temperature, but to define a sharp increase in this parameter by 3 °C.

Its connection is carried out by a serial I2C interface. The measurement of saturation is carried out by methods of measuring changes in light absorption in oxygenated and deoxygenated blood. The physical substantiation of this method lies in the difference in the processes of absorption of saturated and unsaturated blood by oxygen. Therefore, the simplest and most affordable solution would be to use a human finger. In addition, a detectable period between an increase and a decrease in blood oxygenation makes it possible to measure the pulse rate.

The blood pressure non-contact measurement provides it is necessary to use potentially alternative methods, one of which is the method of estimating blood pressure based on the time of passage of a pulse wave obtained using photoplethysmography (PPG) on the wrist and on the finger [22–24] (see Fig. 16.4).

16.3.5 UX-UI Interface

This block provides access to the functions of the robot.

A tablet is used as the main interface, which allows for remote contact with the doctor if necessary. Due to the specifics of the robot application, namely the assumption that patients will use the system from a prone position, the tablet has mechanisms for adjusting the level of inclination. This unit is equipped with a microphone, speaker, and camera to realize feedback. Also, the robot assistant is equipped with a voice recognition system as an alternative UX interface and the ability to comfortably position the tablet.

16.3.6 Data Storing

A database with remote access is used to store patient data. The system contains the following information-personal data (name, address, contact details of the patient

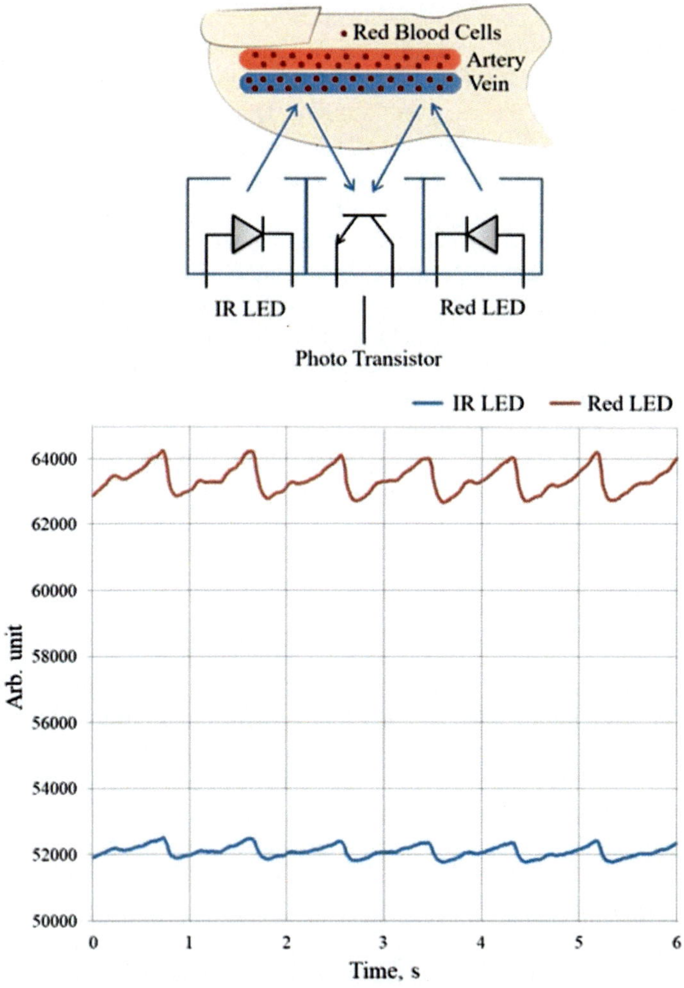

Fig. 16.4 Principle of implementation of the PPG method

and authorized representative), RFID identifier, examination results and indications of daily rounds, as well as personal complaints of the patient.

16.3.7 Control Algorithm

The microcontroller waits for a signal to be sent to the input panel, when the power is turned on. Medical staff can choose one of two work algorithms: the first is if it is necessary to complete a daily survey of the condition of patients and the processing of premises, and the second is only disinfection, for example, of a surgical room or corridors of an institution. Next, healthcare workers need to mark the key points of the movement trajectory on the hospital map. The robot will perform the specified actions when it arrives at the specified locations. During the movement, at intervals of 1 min, the robot checks whether the goal has been reached. If not, then it processes the premises all the way, depending on the presence of a person.

The process of questioning the patient begins with his identification. The robot reads the RFID tag of a person and then proceeds to interrogate the unit for measuring body parameters. An interruption of the optocoupler signal detects the presence of the patient's hand inside the unit, after which the temperature, saturation, and heart rate sensors are turned on. These data are recorded in a database and updated in real time on the attending physician's computer. When the parameter crosses the acceptable threshold of the body's vital activity, the application sends a sound signal notifying about the need for resuscitation.

After completing these steps, the robot returns to the docking station for recharging (Fig. 16.5).

16.4 Conclusion

The article market analyzes in the segment of medical robots that can combine the functions of monitoring the patients state and safe and efficient processing of disinfection. The main aspects of the development of this subject have been identified: the reduction in the cost of existing technologies and the combination of the necessary functionality in a modular system that can be adapted to the needs of a particular medical institution. Based on these observations, the concept of an assistant robot, its 3D model and functional diagram have been developed. The element base selection has been made and the device control algorithm have been developed.

A robotic IoT patient care system for medical institutions has been developed. The assistant provides the following features: daily rounds of patients, disinfection of empty rooms and in the presence of patients and staff in the room, delivery of medicines, tests, and remote survey, examination by a doctor. The robot can measure the following patient parameters: temperature, pressure, saturation, pulse. Due to the specific application of the device, non-contact optical sensors are used to measure all parameters, and all measurement methods are modified to suit operating conditions. The time of continuous battery life of the device is 5 h without recharging.

The developed device allows not only to conduct a daily survey of patients but also allows for effective bactericidal treatment of premises due to the equipping of several room disinfection systems. An ozone generator - which is great for empty rooms and kills over 90% of harmful microorganisms and a closed-type UV recycler - which is suitable for rooms where patients and staff are located.

Due to the modularity of the developed system, the user can configure it at his discretion, depending on the needs of a particular medical institution, which will significantly affect the final cost of the device. Modularity is achieved through the

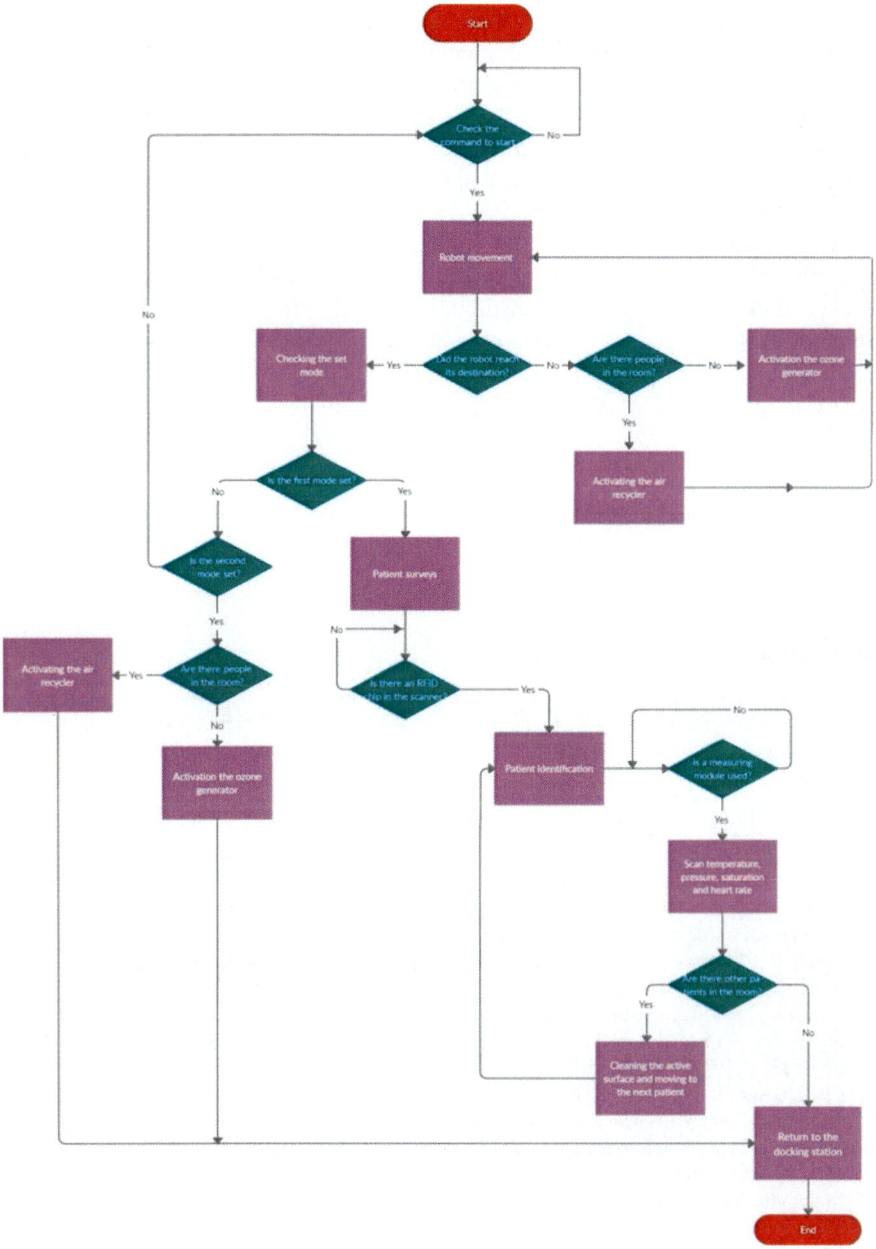

Fig. 16.5 Device operation algorithm

use of a programmable data bus. This will make it possible to use the developed device not only in epidemiological conditions but also as an auxiliary system in non-infectious departments of hospitals to prevent staff overtime.

Among the features of the system, one can also note the ability to build a doctor's round of patients, choosing the optimal trajectory of movement, the ability to independently fix the need for recharging and finding a wireless docking station, UI-UX interface for communication between a doctor and a patient with support for video communication and voice control, real-time transmission data on the state of health of people for comfortable and quick information of the medical database.

This set of functions favorably distinguishes the developed device from competitors both in the range of options provided and in the availability for purchase or manufacture and is an effective means of combating the spread of various infectious diseases.

Acknowledgements V. Chekubasheva, O. Kravchuk, V. Rohovets are grateful to the organizers of the 9th International Symposium "Optics & its applications" and the International Commission for Optics (ICO) for the support which was instrumental to realize their participation in the symposium.

References

1. Ushakov, I.B., Polyakov, A.V., Usov, V.M., Knyazkov, M.M., Motienko, A.I.: Using service robots to counter the SARS-CoV-2 virus spread in enclosed medical premises. J. Medico-Biol. Socio-Psychol. Probl. Safety Emerg. Situations **2**, 104–114 (2021)
2. Rose, D.C., Lyon, J., de Boon, A., et al.: Responsible development of autonomous robotics in agriculture. Nat. Food **2**, 306–309 (2021)
3. Wang, C., Du, D.: Research on logistics autonomous mobile robot system. In: IEEE International Conference on Mechatronics and Automation, pp. 275–280 (2016)
4. Flores Abad, A., Ma, O., Pham, K., Ulrich, S.: A review of space robotics technologies for on-orbit servicing. Prog. Aerosp. Sci. **68** (2014)
5. Simon, P.: Military robotics: latest trends and spatial grasp solutions. Int. J. Adv. Res. Artif. Intell. **4**(4) (2015)
6. Beresnev, V., Chekubasheva, V., Glukhov, O., Kravchuk, O.: Creation of a prototype of a multi-function remote presence robot for physical research in electronics. J. NanoElectron. Phys. **13**(5), 05039 (2021)
7. Chekubasheva, V., Glukhov, O., Kravchuk, O., Rohovets V.: Creation of a remote presence robot based on the TI-RSLK development board. In: Mysko-Krutik, N. (eds.) II International Advanced Study Conference Condensed Matter and Low Physics 2021, p. 229. FOP Brovin O.V., Kharkiv (2021)
8. Majumder, S., Mondal, T., Deen, M.J.: Wearable sensors for remote health monitoring. Sensors **17**(1), 130 (2017)
9. Robot TIMA Homepage, https://robottima.com/en/home/
10. UBTech Robotics, CRUZR, https://www.ubtrobot.com/products/cruzr?ls=en
11. UVD Robots revolutionizing disinfection, https://uvd.blue-ocean-robotics.com/
12. Xia, W., Huang, X.: Rehabilitation from COVID-19. CRC Press, Boca-Raton (2021)
13. Sakphrom, S., Limpiti, T., Funsian, K., Chandhaket, S., Haiges, R., Thinsurat, K.: Intelligent medical system with low-cost wearable monitoring devices to measure basic vital signals of admitted patients. Micromachines **12**(8), 918 (2021)

14. https://www.robotis.us/turtlebot-3-waffle-pi/
15. Benini, A., Mancini, A., Marinelli, A., Longhi, S.: A Biased Extended Kalman Filter for indoor localization of a mobile agent using low-cost IMU and UWB wireless sensor network. In: IFAC Proceedings **45**(22), 735–740 (2012)
16. Karlsson, N., Bernardo, E., Ostrowski, J., Goncalves, L., Pirjanian, P., Munich, M.: The vSLAM Algorithm for Robust Localization and Mapping. In: Proceedings of the 2005 IEEE International Conference on Robotics and Automation, pp. 24–29, Barcelona, ICRA (2005)
17. Xu, Y., Cao, J., Shmaliy, Y.S., et al.: Distributed Kalman filter for UWB/INS integrated pedestrian localization under colored measurement noise. Satell. Navig. **2**, 22 (2021)
18. Dovbnenko, O., Kolesnik, I.: Development of ultraviolet bactericidal recirculator for air purification of livestock buildings. Mech. Electrif. Agric. 149–155 (2019)
19. Grignani, E., Mansi, A., Cabella, R., Castellano, P., Tirabasso, A., Sisto, R., Spagnoli, M., Fabrizi, G., Frigerio, F., Tranfo, G.: Safe and effective use of ozone as air and surface disinfectant in the conjuncture of Covid-19. Gases **1**(1), 19–32 (2021)
20. Amann, M.: Health risks of ozone from long-range transboundary air pollution. WHO Regional Office Europe, Copenhagen (2008)
21. Kovalenko O., Pilshchikova Y., Guseva E. Analysis of some trends of the mathematical methods application to support decisions making and implementation in the creation, testing and use of ultraviolet luminiscent lamps. Caspian J. Control High Technol. **4**(40) (2017)
22. Lochner, C., Khan, Y., Pierre, A., et al.: All-organic optoelectronic sensor for pulse oximetry. Nat. Commun. **5**, 5745 (2014)
23. Lazazzera, R., Belhaj, Y., Carrault, G.: A new wearable device for blood pressure estimation using photoplethysmogram. Sensors **19**(11), 2557 (2019)
24. Tseng, H.W., et al.: A method of measuring sleep quality by using PPG. In: IEEE International Conference on Consumer Electronics-Taiwan (ICCE-TW), pp. 1–2 (2016)